THE NEW ROSETTA TARGETS

ASTROPHYSICS AND SPACE SCIENCE LIBRARY

VOLUME 311

THE NEW ROSETTA TARGETS

Observations, Simulations and Instrument Performances

Edited by

LUIGI COLANGELI

*INAF – Osservatorio Astronomico di Capodimonte,
Napoli, Italy*

ELENA MAZZOTTA EPIFANI

*INAF – Osservatorio Astronomico di Capodimonte,
Napoli, Italy*

and

PASQUALE PALUMBO

*Università "Parthenope",
Napoli, Italy*

KLUWER ACADEMIC PUBLISHERS

DORDRECHT / BOSTON / LONDON

A C.I.P. Catalogue record for this book is available from the Library of Congress.

ISBN 1-4020-2572-6 (HB)
ISBN 1-4020-2573-4 (e-book)

Published by Kluwer Academic Publishers,
P.O. Box 17, 3300 AA Dordrecht, The Netherlands.

Sold and distributed in North, Central and South America
by Kluwer Academic Publishers,
101 Philip Drive, Norwell, MA 02061, U.S.A.

In all other countries, sold and distributed
by Kluwer Academic Publishers,
P.O. Box 322, 3300 AH Dordrecht, The Netherlands.

Cover image: "Capri" – Albert Bierstadt 1857, oil on canvas,
Tarzoli Gallery, San Rafael, California, U.S.A.

Printed on acid-free paper

Contents

Foreword ix

Acknowledgments xi

Discovery, observations and investigations of comet 67P/Churyumov–
    Gerasimenko in Kyiv 1
K. Churyumov

Monitoring Comet 67P/Churyumov–Gerasimenko from ESO in 2003 15
R. Schulz, J.A. Stüwe and H. Böhnhardt

CO and dust productions in 67P/Churyumov–Gerasimenko at 3 AU
    post–perihelion 25
D. Bockelée-Morvan, R. Moreno, N. Biver, J. Crovisier,
J.-F. Crifo, M. Fulle and M. Grewing

The Dust Activity of Comet 67P/Churyumov–Gerasimenko 37
M. Weiler, J. Knollenberg and H. Rauer

Observations of Comet 67P/Churyumov-Gerasimenko with the
    International Ultraviolet Explorer at Perihelion in 1982 47
P. Feldman, M. A'Hearn and M. Festou

Observations of the new Rosetta targets 53
C. Barbieri, S. Fornasier, G. Cremonese, I. Bertini, M. Fulle and
A. Magazzú

Water production rate of comet 67P/Churyumov–Gerasimenko 61
J.T.T. Mäkinen

Rosetta Asteroid Candidates   69
M.A. Barucci, M. Fulchignoni, I. Belskaya, P. Vernazza, E. Dotto and M. Birlan

Characterizing 21 Lutetia with its reflectance spectra   79
V.V. Busarev, V.V. Bochkov, V.V. Prokof'eva and M.N. Taran

Detection of Parent molecules in Comets using UV and visible spectroscopy   85
W.M. Jackson and A. Scodinu

Grain sizes of ejected comet dust. Condensed Dust Analogs, Interplanetary
   Dust Particles and Meteors   97
F.J.M. Rietmeijer and J.A. Nuth III

Physical Properties of Cometary Dust. The Case
   of 67P/Churyumov-Gerasimenko   111
A.C. Levasseur-Regourd, E. Hadamcik, J. Lasue, J.B. Renard and J.C. Worms

Physical Model of the coma of Comet 67P/Churyumov–Gerasimenko   119
J.F. Crifo, G.A. Lukyanov, V.V. Zakharov and A.V. Rodionov

The Dust Environment of Comet 67P/Churyumov–Gerasimenko   131
M. Fulle, C. Barbieri, G. Cremonese, H. Rauer, M. Weiler, G. Milani and R. Ligustri

Modelling the Environment of 67P/Churyumov–Gerasimenko   143
J. Agarwal, M. Müller and E. Grün

Plasma environment of comet Churyumov-Gerasimenko – 3D hybrid
   code simulations   153
T. Bagdonat, U. Motschmann, K.-H. Glassmeier and E. Kührt

Modelling of the structure and light scattering of individual dust particles   167
I. Bertini, N. Thomas and C. Barbieri

Gas and Dust Activity of the Nucleus of 67P/Churyumov-Gerasimenko:
   Some Preliminary Results   177
M.T. Capria, A. Coradini, M.C. De Sanctis and M. Fulle

Big Particle Emission from Comet 67P/Churyumov-Gerasimenko   185

E. Grün and J. Agarwal

The solar wind removal of volatiles from comets – perspectives
  on the Rosetta mission   197
R. Lundin and H. Nilsson

Non-resonant gravitational motion and equilibrium stability of Rosetta   211
E. Mysen

VIRTIS experiment at Churyumov – Gerasimenko comet, new Rosetta
  Target   223
A. Coradini, F. Capaccioni, G. Filacchione, G. Magni, E. Ammannito,
M. T. Capria, G. Piccioni, P. Drossart and G. Arnold

Consert experiment: description and performances in view
  of the new target   237
W. Kofman, A. Herique, J–P. Goutail and the CONSERT team

Rosina's Scientific Perspective at Comet Churyumov-Gerasimenko   257
K. Altwegg, A. Jäckel, H. Balsiger, E. Arijs, J.J. Berthelier,
S. Fuselier, F. Gliem, T. Gombosi, A. Korth and H. Rème

The GIADA experiment for the Rosetta mission   271
L. Colangeli, V. Della Corte, F. Esposito, E. Mazzotta Epifani,
E. Palomba, J.J. Lopez–Moreno, J. Rodriguez, R. Morales,
A. Lopez–Jimenez, M. Herranz, F. Moreno, P. Palumbo, A. Rotundi,
M. Cosi and The International GIADA Consortium

Implications of the new target comet on science operations
  for the Rosetta Lander   281
J. Biele and S. Ulamec

First contact with a comet surface: Rosetta Lander simulations   289
M. Hilchenbach, H. Rosenbauer and B. Chares

The Rosetta Lander experiment SESAME and the new target comet
  67P/Churyumov–Gerasimenko   297
K.J. Seidensticker, K. Thiel, A. Péter, W. Schmidt, H.-H. Fischer,
D. Madlener, S. Schieke and R. Trautner

Foreword

This volume collects papers presented, as invited and contributed talks or posters, at the workshop on "The NEW Rosetta targets. Observations, simulations and instrument performances", which was held in Capri on October 13-15,2003. More than 100 scientists covering different fields, such as optical and radio astronomy, laboratory experiments and modelling of comet physics and processes, as well as several Principal Investigators of the instruments on board Rosetta, participated to this highly interdisciplinary workshop.

The Rosetta mission was programmed for launch in January 2003 towards the short period comet 46P/Wirtanen and the asteroids 140 Siwa and 4979 Otawara. However, due to problems with the Ariane V launcher, the launch was postponed and the European Space Agency had to identify new mission targets, suitable for a launch window at the end of February 2004. The short period comet 67P/Churyumov-Gerasimenko was chosen as primary target for the new baseline, together with one or two asteroids to be selected, depending on the Δv available after the Rosetta probe interplanetary orbit insertion. After the successful launch of the mission on March 2nd, 2004, the new baseline foresees a double fly-by with asteroids 21 Lutetia and 2867 Steins, on the way towards the rendezvous with the primary target, that will be reached in 2014.

The papers included in this volume cover different complementary fields: observations of the new Rosetta targets, laboratory experiments, theoretical simulation of comet environment and processes, performances of the experiments on board Rosetta spacecraft, also in view of the new mission target(s).

After the postponement of the mission launch, comet 67P/Churyumov-Gerasimenko was the subject of an intense observation campaign, between February and June 2003. A dedicated monitoring was conducted from optical to radio wavelengths, in order to determine characteristics of nucleus, coma and tails. The main result obtained by the observations is that the comet is more active than the previous target, with jets and structures in the coma, and probably active fractions of its surface; all this provides an exciting scenario for the orbiter and lander experiments. The comet was intensely observed after its perihelion passage on August 2002, revealing significant asymmetries with respect to pre-perihelion observations during previous passages. Since Rosetta is planned to encounter the comet and start to follow it before perihelion, results presented in the workshop enhance the need to further improve our knowledge on this comet, being prepared to the next perihelion passage in 2009.

Continuing strong efforts are placed in laboratory experiments and theoretical models aimed at simulating cometary physical processes. The implication for the Rosetta mission is to describe the gas, dust and plasma environment to be afforded by the probe during the operation phases, also based on comparisons with the former mission target. Results presented at the workshop indicate, in particular, that the average coma environment of 67P/Churyumov-Gerasimenko is probably more benign for the Rosetta spacecraft than that of 46P/Wirtanen, but it could probably be more severe inside dust jets. Further efforts are needed to simulate at best the expected gas, dust and plasma environment.

The experiments on board Rosetta, both on the orbiter and on the lander, were designed and built in view of a rendezvous with the former target, 46P/Wirtanen. The selection of 67P/Churyumov-Gerasimenko determined the need to reconsider the instrument performances and the operation scenario in view of the new expected gas, dust and plasma environment in the coma and of the nucleus properties. The papers concerning orbiter instruments and lander analyse the compatibility with the new expected environment. The overall result is that the new comet is suitable to achieve the scientific objectives of the mission, i.e. the exploration of primitive bodies of the Solar System.

In conclusion, the papers collected in this volume provide a nice photograph of the different scientific and technical aspects connected to the Rosetta mission, up today. A 10-years wait divides us from the real operations around 67P/Churyumov-Gerasimenko. It is mandatory, however, that during this time astronomical observations, laboratory experiments and theoretical models dedicated to comets shall continue to progress. The aims are both to increase the capability to understand the physics and chemistry of comets and to better prepare for the "quantum jump" that a fruitful use of the Rosetta data shall provide.

L. Colangeli, E. Mazzotta Epifani, P. Palumbo
Napoli, Italy, April 2004

Acknowledgments

The success of the workshop, which was organised under the auspices of the Giunta Regionale della Campania, the Osservatorio Astronomico di Capodimonte and Università degli Studi di Napoli "Parthenope", has been possible thanks to the contributions that several sponsors have provided to the Local Organising Committee. The Research and Scientific Support Department of the European Space Agency has been invaluable in supporting most of the logistics expenses. The Scientific Directorate of the European Space Agency kindly provided financial support to allow young and extra-European scientists to participate to the workshop. ALENIA Spazio and ASI (Italian Space Agency) have also contributed, both financially and with man power, to the success of the conference. A special thank is due to the winery "Planeta", (Sicily). We thank the Osservatorio Astronomico di Capodimonte and Università Parthenope for the continuous help in man power and logistics provided before, during and after the workshop: a special and warm thank is due to Mimma Lauria and Maria Teresa Fulco for their great help during the workshop.

DISCOVERY, OBSERVATIONS AND INVESTIGATIONS OF COMET 67P/CHURYUMOV–GERASIMENKO IN KYIV

Klim Churyumov
Kyiv Shevchenko National University, Kyiv, Ukraine

Abstract   Rosetta, an European space vehicle 15 years in development, will head for the short period comet 67P/Churyumov-Gerasimenko in February, 2004. In September 1969 S.Gerasimenko and myself went to the Alma-Ata Astrophysical Institute to conduct a survey of short period and new comets. Later that month, I examined an exposure of comet 32P/Comas Sola made on September 11.92 UT, 1969, and found a cometary object near the center of the plate which I assumed was the expected short period comet 32P/Comas Sola. Later explorations at Kyiv University revealed that this comet's position was $1.8°$ from the predicated calculations of comet 32P. It was a new comet. The comet had an apparent magnitude of 13 and a faint tail about 1 arcmin in length at position angle 280 degrees. On the basis of the observations of comet 67P obtained in Nizhny Arkhyz with the help of the 6- BTA reflector of the Special Astrophysical Observatory (SAO) of the Russian Academy of Sciences (RAS) some physical parameters of its comet plasma tail (coefficients of diffusion $D_{||}$, $D_{\perp}$ and induction of magnetic field B) were determined (Jan. 12.105, 1983 UT: $D_{||} = 5.07 \times 10^{14} \div 1.21 \times 10^{15}$ cm^2/s, $D_{\perp} = 5.73 \times 10^{13} \div 1.37 \times 10^{14}$ cm^2/s, B=46$\div$111 nT; Jan. 13.124, 1983 UT: $D_{||} = 4.67 \times 10^{14} \div 1.14 \times 10^{15}$ cm^2/s, $D_{\perp} = 4.30 \times 10^{13} \div 1.05 \times 10^{14}$ cm^2/s, B=55$\div$134 nT). Other results of exploration of comet 67P are discussed.

Keywords: Rosetta, comet 67P/Churyumov-Gerasimenko, discovery, orbit, evolution, activity, plasma tail, magnetic field, ROMAP

1.   Discovery and the first observations of comet 67P/Churyumov-Gerasimenko in 1969-1970

During August and September 1969 the author took part in the third Kyiv University astronomical expedition to the Alma-Ata Astrophysical Institute (Churyumov and Gerasimenko, 1972) together with post graduated student Svetlana Ivanivna Gerasimenko and laboratory assistant Chirkova Lyudmila Mykolayivna.   The purpose of the expedition was to carry out visual and photographic searches for new comets in the morning and evening Everhart zones and also to make photographic observations of the well-known short-

1

L. Colangeli et al. (eds.), The New ROSETTA Targets, 1–13.
© 2004 *Kluwer Academic Publishers. Printed in the Netherlands.*

Table 1.   Orbital parameters of comet 67P/Churyumov-Gerasimenko in 1969

T = 1969 Sept. 11.03770 ET	Epoch = 1969 Sept. 16.0 ET
$\omega_{(1950.0)} = 11.°20017 \pm 0.00024$	e=0.633018±0.000000
$\Omega_{(1950.0)} = 50.35917 \pm 0.00000$	a=3.501454 ± 0.000001 A.U.
$i_{(1950.0)} = 7.14565 \pm 0.00007$	n^o=0.150428986±0.00000004
$q = 1.284970 A.U.$	P=6.552 y.

period comets 4P/Faye (1969 VI), 32P/Comas Sola (1969 VIII), 45/PHonda-Mrkos-Pajdusakova (1969 V) and the two new comets Kohoutek (C/1969 O1-A =1969b) and Fujikawa (C/1969 P1 = 1969 VIII). The observations were made with a 50-cm $f/2.4$ Maksutov telescope and a 17-cm $f/1$ Schmidt camera. Altogether we took about 100 plates suitable for integral photometry and the determination of exact positions of the above-mentioned comets.

Still in Alma-Ata, I noted on September 20 a cometary object of magnitude 13 on a plate taken on September 11 for P/Comas Sola. Back in Kyiv on October 22 Svetlana Gerasimenko and myself found that the object was 1.8° from the position of P/Comas Sola given in the ephemeris. Then we saw P/Comas Sola close to its ephemeris position, suggesting that the object we had noted was a new comet. Examination of other plates for P/Comas Sola - two on September 9 (Fig. 1) and two on September 21 - immediately revealed the new object, and although it was near the edge, it still had its cometary appearance and showed motion among the stars. Professor Sergey Vsekhsvyatskij cabled news of our discovery to the IAU Central Bureau for Astronomical Telegrams. The new comet was given the preliminary designation 1969h, and later the final designation C/1969 R1. Now it has constant designation 67P in the Catalogue of cometary orbits (Marsden and Williams, 1999).

On the basis of the first exact positions, reduced by N. A. Shmakova (Leningrad) from our measurements, Brian Marsden (Marsden, 1969) calculated six ephemerides from two parabolic and four elliptical orbits. On October 31, 1969 the new comet was photographed by Scovil (U.S.A.). This observation, and then observations by T. Seki (Japan), E. Roemer (U.S.A.), and B. Milet (France), closely confirmed one of the elliptical orbits. Thus the new comet proved to be one of members of the Jupiter family of short-period comets. Later on the orbital parameters of the elliptical orbit of comet C/1969 R1 were determined (Belyaev and Churyumov, 1982a). They are listed in Table 1.

In Table 2 the results of exploration of evolution of the comet orbit during the time interval 1800-2000 (Belyaev and Churyumov, 1982b) are given. Data in Table 2 show that the comet has an interesting unusual history of its orbital motion. Up to 1840 its perihelion distance was at 2.2 AU and the comet was then unobservable from Earth. That year there was an encounter with Jupiter

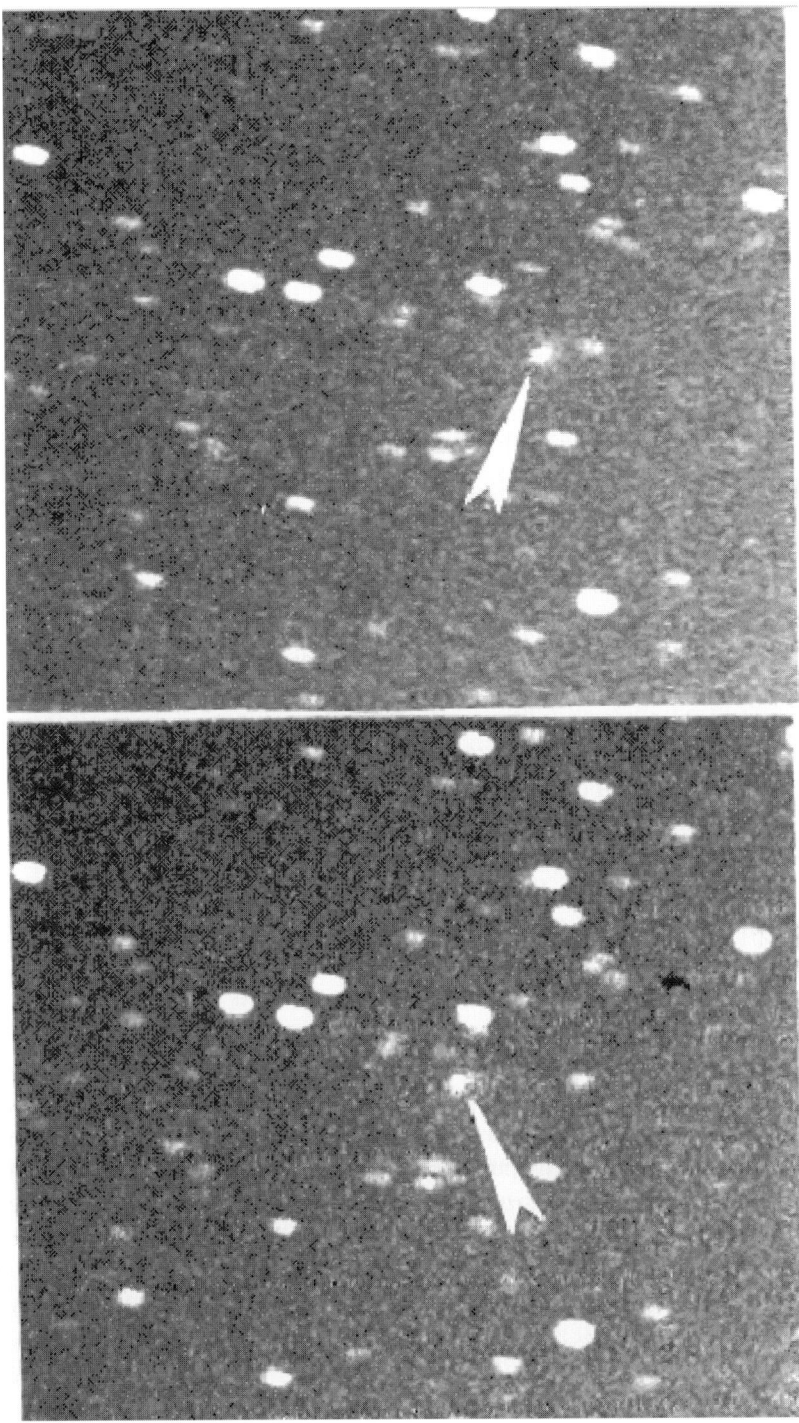

Figure 1. Discovery of comet 67P/Churyumov-Gerasimenko on a couple of plates, obtained on Sept.9, 1969

Table 2. Evolution of the comet 67P/Churyumov-Gerasimenko orbit during 1800-2000. T_s is the date of comet encounter with Jupiter, Δ_{min} is the distance from Jupiter, π is the longitude of perihelion, Ω is the longitude of ascending node, i is the inclination of orbit, e is the eccentricity, q is the perihelion distance, Q is the aphelion distance, P is the period of revolution around the Sun.

T_s	$\Delta_{min}(AU)$	π	Ω	i	e	$q(AU)$	$Q(AU)$	P (y)
1800 Jan. 25		77°	57°	25°	0.43	2.21	5.49	7.56
1817 Mar. 10	1.48							
1825 June. 23		77°	57°	25°	0.41	2.29	5.53	7.74
1840 Sept. 17	0.29							
1850 Aug. 31		88°	55°	24°	0.35	2.81	5.82	8.97
1875 Apr. 22		88°	55°	24°	0.35	2.80	5.85	8.99
1876 Mar. 10	1.09							
1900 Apr. 11		90°	54°	23°	0.34	2.93	5.94	9.35
1923 Oct. 9	0.96							
1925 Sept. 7		85°	53°	23°	0.37	2.72	5.89	8.92
1928 Apr. 27	1.86							
1950 Nov. 15		84°	52°	23°	0.36	2.75	5.90	9.01
1959 Feb. 4.3	0.052							
1969 Sept. 16		62°	50°	7°	0.63	1.28	5.72	6.55
1975 July 7		62°	50°	7°	0.63	1.30	5.73	6.59
2000 Jan. 17		62°	50°	7°	0.63	1.29	5.73	6.58

at 0.29 AU and the orbit shifted outwards to a perihelion distance of 2.9 AU. From there the perihelion distance slowly decreased further to 2.75 AU, from which a close Jupiter encounter moved the comet into an orbit with perihelion at just 1.28 AU.

The most remarkable peculiarities of orbital evolution of 67P is this very close encounter of the comet with Jupiter on Feb. 4.3, 1959 which was characterized by a Δ_{min} = 0.052 AU. Thanks to this encounter all the orbital elements were essentially changed and the comet could be discovered with the help of terrestrial telescopes, through 1.5 revolution after this encounter. From the beginning of November 1969 scientists from the Astronomy Department of Kyiv University of Taras Shevchenko (Vsekhsvyatskij, Gerasimenko, Afanasiev) systematically observed comet C/1969 R1 at mountain observatories (Alma-Ata, Byurakan). The purpose of these observations was to obtain a continuous photometric series (which is very important in the case of periodic comets) to determine integral magnitudes of the comet, to study the comet's physical structure and to obtain accurate positions. During this period visibility conditions made it possible to photograph the comet at low zenith distances (less than 25 ˚), which is essential for the photography of faint, extended objects. Most of the material suitable for photometric work was obtained with the 20-cm and 50-cm Schmidt cameras at the Byurakan Astrophysical Observatory (observers Vsekhsvyatskij, Afanasiev) and the 50-cm Maksutov telescope at the Alma-Ata Astrophysical Institute (observers Afanasiev, Gerasimenko, Churyumov). During September 1969-March 1970 more than 50 plates were taken on 34 nights. About 45 plates were selected to determine nuclear magnitudes of the comet. Integral photographic magnitudes were obtained from nine plates taken at times of good atmospheric transparency and giving reliable extra focal standards.

On most of the plates the comet had a strongly pronounced nucleus 0.06 mm (50-cm Schmidt) and 0.10 mm (50-cm Maksutov telescope) in diameter. For our microphotometric measurements we used a circular diaphragm of diameter 0.12 mm (corresponding to 1 to 2×10^4 km at the comet). We obtained extra focal exposures of star clusters (Coma Berenices, the Pleiades, NGC 2632, NGC 1647, NGC 1628, NGC 2264) and the North Polar Sequence as standard stars reference.

From Kodak O-aO and ORWO ZU-2 plates (without filters) we reduced the $B-$magnitudes of the standard clusters to the international photographic system in the following way:
$$m_{pg} = -0.18 + 1.09(B - V) + V.$$
On the basis of photographic observations made during 30 nights (from Sept. 10, 1969 to Mar. 6, 1970) we determined m_{pg} for the photometric nucleus of comet C/1969 R1 and also the magnitudes h_{10} reduced to unit distances from the Earth and Sun (Churyumov and Gerasimenko, 1972).

Analysis of obtained data shows that change of the photometric nucleus brightness of comet C/1969 R1 have a tendency for a 27-day recurrence period in the comet outbursts: Sept. 22-Nov. 16 (55 days), Nov. 16-Dec. 14 (29 days). Dec. 14-Feb. 7 (54 days), Feb. 7-Mar. 5 (27 days). We also note a 27-day recurrence period in the minima: Nov. 18, Dec. 15, Feb. 8.

From nine plates we have determined the absolute integral magnitude and the photometric parameters H_y and n which were found by least squares:
$$H_y = 11.91 \pm 0.54, n = 4.0 \pm 0.8.$$
The large residuals are an indication of the high activity in the comet. Over the entire period of our observations the comet had a narrow, straight tail, probably of type I. Its length ranged from 1' to 11'. On November 16 Burnasheva (Crimean Astrophysical Observatory) noted that there was a fan-like tail. Some characteristics related to the structure of the comet are listed in Table 3. The tail axis deviated from the prolonged radius vector by up to $\pm$ 20°, which is probably a result for strong interaction between the comet tail plasma and the solar wind.

On November 16 a spectrogram of the comet was obtained by Lipovetskij with the 100-cm Schmidt camera (1.5° prism) at Byurakan. Since the dispersion was very low (1800 Å/mm at H_γ), only the continuous spectrum of the head and tail was noted. The emissions of CN, C_2 and apparently C_3 are visible on a contact print. Only a single emission becomes visible in the tail (perhaps CO^+).

Table 3. Coma diameter (*d*), position angle (*p.a.*) and length (*s*) of the tail

1969/1970 UT	*d*	*p.a.*	*s*
Nov. 16.90	1.52'	298.24°	2.4'
Nov. 16.93	1.43'	300.90°	2.4'
Nov. 16.95	0.99'	297.05°	6.3'
Nov. 17.91	1.72'	295.8°	10.6'
Nov. 17.93	1.28'	297.24°	3.5'
Nov. 19.08	1.11'	292.20°	5.5'
Dec. 6.09	0.55'	306.54°	4.5'
Dec. 6.98	1.15'	288.70°	4.1'
Dec. 7.99	1.46'	301.01°	9.4'
Dec. 8.11	1.02'	300.37°	11.2'
Dec. 10.05	0.60'	301.02°	1.7'
Dec. 11.08	0.42'	295.50°	7.4'
Dec. 16.08	1.48'	303.89°	3.4'
Dec. 19.08	0.65'	305.37°	3.8'
Feb. 6.86	1.87'	294.92°	6.0'
Feb. 7.94	2.03'	297.40°	6.3'
Feb. 8.91	2.18'	300.66°	7.2'
Feb. 12.92	1.48'	279.04°	5.1'
Mar. 1.83	0.24'	293.62°	4.2'

The comet has been seen in 1969/70, 1976, 1982/83, 1989, 1996 and 2002/ 2003. The comet is unusually active for a short period object and has a coma and often narrow tail even far from perihelion, which are almost certainly a result of the comet's big decrease in perihelion distance. During the 2002/2003 apparition the tail has been as long as 10 arcminutes, with a stellar central condensation in a faint extended coma. Even 7 months after perihelion the tail continues to be very well developed.

The substantial variations in the nuclear and integral head magnitudes of comet C/1969 R1, the structural changes in the head, and the presence of the long tail are all evidences of a high cometary activity.

2. Connection of brightness variations of 67P/Churyumov-Gerasimenko with the solar activity in 1982-1983

During the comet's apparition in 1982/1981 many observation data (more than 330 magnitude estimates) were obtained. This allowed to construct a detailed comet's light curve and to thoroughly study its peculiarities. The comet was active, and during the six months observations (late August, 1982- early March, 1983) it had numerous outbursts (more than 16) and variations of its integral magnitudes. At that time the mutual space dislocation of the Earth, the Sun and the comet during the whole period of observations was exceptionally favorable from the point of view of the study of the solar activity influence upon the comet's light curve. A temporal shift preconditioned by the difference in the heliographic longitudes of the Earth and the comet did not exceed 2.3 of a day, and during October 25, 1982 – February 5, 1983 it was less than one day. The solar activity, visible from the Earth, insubstantially changed when observed from the comet. The solar activity during the observation period was situated on the decrease branch of the 11-year solar activity cycle when most stable high speed corpuscular streams were being formed. The comparison of comet 67P/Churyumov-Gerasimenko's light curve and the curve of the solar activity indices changes that are reduced to the comet's center shows that the variations of the comet's brightness correlate rather well with the changes of the solar indices.

A correlation coefficient R_{mW} between the heliocentric magnitude m_Δ of the comet and the Wolf number W is calculated by the formula:

$$R_{mW} = \frac{N_{mW} N - N_W N_m}{\sqrt{\bar{N}_W \bar{N}_{\bar{W}} \bar{N}_m \bar{N}_{\bar{m}}}}$$

where N_{mW} is the number of intervals with the maxima of W and with the maxima of m_Δ, N_W is the number of intervals with the maxima of W and without the maxima of m_Δ, N_m is the number of intervals with the maxima of m_Δ and without the maxima of W, N is number of intervals without the maxima of W and m_Δ , $\bar{N}_W$ is tha total number of intervals with the maxima of W, $\bar{N}_{\bar{W}}$ is the total number of intervals without the maxima of W, $\bar{N}_m$ is the number of intervals with the maxima of m_Δ, $\bar{N}_{\bar{m}}$ is the number of intervals without the maxima of m_Δ.

Also a coupling coefficient K was calculated by the formula:

$$K_{AB} = \frac{m/n - p}{m/n + p - 2m/np}$$

where m is the number of coincidence of the values A and B, p is the probability of an occasional coincidence of values A and B, n is the total number of intervals on which the event duration (T) is divided.

Values of R, K and D (probability of an occasional apparition of the value k not equal to 0) are given in Table 4. It is evident that the correlation between the comet's brightness maxima and those of solar activity indices is statistically meaningful. It indicates that the growing spot total area increases the probability of the growth in the comet's brightness, but an increase in m_Δ may not be accompanied by an increase of S. The analogous connection with the Wolf numbers is less obvious and the maximum of m_Δ is accompanied by the maximum of W with more probability than the maximum of W is by the brightness maximum.

Table 4. Values of R (correlation coefficient), K (coupling coefficient) and D (probability of an occasional apparition of the value k not equal to 0. m_Δ is the total heliocentric magnitude of the comet, W is Wolf number, S is the total spot area

Comparable values	R	K	D
m_Δ, W	0.51 ± 0.09	0.72	4.1
W, m_Δ		0.58	3.3
m_Δ, S	0.51 ± 0.09	0.46	2.5
S, m_Δ		0.79	4.6

3.   Parameters of the magnetic field of the comet 67P/Churyumov-Gerasimenko (1982 VIII) plasma tail

The original plates with the image of comet P/Churyumov-Gerasimenko were obtained by I.D.Karachentsev and K.I.Churyumov on January 12.105 UT and 13.124 UT, 1983 with a 6-m telescope BTA of the Special Astrophysical Observatory of Russia's Academy of Sciences at Mount Pastukhov. These plates are the first observations of a comet with one of the biggest ground based telescope. The BTA is located at the gorge of Seven Brooks at the height of 2070 meters. The main mirror diameter is 6.05 meters, the focal distance 24 meters. The Ritchy cassette with a corrector and a field lens was used. The pictures of the comet Churyumov-Gerasimenko were obtained on emulsion Kodak IIaO with hypersensitivity in H_2. For quantitative estimation of some physical parameters of the plasma tail of comet Churyumov-Gerasimenko the diffusion model (Nazarchuk and Shul'man,1968) was used.

The diffusion model permits to analyze the photometric structure of cometary plasma tails. The diffusion model is based on the following assumptions: 1) the cometary nucleus with its surroundings is considered a point source of matter. This assumption is acceptable because of the small size of a cometary nucleus as compared with its tail; 2) the Green function for an instantaneous source is used. It is assumed that the center of mass of any instantaneously emitted package of particles moves with an uniform acceleration along the axis of the comet's tail, the outflow of matter has begun infinitely long ago, and the source power C is constant. A cometary ion is also considered to acquire random momenta from inhomogeneities of self-consistent fields which move through the tail; i.e., the process of the interaction between cometary ions and the solar wind is assumed to be macrostochastic. Then the motion of a cometary ion is a superposition of diffusion and transport to the tail. In this case, the Green function has the form of an anisotropic function for exponentially disappearing particles:

$$G = \frac{1}{4\pi t^* \sqrt{D_{||}^* D_\perp}} \times \exp\left(-\frac{\left(x - a\,(t^*)^2/2\right)^2}{4D_{||}^* t^*} - \frac{y^2}{4D_\perp t^*} - \frac{t^*}{\tau}\right) \tag{1}$$

where t^* is the age of the particle package, $D_{||}^* = D_{||}\cos^2\beta + D_\perp \sin^2\beta$ is the coefficient of diffusion along the comet's tail in the plane of the sky, $D_{||}$ and $D_\perp$ are the coefficients of longitudinal and transversal diffusion, respectively, τ is the mean lifetime of glowing particles, and β is the angle between the tail axis and its projection onto the plane of the sky. Here, the x-axis is directed along the comet's tail (antisunward), and the y-axis is perpendicular to it. The chemical composition of the glowing matter is assumed to be uniform in the tail. Then the surface brightness I is proportional to the surface density $n(x, y)$ of the glowing particles

$I=kn(x,y)$

Here

$$n\,(x, y) = \frac{C}{4\pi\sqrt{D_{||}^* D_\perp}} \times \int_0^\infty \exp\left(-\frac{\left(x - \frac{at^2}{2}\right)^2}{4D_{||}^* t^*} - \frac{y^2}{4D_\perp t^*} - \frac{t^*}{\tau}\right) \frac{dt}{t} \tag{2}$$

For convenience, we introduce the dimensionless coordinates

$$X = \frac{x}{L_{||}}, \quad Y = \frac{y}{L_\perp} \tag{3}$$

and the dimensionless quantity

$$\theta = \frac{t}{2\tau}. \tag{4}$$

Here, the longitudinal $L_{||}$ and the transverse $L_\perp$ scale are chosen as follows:

$$L_{||} = 2\sqrt{D_{||}^* \tau}, \quad L_\perp = 2\sqrt{D_\perp \tau} \tag{5}$$

We also introduce a dimensionless parameter Γ that determines the acceleration a:

$$\Gamma = a\sqrt{\frac{\tau^3}{D_{||}^*}}\cos\beta. \tag{6}$$

Then the surface brightness can be represented as (Nazarchuk , 1969):

$$I(X,Y) = k \frac{C}{4\pi \sqrt{D_{\parallel}^* D_{\perp}}} \times \int\limits_0^\infty \exp\left(-\frac{(X - \Gamma\theta^2)^2 + Y^2}{\theta} - \theta\right) \frac{d\theta}{\theta}. \quad (7)$$

As seen from Eq. (7), the surface brightness depends on one significant parameter Γ and three scale factors $L_{\parallel}$, $L_{\perp}$ and $k\frac{C}{4\pi\sqrt{D_{\parallel}^* D_{\perp}}}$. The problem is in the determination of these values from the surface-brightness distribution known from observations. As the diffusion model is used, the longitudinal and transverse photometric cross sections of the cometary plasma tail are fitted with those calculated theoretically through the selection of the above parameters. Eq. (7) for the surface brightness contains an improper integral, whose numerical computation requires a large volume of computation. This paper presents the results of a study of the comet 67P/Churyumov-Gerasimenko plasma tail for January 12 and 13, 1983, obtained by comparison with the improved theoretical surface-brightness values. With the use of the logarithmic, the theoretical law of the brightness decrease can be represented as the sum of a constant and a family of curves:

$$-2.5 \log I = const - 2.5 \log \Phi(X, Y, \Gamma) \quad (8)$$

where

$$\Phi = \int\limits_0^\infty \exp\left(-\frac{(X - \Gamma\theta^2)^2 + Y^2}{\theta} - \theta\right) \frac{d\theta}{\theta}$$

Relative photometry was used to compare the observed cross sections and those calculated theoretically from the diffusion model. Although absolute photometry and reliable spectral data are needed for transition from the surface brightness to the surface density, relative photometry is sufficient to determine the parameters Γ, $L_{\parallel}$ and $L_{\perp}$; they can be obtained without knowing absolute values of the surface density of particles (Nazarchuk and Shul'man, 1968). Photometry of the comet's tail was performed along its axis and in perpendicular directions (several cross sections). The magnetic induction was estimated by the formula:

$$B = 2 \times 10^{11} \frac{T}{D_{\parallel}} L_{\parallel} / (L_{\perp} \cos\beta) [nT] \quad (9)$$

The temperature of the cometary plasma was assumed to be 5×10^5 and 2×10^6 K for the minimum and the maximum estimate of the magnetic induction, respectively (Wegmann, et al., 1987; Galeev, 1989).

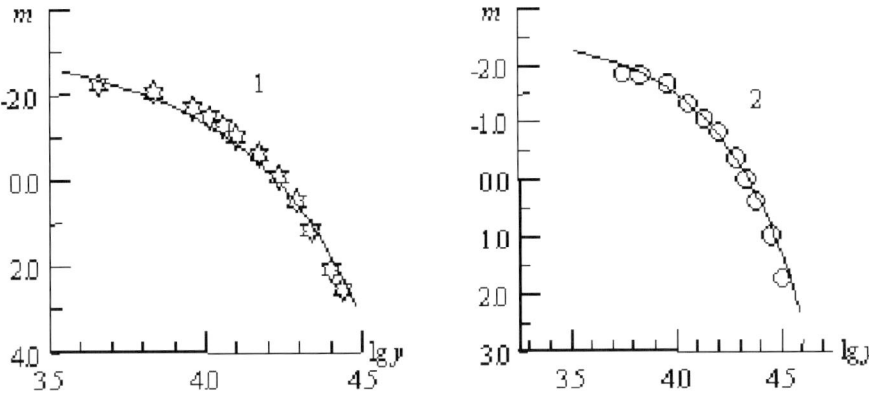

Figure 2. Two cross photometric profiles of the plasma tail of comet 67P/Churyumov-Gerasimenko on Jan. 12,1983

Table 5. Parameters of the diffusion model for the plasma tail of comet 67P/Churyumov-Gerasimenko. Γ is the dimensionless parameter determining the acceleration a; $L_{||}$ and $L_{\perp}$ are the longitudinal and transvers scale, respectively.

Date, UT	Γ	$L_{||}$ (km)	$L_{\perp}$ (km)
1983, Jan. 12.105	14	$2.2 \cdot 10^5$	$2.21 \cdot 10^4$
1983, Jan. 13.124	15	$2.1 \cdot 10^5$	$1.48 \cdot 10^4$

Fig. 2 shows two cross photometric profile of the plasma tail of comet P/Churyumov-Gerasimenko on Jan. 12,1983. Here y is the distance of cross profile from the tail axis (km). Fig. 2 also shows respective theoretical profiles (solid curves), which have been selected from the theoretical families of curves $\Phi(logx, \Gamma)$ and $\Phi (x_i, logy, \Gamma)$, where x_i are the dimensionless distances of cross photometric profiles from the nucleus and Γ is the dimensionless parameter defined in Eq. 6.

Comparison of the theoretical and observed photometric profiles made it possible to determine parameters of the diffusion model ($L_{||} = 2\sqrt{D_{||}\tau}$ and $L_{\perp} = 2\sqrt{D_{\perp}\tau}$ are longitudinal and cross characteristic scales) for plasma tails of P/Churyumov-Gerasimenko comet. Their values are given in the Table 5.

Estimates of the physical parameters are given in Table 6. Changes of magnetic induction B in the plasma tails of comet 67P during Jan. 10-16, 1983 are shown in Fig.3. These changes of B were approximated by the formula derived by Natalia Shabas for the plasma tails of comet 1P/Halley (Shabas, 1999):

12

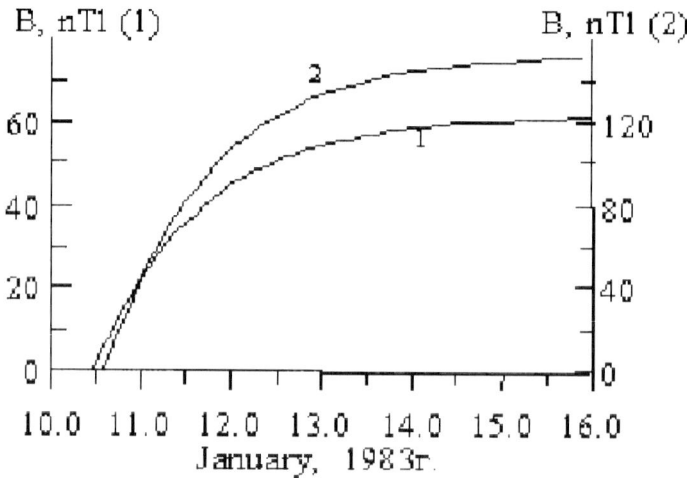

Figure 3.   Two models of change of magnetic field in the plasma tail of comet
67P/Churyumov-Gerasimenko from 10 to 16 Jan.,1983

Table 6.   Physical parameters of the plasma tail of comet 67P/Churyumov-Gerasimenko. $D_{||}$
and $D_{\perp}$ are the longitudinal and transverse diffusion coefficients. B is the magnetic induction

Date, UT	$D_{||}$ (cm/s)	$D_{\perp}$ (cm/s)	B (nTl)
1983, Jan.12.105	5.07×10^{14}	5.73×10^{13}	$46 \div 111$
1983, Jan.12.105	1.21×10^{15}	1.37×10^{14}	
1983, Jan.13.124	4.67×10^{14}	4.30×10^{13}	$55 \div 134$
1983, Jan.13.124	1.14×10^{15}	1.05×10^{14}	

$$B = B_f \left(1 - e^{-\Delta t/\tau^*}\right) \tag{10}$$

The obtained upper estimates of induction of the magnetic field B≅111 nT
for Jan. 12,1983 and B≅134 nT for Jan. 13,1983 probably exceed real values
of B in the cometary plasma tail. However good coincidence of the theoretical
and observed data seems to be proof of the plasma nature of the comet tail in
question. Moreover, the comet tail looks rather narrow and straight without a
noticeable expansion that may be a proof in favor of the high magnetic field
that keeps the cometary plasma in a narrow cylinder. The tail shape that practically did not change during the day makes it more probable to consider this tail
to be a strongly magnetized plasma jet. I think that this peculiarity of magnetic
fields in plasma tail of comet 67P is tight connected with the magnetic properties of the surface layers of the cometary nucleus. I hope that this problem
will be successful solved with the help of the device ROMAP installed on the
ROSETTA Lander when it will land on the comet 67P nucleus in 2014.

References

Belyaev, N.A., Churyumov, K.I.(1982a). Unit of three apparitions of periodic comet Churyumov-Gerasimenko. *Kyiv Kometny Tsircular*, 296:2.

Belyaev, N.A., Churyumov, K.I.(1982b). Evolution of orbit of short period comet Churyumov-Gerasimenko. *Kyiv Kometny Tsircular*, 296:3.

Churyumov, K.I. and Gerasimenko, S.I.(1972). Physical observations of the short-period comet 1969 IV. *The motion, evolution of orbits, and origin of comets.*, 27–34. Proc. of Symposium 45 of the IAU. Chebotarev et al. (eds.)

Galeev, A.A. (1989). Plasma processes in the outer coma. *Astroph.and Space Sci. Library.*, 167:1145–1170.

Marsden, B.G. (1969). Comet Churyumov–Gerasimenko (1969) *IAU Circ.* Nos 2181, 2187.

Marsden, B.G. and Williams G.V. (1999). Catalogue of cometary orbits. *13th ed.*, 127

Nazarchuk, G.K., Shul'man L.M. (1968). Diffusion model of cometary tails. *Problemy kosmicheskoy fiziki*, 3:11-16

Nazarchuk, G.K., (1969). Analysis of surface-brightness distribution in the tail of comet 1956h. *Physics of comets*, Kyiv, Naukova Dumka publishing, 77–99

Shabas N.L. (1999). Physical parameters of the plasma tail of comet 67P/Churyumov–Gerasimenko. *Visnyk of Kyiv Univ. Astronomy*, 35:80-83

Wegmann R., Schmidt H.U., Hubner W.F., Boice D.C. (1987). Cometary MHD and Chemistry. *Astron. Astrophys.*, 187:339–350.

MONITORING COMET 67P/CHURYUMOV–GERASIMENKO FROM ESO IN 2003

Rita Schulz
ESA Research and Scientific Support Department, ESTEC, Noordwijk, The Netherlands
Rita.Schulz@rssd.esa.int

Joachim A. Stüwe
Leiden Observatory, Leiden, The Netherlands
stuwe@strw.leidenuniv.nl

Hermann Böhnhardt
MPI for Astronomy, Heidelberg, Germany
hboehnha@mpia-hd.mpg.de

Abstract   A new suitable target comet had to be found for the Rosetta mission after its launch planned for mid-January 2003 was cancelled.   As soon as 67P/Churyumov-Gerasimenko became a likely new target for Rosetta, an extensive monitoring of this comet started from ESO with the first observation on 11 February 2003. Therefore, a postperihelion characterization of this comet is now existing, which covers its evolution along the orbit between 2.29 AU and 3.22 AU. Broad-band BVRI images and low resolution long-slit spectra were obtained for morphological, colour and compositional analysis of the coma and for studying the comet's activity along the orbit. Various observational techniques have been used to collect data suitable for the analysis of long-term as well as short-term variability, from which conclusions on several nucleus properties can be drawn. Here an overview of the existing observations is given and the first results of the analysis of some of the data are presented.

Keywords: Comets, Rosetta, 67P/Churyumov-Gerasimenko

Introduction

The Rosetta mission was originally scheduled to launch to comet 46P/Wirtanen in January 2003. However, as the launch had to be delayed owing to unforeseen problems with the launch vehicle, the original launch window closed and alternative mission opportunities had to be found. Comet 67P/Churyumov-

15

L. Colangeli et al. (eds.), The New ROSETTA Targets, 15–24.
© 2004 *Kluwer Academic Publishers. Printed in the Netherlands.*

Gerasimenko has now been identified as the new Rosetta target and the mission
has been scheduled for launch in February 2004. However, before this decision
could be taken, a number of basic properties of the new target comet had to be
determined to ensure that the rendezvous and the landing on the nucleus will
be successful. The design of the scientific instruments on board the Orbiter
as well as the design of the landing scenario had been optimized to the characteristics of comet 46P/Wirtanen, as determined from ground. Consequently,
the properties of the new target comet, particularly its size and activity, had
to be within the range that is required to ensure the feasiblity of the mission
and an optimal scientific return. Comet 67P/Churyumov-Gerasimenko was
therefore extensively observed between February and June 2003. A first measurement of the nucleus size was obtained with HST in March 2003 (Lamy et
al. 2003). It resulted in a value for the effective radius of about 2 km (Lamy
et al. 2003), hence the nucleus of 67P/Churyumov-Gerasimenko is larger than
that of 46P/Wirtanen (0.6 km; Lamy et al. 1998; Boehnhardt et al. 2002).
In addition to the nucleus size, a number of other properties of 67P/Churyumov-
Gerasimenko had to be known to be able to decide whether this comet is
indeed a suitable target for Rosetta. A dedicated monitoring was therefore
conducted with ESO telescopes, whereby observational strategies were used
which allow to determine important characteristics of comet 67P/Churyumov-
Gerasimenko.

1.   Observations

Comet 67P/Churyumov-Gerasimenko was monitored from the European
Southern Observatory between 11 February 2003 and 26 June 2003. The comet
was observed postperihelion while moving outbound at heliocentric distances
between 2.295 AU and 3.224 AU. We obtained broad-band images in B,V,R,I
and low resolution long-slit spectra to carry out a morphological and compositional coma analysis as well as to derive a number of basic nucleus properties.
Out to r = 2.6 AU, the 3.6-m telescope on La Silla was used, whereas all observations beyond r = 2.8 AU were taken with the VLT (Antu) at Paranal. An
overview of the observations is given in Table 1.

We aimed at collecting data, which are suitable for the determination of
a number of properties of the target comet, namely, astrometric position, activity as a function of heliocentric distance, long- and short-term variability,
rotational properties of the nucleus, coma morphology and colour, production
rates of various gas coma species (CN, C_2, C_3, NH_2) and Afρ values of the
dust coma. Various observational strategies have therefore been used. The exposure times and sequence, in which images and spectra were obatined, were
adjusted such that the requirements for the individual tasks were fulfilled (e.g.

Table 1. Overview of Observations

Date (2003)	Telescope/ Instrument	Observations	r (AU)	Δ (AU)	phase °
11 Feb	3.6-m EFOSC2	BVRI images spectra	2.30	1.40	13.4
09 Mar	3.6-m EFOSC2	BVRI images spectra	2.49	1.51	4.4
30 Apr	VLT Antu FORS1	BVRI images spectra	2.86	2.24	17.9
01 May	VLT Antu FORS1	BVRI images spectra	2.86	2.24	18.0
02 May	VLT Antu FORS1	BVRI images spectra	2.87	2.26	18.1
03 May	VLT Antu FORS1	BVRI images spectra	2.87	2.27	18.3
04 May	VLT Antu FORS1	BVRI images spectra	2.88	2.29	18.4
05 May	VLT Antu FORS1	BVRI images spectra	2.89	2.31	18.5
03/04 Jun	VLT Antu FORS1	R images	3.08	2.89	19.2
19/20 Jun	VLT Antu FORS1	R images	3.18	3.21	18.3
23/26 Jun	VLT Antu FORS1	R images	3.21 -3.22	3.29-3.35	18.0-17.7

some observations needed special seeing conditions, and/or low airmass, while for other tasks special sequencing of the observations was required).

2.   Data Reduction

All reduction processes were carried out with the aid of ESO's *Munich Image Data Analysis System* MIDAS. We performed bias subtraction and flat fielding on all frames. The EFOSC2 data (i.e. the observations taken in February & March) were prepared using the standard MIDAS routines. In particular the reduction procedures of the MIDAS context "long" were applied to the spectrum frames. The FORS1 observations were reduced by sending them through the *FORS Pipeline* (Bogun, 2001), either at the home institute, for the observations obtained in visitors mode in May 2003, or "automatically" by the *VLT Dataflow system* (DFS 2001), in case of the service mode observations in June 2003. Wavelength calibration of all spectra was done via HeAr-frames. To facilitate a calibration of the broad-band images, appropriate fields from the list of Landoldt (1992) were observed at different airmasses each night. Because of the long slit of the EFOSC2 (5 arcmin, see Patat 1999) as well as the FORS1 (6.8 arcmin, see Szeifert, 2002) instrument, the sky contamination of the spectra could be determined from the spectrum frames themselves. The extinction correction was performed assuming standard atmospheric conditions described by Tüg (1977). For the flux calibrations observations of the spectrophotometric standard Feige 56 (EFOSC2) and Feige 66 & EG 274 (FORS1) taken on the same nights as the comet frames were used.

3.   Data Analysis

Image Processing

Comet 67P/Churyumov-Gerasimenko showed indications for the presence of distinct features already in the unprocessed images. To enhance these structures (and to search for weaker features) the two-dimensional coma profile was approximated by its azimutal average and this average was then subtracted from the observed brightness distribution (see e.g. Schulz, 1991, for a description of this and other structure enhancement methods). The applied enhancement technique also yields the azimutaly averaged (one-dimensional) coma profile as a by-product, which can be used for further analysis.

Spectra Processing

The flux calibrated spectra for each observational run, obtained with the same slit orientation, were co-aligned and an average spectrum as well as its r.m.s. was computed. From these spectra spatial profiles were extracted by averaging over certain wavelength bands. These profiles were used for an analysis as described in Kolokolova et al. (2003) and the results of this analysis will be published elsewhere. Furthermore the two dimensional spectra were added up along the slit over the innermost 5" to obtain the wavelength dependence of the cometary flux in the inner coma. To the resulting one dimensional spectra (of S/N > 100) a "reddened" spectrum of a solar analog star was fitted, while excluding regions of cometary emission, resulting in an approximation of the cometary continuum flux. Subtraction of this continuum from the measured flux distribution then yields the flux of the cometary emission. From these emission spectra production rates (or their upper limits, in case of non-detections) were computed using the *Vectorial Model* described by Festou (1981).

Distinct Coma Features

Distinct coma features were present in the BVRI images during all observational runs. Figure 1 exemplifies the evolution of the coma features between February and June 2003. Up to 4 distinct features can be identified in the structurally enhanced coma images. The comet exhibited an anti-tail during the observations in May and June 2003. The results of the first morphological analysis of the observations is presented in Boehnhardt et al. (2003).

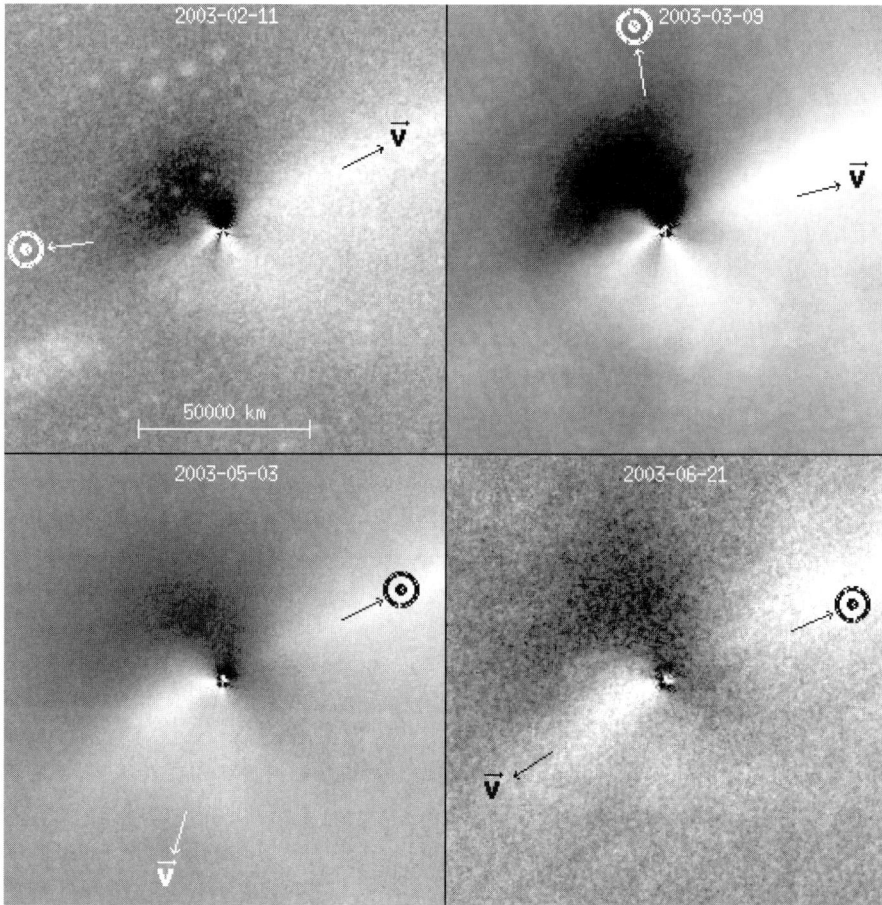

Figure 1. Structurally enhanced broad-band R images showing the evolution of the distinct features in the coma of comet 67P/Churyumov-Gerasimenko between February and June 2003. The projected direction of the Sun, ⊙ and of the movement of the comet $\vec{v}$ are indicated.

Coma Profiles

Figure 2 shows one-dimensional profiles that were constructed from the photometrically calibrated broad-band images by azimuthal averaging around the optocenter. For a steady-state dust coma these profiles should be straight lines in double logarithmic representation. Their slopes should range between m = -1 for a sherically symmetric coma and m = -1.5 representing the limiting case for distortion by solar radiation pressure (Jewitt and Meech, 1987). However, it is obvious from Figure 2 that the profiles cannot be sufficiently represented by a single straight line. To get a first estimate of the particle distribution in the coma and for a first crude analysis of the coma evolution with increasing heliocentric distance, r_h, we have therefore fitted straight lines to the profiles between 2000 km $\leq \rho \leq$ 10000 km, and derived an average slope for each of the four observing windows. These slopes are flatter than m = -1 already in February 2003 and become even flatter with increasing r_h. The average slopes are: -0.86 ± 0.02 (February), -0.75 ± 0.02 (March), -0.52 ± 0.01 (May), -0.44 ± 0.02 (June). The determined slopes are in disagreement with the fountain model. Further detailed analysis of the overall shape of the coma is therefore absolutely necessary.

Colour Determination

Colour indices of the coma were derived from the images obtained in B, V, R and I. The radial colour profiles of the coma are presented in Figure 3. They show constant values (within the error margins) as a function of the projected distance to the nucleus. Owing to the low brightness of the coma, the determined values have rather large uncertainties, particularly at larger distances to the nucleus. The coma became redder with increasing heliocentric distance which is expected because the gas production decreases.

Production Rates

The production rates of a number of gas species as well as of the dust (in $Af\rho$) have been determined from the spectrophotometric observations. The derived production rates and upper limits are given in Table 2. Comet 67P/Churyumov-Gerasimenko showed a major drop of activity between r_h = 2.5 AU and r_h = 2.9 AU. Between these two positions the CN production rate decreased by about one third while the $Af\rho$ value decreased by a factor of 2. When comparing the measurements obtained of comet 67P/Churyumov-Gerasimenko and comet 46P/Wirtanen at similar heliocentric distances it shows that before the major drop of activity occured, the $Af\rho$ value of 67P/Churyumov-Gerasimenko was about twice as high as that of comet 46P/Wirtanen, whereas it was comparably low at around 2.9 AU (see Schulz et al., 1998). Note that 67P/Churyumov-Gerasimenko was observed along its post-perihelion orbit, while 46P/Wirtanen was covered pre-perihelion. During their post-perihelion phase comets usually show higher activities than at the same pre-perihelion distance to the Sun. It is therefore likely that comet 67P/Churyumov-Gerasimenko will be of the similar activity as comet 46P/Wirtanen at around r_h = 3 AU pre-perihelion, i.e. at the time of landing on the nucleus.

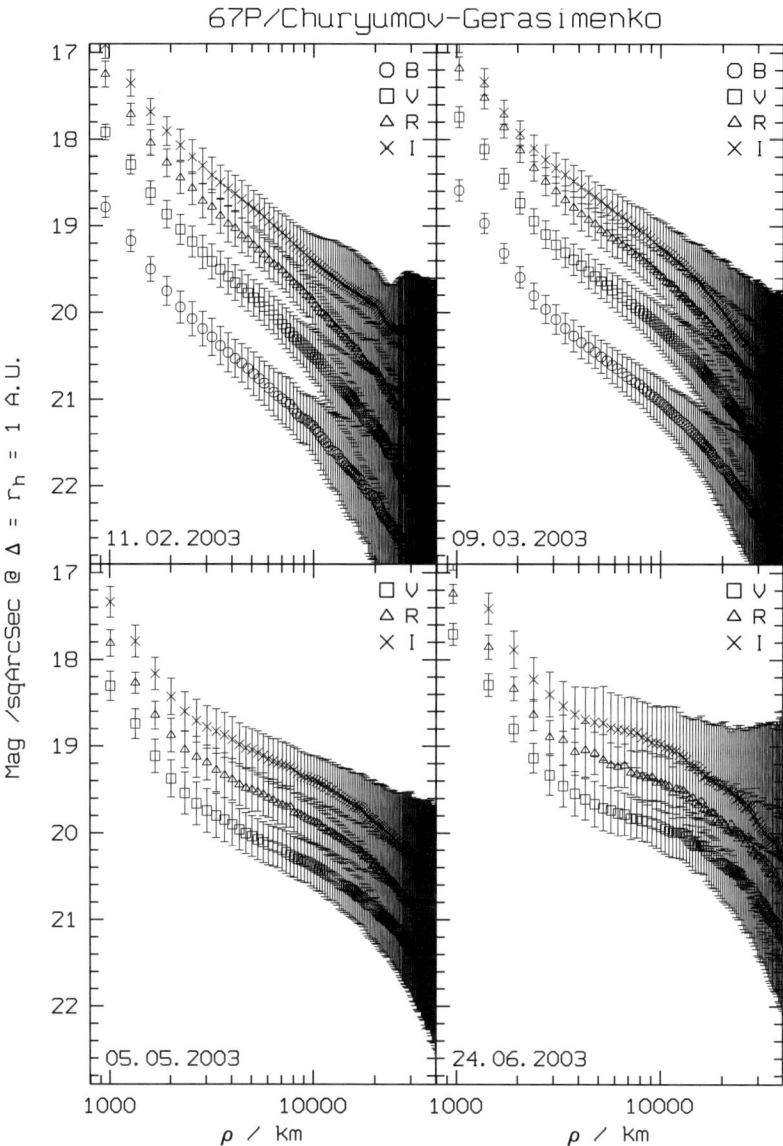

Figure 2. Azimuthally averaged surface brightness, Mag, as a function of projected nucleus distance, ρ. The brightness values were normalized to a geocentric and helioncentric distance of 1 AU, respectively. The error bars indicate 1 σ errors.

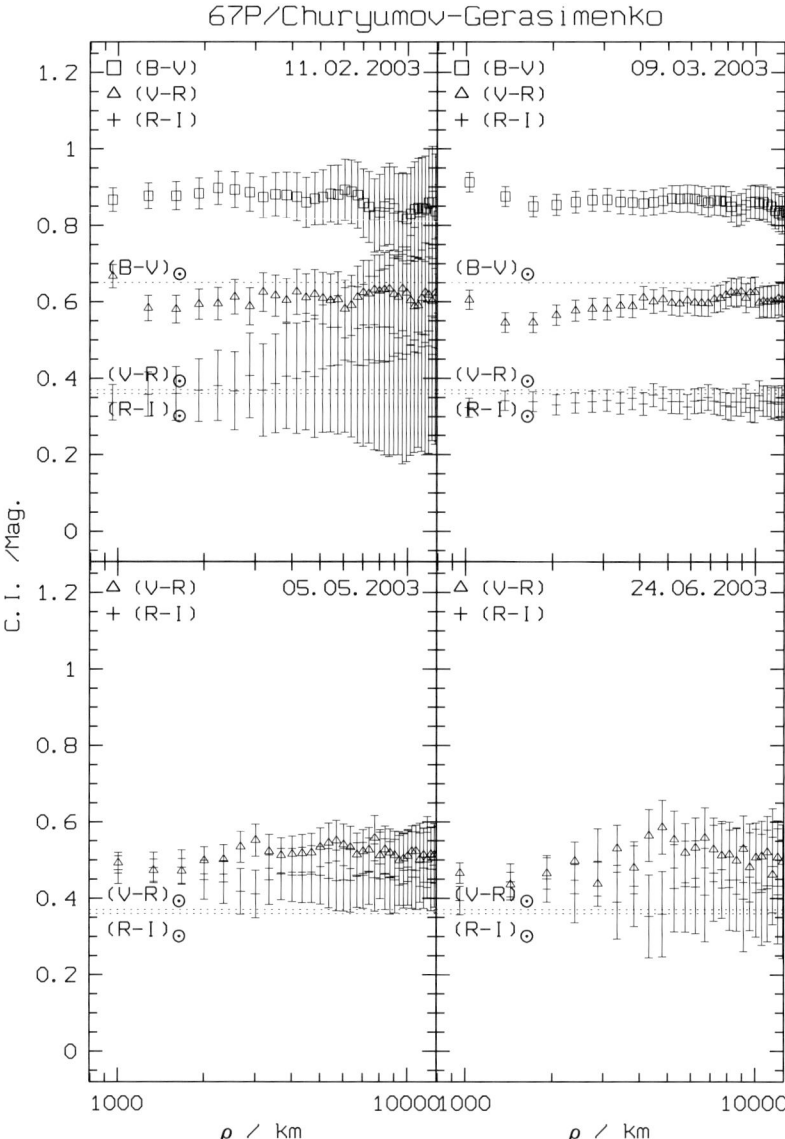

Figure 3. The colour index, C.I., of the coma as a function of projected nucleus distance, ρ. The respective values for the Sun are indicated as a straight lines. The error bars indicate 1 σ errors.

Table 2. Gas Production Rates and *Afρ* Values of 67P/Churyumov-Gerasimenko

Date (2003)	r (AU)	Q(CN) $10^{23}s^{-1}$	Q(C$_2$) $10^{23}s^{-1}$	Q(C$_3$) $10^{23}s^{-1}$	Q(NH$_2$) $10^{23}s^{-1}$	*Afρ* cm
11 Feb	2.3	24.2 ± 3.4	< 27.9	< 0.87	< 116.7	31 ± 5
09 Mar	2.5	24.7 ± 5.5	< 21.9	< 1.05	< 82.3	39 ± 10
30 Apr - 04 May	2.9	15.8 ± 5.1	< 3.13	< 0.09	< 14.4	16 ± 3

4. The new Rosetta Target Comet

The coma morphology of comet 67P/Churyumov-Gersimenko is similar to that of comet 1P/Halley, in that it shows distinct coma features such as jets. The Giotto images of comet 1P/Halley showed that material evaporated off certain areas of Halley's nucleus in form of jets, which led to the concept of a comet nucleus being divided in active and inactive areas. Monitoring of 67P/Churyumov-Gerasimenko will therefore provide the information needed to verify whether this concept is correct and could be generally applied to comets that show jet structures in the coma. The detailed knowledge we will gain on the formation of distinct coma features will lead to major progress in the interpretation of such features not only for the Rosetta target comet, but also for other comets.

The orbit of 67P/Churyumov-Gerasimenko allows Rosetta to not only monitor its pre-perihelion evolution, but to also continue through perihelion and along a significant part of its post-perihelion phase. This is particularly important because at perihelion the light curve of 67P/Churyumov-Gerasimenko clearly shows a certain asymmetry usually attributed to a seasonal effect. Such an effect has generally been assumed to result from an active area of the nucleus, which was in shadow during the approach of the comet, and is exposed to sunlight for the first time at or shortly after perihelion passage. Therefore, for this comet Rosetta will be able to confirm whether the interpretation of a seasonal effect is indeed correct. If it is, Rosetta will not only study how the activity of a comet nucleus develops at large heliocentric distance, hence at low solar radiation, but also how such a process functions for high solar radiation close to the Sun. In conclusion, comet 67P/Churyumov-Gerasimenko is a very interesting target for the Rosetta mission. The detailed study of this target will increase our knowledge of a number of physical and chemical processes on the comet nucleus and the near-nucleus environment that are very relevant also for the interpretation of remote-sensing observations of other comets, which cannot be visited by any spacecraft.

Acknowledgments

We wish to thank the Director General of the European Southern Observatory for granting telescope time at the VLT on very short notice and the ESO staff at Headquartes, La Silla and Paranal for their support.

References

Boehnhardt H., Delahodde C., Sekiguchi T., Tozzi G.P., Amestica R., Hainaut O., Spyromilio J., Tarenghi M., West R.M., Schulz R., Schwehm G. (2002). VLT observations of comet 46P/Wirtanen. *A&A*, 387: 1107-1113

Boehnhardt H., Stüwe J. A., and Schulz R. (2003). Dust Coma and Tail Structures on comet 67P/CG. *Presentation at Capri Rosetta Workshop*

Bogun S. (2001) FORS-Pipeline and Qualtity Control User's Manual, European Southern Observatory, VLT-MAN-ESO-19500-1771

DFS Pipeline & Quality Control - User's Manual, European Southern Observatory, VLT-MAN-ESO-19500-1619

ESO-MIDAS Users Guide, MIDAS Release 98NOV, European Southern Observatory, MID-MAN-ESO-11000-0002

Festou M. (1981). The density distribution of neutral compounds in cometary atmospheres. I – Models and equations. *A&A*, 95: 69–79

Jewitt D. G. and Meech K. J. (1987). Surface brightness profiles of 10 comets. *ApJ*, 317: 992–1001

Kolokolova L., Lara L. M., Schulz R., Stüwe J. A. and Tozzi G. P. (2003). Color of an ensemble of particles with a wide power–law size distribution: application to observations of comet Hale–Bopp at 3 AU. *Journal of Quantitative Spectroscopy & Radiative Trasfer*, 79–80: 861–871

Lamy P., Toth I., Jorda L., Weaver H. A., and A'Hearn M. F. (1998). The nucleus and inner coma of Comet 46P/Wirtanen. *A&A*, 335: L25–L29

Lamy P., Toth I., Weaver H. A., Jorda L., and Kaasalainen M. (2003). The nucleus of comet 67P/Churyumov–Gerasimenko, the New Target of the Rosetta mission. *Bull. Am. Astron. Soc.*, 35: 970

Landoldt, A. U. (1992). UBVRI photometric standard stars in the magnitude range $11.5 < V < 16.0$ around the celestial equator. *AJ*, 104: 340–491

Patat, F. (1999). EFOSC2 User's Manual, European Southern Observatory, LSO-MAN-ESO-36100-0004

Schulz, R. (1991). in *Proceedings of the 3_{rd} ESO/ST-ECF Data Analysis Workshop*, P.J. Grosbol & R.M. Warmels Eds.

Schulz, R., Arpigny C., Manfroid J., Stüwe J.A., Tozzi G.P. Rembor K., Cremonese G., and Peschke S. (1998). Spectral evolution of Rosetta Target comet 46P/Wirtanen. *A&A*, 335:, L46–L49

Szeifert, T. (2002) FORS1+2 User Manual, European Southern Observatory, VLT-MAN-ESO-13100-1543

Tüg, W. (1977). Vertical extinction on La Silla. *The Messenger (ESO)*, 11, 7–8

CO AND DUST PRODUCTIONS IN 67P/CHURYUMOV–GERASIMENKO AT 3 AU POST–PERIHELION

D. Bockelée-Morvan, R. Moreno, N. Biver, J. Crovisier
LESIA, Observatoire de Paris, 5 place Jules Janssen, F-92195, France
dominique.bockelee@obspm.fr

J.-F. Crifo
Service d'Aéronomie, BP3, F-91371, Verrières Le Buisson Cedex, France

M. Fulle
Osservatorio Astronomico di Trieste, Via Tiepolo 11, I-34131, Trieste, Italy

M. Grewing
Institut de radioastronomie millimétrique, 300 rue de la Piscine, Domaine universitaire, F-38406, St-Martin-d'Hères, France

Abstract   Following the selection of comet 67P/Churyumov-Gerasimenko as the new target of the *Rosetta/ESA* mission, observations of this comet were performed at the 30-m telescope of the Institut de radioastronomie millimétrique in May–June 2003. Observations focussed on the CO $J(2–1)$ and $J(1–0)$ lines and dust thermal continuum at 1.2 mm in order to constrain the CO and dust productions at 3 AU from the Sun. Neither the CO lines, nor the continuum emission were detected. The derived 3–σ upper limit on the CO production rate is $Q_{CO} < 9.2 \times 10^{26}$ molecules s^{-1}. That derived for the dust production rate is strongly model dependent, and at best is $Q_{dust} < 36$ kg s^{-1}. The first synthetic CO line profiles computed with a 3-D gas-dynamical simulation of a gas coma are presented.

Keywords: 67P/Churyumov-Gerasimenko, CO, dust, radio observations, Rosetta

Introduction

In spring 2003, the European Space Agency selected comet 67P/Churyumov–Gerasimenko (hereafter designated as 67P) as the new target of the *Rosetta*

L. Colangeli et al. (eds.), The New ROSETTA Targets, 25–36.
© 2004 *Kluwer Academic Publishers. Printed in the Netherlands.*

mission. Strong effort was requested to obtain information on this comet. From analysis of wide field images, where the neck line is visible, it was inferred that 67P released large quantity of 1 cm-sized particles in 2002 at 2 AU pre-perihelion (Fulle et al. 2004a, b). This finding, which implies high gas production, motivated CO and continuum observations at millimetre wavelengths in May–June 2003, when the comet was at 3 AU from the Sun, post-perihelion. Millimetre observations are more sensitive than IR or UV observations to detect CO at large heliocentric distances. Continuum flux radiation in the millimetre spectral range is dominated by thermal radiation of mm-sized particles, which contain a large fraction of the total grain mass. We present here observations which were undertaken with the 30-m telescope of the Institut de radioastronomie millimétrique (IRAM) on Pico Veleta (Spain).

1. CO and HCN

Observations

The observations of the CO lines were carried out on 2 and 3 May UT and from 20 to 24 May UT, 2003, when the comet was at $r_h \sim 3$ AU from the Sun post-perihelion, and at $\Delta = 2.27$–2.70 AU from the Earth. A log of the observations is given in Table 1. Four receivers were used at the same time. The A100 and B100 receivers were tuned to the frequency of the CO $J(1$–$0)$ line (115.271 GHz). The A230 and B230 receivers were tuned to that of CO $J(2$–$1)$ (230.538 GHz). On 24 May, part of the observations were made with the B100 receiver tuned to the HCN $J(1$–$0)$ line at 88.632 GHz. The receivers were connected to the autocorrelator *VESPA* divided in four units of 35 MHz bandwidth and 19.5 kHz spectral resolution, corresponding to a velocity resolution of 0.051 and 0.025 km s^{-1} for the $J(1$–$0)$ and $J(2$–$1)$ lines, respectively. The receivers were also connected to low resolution (0.3 to 1.25 MHz) backends for redundancy and checking purposes. In order to increase the sensitivity of the measurements, the observations were performed using the frequency switching mode with a frequency throw of 6.5 MHz for the A230/B230 receivers, and of 7 MHz for the A100/B100 receivers. The weather conditions were generally good, with typical SSB system temperatures at comet transit ($\sim$65° elevation) in the range 220–240 K (115 GHz) and 320–400 K (230 GHz) in the T_A^* scale.

On each day, the comet observations were preceded by line observations of IRC+10216 using the same instrumental set-up, in order to check that the receivers and spectrometers were operating well. The detection of the strong mesospheric CO lines in the frequency-switch spectra of comet 67P provided an additional check that the system was working well.

The comet was tracked using osculating orbital elements taken from the JPL's Horizons System (http://ssd.jpl.nasa.gov/horizons.html). To check and correct the absolute pointing of the telescope, continuum sources close to the

Table 1. Log of the observations of comet 67P at the IRAM 30-m telescope

Date (UT)	r_h (AU)	Δ (AU)	Obs.	Integ.[a] (min)	Offset[b] (")	r.m.s.[c]
2003 May 2	2.87	2.27	CO $J(1–0)$	190	2.0	17 mK
			CO $J(2–1)$	190	2.0	25 mK
2003 May 3	2.88	2.29	CO $J(1–0)$	170	2.0	18 mK
			CO $J(2–1)$	170	2.0	29 mK
2003 May 20	2.99	2.62	CO $J(1–0)$	400	3.4	14 mK
			CO $J(2–1)$	390	3.4	29 mK
2003 May 21	3.00	2.64	CO $J(1–0)$	580	3.1	12 mK
			CO $J(2–1)$	580	3.1	17 mK
2003 May 22	3.00	2.66	CO $J(1–0)$	460	5.0	10 mK
			CO $J(2–1)$	440	5.0	13 mK
2003 May 23	3.01	2.68	CO $J(1–0)$	500	3.8	11 mK
			CO $J(2–1)$	490	3.8	13 mK
2003 May 24	3.02	2.70	CO $J(1–0)$	90	1.9	27 mK
			CO $J(2–1)$	180	1.9	21 mK
			HCN $J(1–0)$	90	1.9	10 mK
2003 June 13	3.15	3.10	cont. 250 GHz	90	0.0	1.3 mJy
2003 June 16	3.17	3.17	cont. 250 GHz	60	0.0	1.2 mJy

[a] total integration time on source;
[b] pointing error for the B100/B230 receivers;
[c] for lines, r.m.s. in T_A^* for a 19.5 kHz resolution channel.

comet were regularly observed with the B100 receiver tuned at 115 GHz. Because of the faintness of the continuum pointing sources and of atmospheric instabilities, adopted pointing corrections were not always optimum during the 20–24 May period. Pointing errors were evaluated afterwards, and range from 2 to 5" depending of the date, as given in Table 1. The mean offsets for the average 2–3 May and 20–24 May CO observations are 2.6 and 4.3", respectively (Table 2). These offsets include the misalignment (2.4" in azimuth) of the A100/A230 receivers with respect to the B100/B230 receivers.

The half-power beam width (*HPBW*) was evaluated to 21" at 115 GHz and to 10.7" at 230 GHz, from continuum observations of Jupiter performed during the observing run. The main beam efficiencies measured on this planet are 0.75 and 0.55, at 115 and 230 GHz, respectively.

The total integration time on the comet was $\sim$2400 min on each CO line, adding the time spent on each receiver. The CO $J(1–0)$ and $J(2–1)$ lines were not detected. Table 2 summarizes the results of the observations.

Modelling and results on CO production

Models have been developed in order to interpret intensities of molecular millimetric lines in terms of gas production rate (e.g., Crovisier 1987; Biver

Table 2. Summary of CO and HCN observations

Date	r_h (AU)	Δ (AU)	Offset[a] (")	line	Integ.[b] (min)	$\int T_{mB} dv$ [c] (mK km s^{-1})	Q (10^{27} s^{-1})
May 2–3	2.87	2.28	2.6	CO J(1–0)	360	< 14	< 6.0
				CO J(2–1)	360	< 19	< 2.1
May 20–24	3.00	2.66	4.3	CO J(1–0)	2200	< 5.8	< 3.1
				CO J(2–1)	2080	< 7.2	< 1.0
May 24	3.02	2.70	2.0	HCN J(1–0)	90	< 22	< 0.016

[a] includes 2.4" misalignment between 230 and B230 receivers;
[b] total integration time on source;
[c] 3–σ upper limit over –1.0 to 0.6 km s^{-1} velocity interval.

et al. 1999a). These models follow a two-step process. First, the evolution of the population of the rotational levels within the ground vibrational state is computed as the molecules expand in the coma along a radial direction. The excitation model takes into account the main excitation processes, which are collisions, excitation of the infrared bands by the Sun radiation and, for comets at heliocentric distances larger than 3 AU, excitation by the 3 K cosmic background (Biver 1997). For CO under optically thin conditions, which applies here, the input parameters are the gas temperature T and the expansion velocity v_{exp} throughout the coma, the neutral–neutral cross-sections σ_c and the density distribution of the collision partners. These two last parameters control the size of the inner coma region where the populations are at local thermal equilibrium before evolving towards fluorescence equilibrium. Collisions between CO and electrons have no significant impact on CO excitation, because the dipole moment of CO is small (Biver et al. 1999a). In the second step, the radiative transfer equation is solved and provides the line intensity within the antenna beam, to be compared to the observations. A synthetic line profile can also be computed, taking into account thermal broadening. In the current state of the model, a Maxwellian thermal velocity distribution defined by the temperature is used, though this does not apply when the flow becomes free-molecular.

In most interpretations, the molecular density of the considered species (here CO) and collision partner (H_2O and/or CO) is described by a Haser law, the gas temperature and velocity are assumed to be constant throughout the coma, and the coma is isotropic. The input parararameters are constrained by observations (e.g., v_{exp} from the line profile) or guessed (see, e.g., Biver et al. 1999a). A gas production rate is then derived.

A more advanced approach was used here, taking advantage of gas dynamical 3-D models described in Crifo et al. (2004). This approach was motivated by: i) the absence of observational constraints on the physical parameters of a rarefied coma as that of 67P at 3 AU from the Sun; ii) the need to prepare the

Table 3. Model parameters

Gas model	Q_{CO} (s^{-1})	Q_{CO} day/night	Q_{H_2O} (s^{-1})	Nucleus
1	10^{27}	$\sim 1^{a}$	0	2-km sphere
2	10^{27}	19^{b}	0	2-km sphere
3	5.45×10^{26}	$\sim 1^{a}$	4×10^{26}	2-km sphere
4	5.09×10^{26}	19^{b}	4×10^{26}	2-km sphere
5	10^{27}	1^{a}	2.5×10^{26}	"starfish"

[a] corresponds to asymmetry parameter $a_0 = 0.9$ defined in Crifo et al. (2004);
[b] corresponds to asymmetry parameter $a_0 = 0.1$.

exploitation of the spectral data that will be acquired by the *MIRO* microwave experiment on *Rosetta*; iii) last but not least, the concern to provide an accurate constraint on CO production in 67P at 3 AU from the Sun.

Gas dynamical 3-D calculations were performed for 67P at 3 AU from the Sun for various CO/H_2O gas production models and nucleus shapes (Crifo et al. 2004). Five models are considered here, which parameters are given in Table 3. CO production from the surface is either almost uniform (models 1, 3 and 5) or strongly asymmetric (models 2 and 4). For the mixed CO/H_2O models (models 3–5), H_2O ice is distributed uniformly on the surface, but sublimates according to the solar illumination. For models 1–4, the nucleus is a 2 km radius sphere. In model 5, the nucleus has a "starfish" analytical shape (it mimics the shape proposed by Lamy et al. (2003) for the nucleus of 67P). Models 1 and 5 are presented in details in Crifo et al. (2004).

The gas dynamical calculations provide the distributions of T, v_{exp}, CO and H_2O (for CO-H_2O mixtures) densities (n_{CO}, n_{H_2O}) distributions in the coma out to a distance $D \sim 10^5$ km. However, above some much closer distance in the coma, the flow becomes practically free-molecular, and the gas velocity distribution ceases to be Maxwellian. One can then define a "parallel temperature" $T_\parallel$ and a "perpendicular temperature" $T_\perp$ which differ. $T_\parallel$ and the velocity vector freeze-out, while $T_\perp$ decreases to zero. These freezing-out are not present in the gas dynamical solutions. Direct Simulation Monte Carlo (DSMC) calculations performed up to 40 km from the nucleus for model 1 show excellent agreement with the fluid approach for T and v_{exp} in this region (Crifo et al. 2004), but reveal that the freezing-out of $T_\parallel$ proceeds above 2–4 km above the surface. Thus, for this model, we present results assuming frozen T and velocity vector beyond $D = 40$ km. For models 1 and 2, we also present calculations with the full gas dynamical results ($D = 80\,000$ km). Table 4 gives the terminal temperature and expansion velocity in the Sun and anti-Sun directions for the physical models considered here.

The resulting 3-D distributions of T, v_{exp}, n are used to compute the 3-D distribution of the populations of the CO rotational levels, and then synthetic

Table 4. Model results for the CO $J(2–1)$ line

Gas model	D^a (km)	v_{exp}^b (km s^{-1})		T^b (K)		$\int T_{mB}dv^c$ (mK km s^{-1})	n_u^d	N^e 10^{31}
		Sun	anti-Sun	Sun	anti-Sun			
1	40	0.74	0.38	15	4	9.2	0.22	3.28
	80 000	0.77	0.45	40	71	9.0	0.22	3.13
2	40	0.74	0.38	14	4	9.1	0.26	2.77
	80 000	0.77	0.45	40	74	7.7	0.22	2.68
3	40	0.77	0.51	14	7	4.3	0.25	1.34
4	40	0.76	0.51	14	7	4.7	0.26	1.38
5	40	0.76	0.52	12	6	7.5	0.23	2.53

^{a} distance above which the velocity vector and T are frozen;
^{b} terminal velocity and temperature;
^{c} line area of the CO $J(2–1)$ line for the IRAM observing conditions on May 20–24, 2003, but with a beam offset equal to 0. For models 1–4, the Earth-comet-Sun angle (i.e., phase angle) is taken equal to $0°$; results with phase angle = $19°$ (corresponding to 67P in May 2003) are similar within 1%. For model 5, the line of sight is along the smallest nucleus axis, and the phase angle is $45°$;
^{d} mean population of the $J = 2$ CO rotational level probed by the IRAM beam;
^{e} number of molecules probed by the IRAM beam; see other parameters in footnote c).

line profiles. The time-dependent excitation model is run along radial rays, which assumes implicitly that the molecules flow outwards radially (and this is indeed almost the case). The total CO–CO cross-section for collisional de-excitation is taken equal to 2×10^{-14} cm^2. That for CO–H$_2$O is 5×10^{-14} cm^2. These cross-sections are poorly known.

The computed CO $J(2–1)$ line intensities for the observing conditions at IRAM on 20–24 May 2003 are given in Table 4. When scaled to the same CO production rate, all models give consistent results within 20%, with most of them within 8%. There are two reasons for that. First, the molecules probed by the IRAM beam ($\sim 10\,300$ km radius) have about the same excitation state for all models (see the mean population n_u of the $J = 2$ state in Table 4). These molecules have not reached yet fluorescence equilibrium (where $n_u = 0.15$) but are at altitudes where collisional excitation is not significant. Second, the number of molecules N probed by the IRAM beam does not vary much with the model (see Table 4). The derived 3–σ Q_{CO} upper limit, taking into account the beam offset, is 10^{27} s^{-1} for the 20–24 May period (Table 2), and 9.2×10^{26} s^{-1} when including the 2–3 May observations. The upper limit derived from the $J(1–0)$ line is three times higher (Table 2).

Synthetic line profiles are shown in Fig. 1. The two spectra in Fig. 1A (model 1, with $D = 40$ km and $D = 80\,000$ km) illustrate the effect of terminal velocity and temperature on the line shape. The spectra for model 1 (Fig. 1A) present a strong peak towards positive velocities, because the coma is more dense on the night side. In contrast, for model 2, the CO coma is more dense in the day side: on the corresponding spectrum (Fig. 1B), the strong peak is at a negative velocity. The features seen at 0.15–0.20 km s^{-1} on these spectra are real and trace the relatively dense conical structure (weak shock) formed in the night side coma, near the terminator, to match the supersonic flows originating from the dayside nucleus surface to that originating from the night-side nucleus surface.

Fig. 1C shows the synthetic line profile for the "starfish" nucleus, with the line of sight along the smallest nucleus axis (see Fig 1 of Crifo et al. 2004; the phase angle is 45°). In this case, the coma is structured by the nucleus topography. These synthetic profiles prefigure the spectra which will be obtained by the *MIRO* experiment onboard *Rosetta*.

Upper limit on HCN production

The observations of HCN made on 24 May 2003 were analysed with the excitation model presented in Biver et al. (1999a), in the simple case of an isotropic HCN coma with v_{exp} = 0.6 km s^{-1} and T = 15 K. The derived upper limit on the HCN production rate is $Q_{HCN} < 1.6 \times 10^{25}$ s^{-1}, and is ten times larger that the CN production rate determined by Schulz et al. (2004) for beginning of May, from observations at the European Southern Observatory.

2.   Dust

Observations

The continuum observations of comet 67P were performed on 13 and 16 June 2003. We used the Max Planck Institut für Radioastronomie bolometer array with 117 pixels (*MAMBO-2*). Pixels are separated by $\sim$ 20", and their half-power beam width is 11". The effective frequency is close to 250 GHz, and the total bandwidth is about 60 GHz. Observations of comet 67P were performed using the ON–OFF mode, where the subreflector of the telescope is alternatively looking at the target and at a position 32" away in azimuth with a rate of 2 Hz. This observing technique subtracts most of the atmospheric emission. Some weak atmospheric continuum residuals remain, due to fast temporal fluctuations of the terrestrial atmosphere. The multi-beam observations allow us to measure these residuals from the channels adjacent to the central channel.

All bolometric measurements were obtained by observing remotely from Grenoble (France). Flux calibration was determined through observations of Jupiter, Uranus, and Mars. The pointing and focus of the telescope were achieved by observing the quasar 3C273. Sky observations at several elevations (skydip) permitted us to measure the zenithal atmospheric opacity. Observations were performed in average weather conditions (zenithal sky opacity

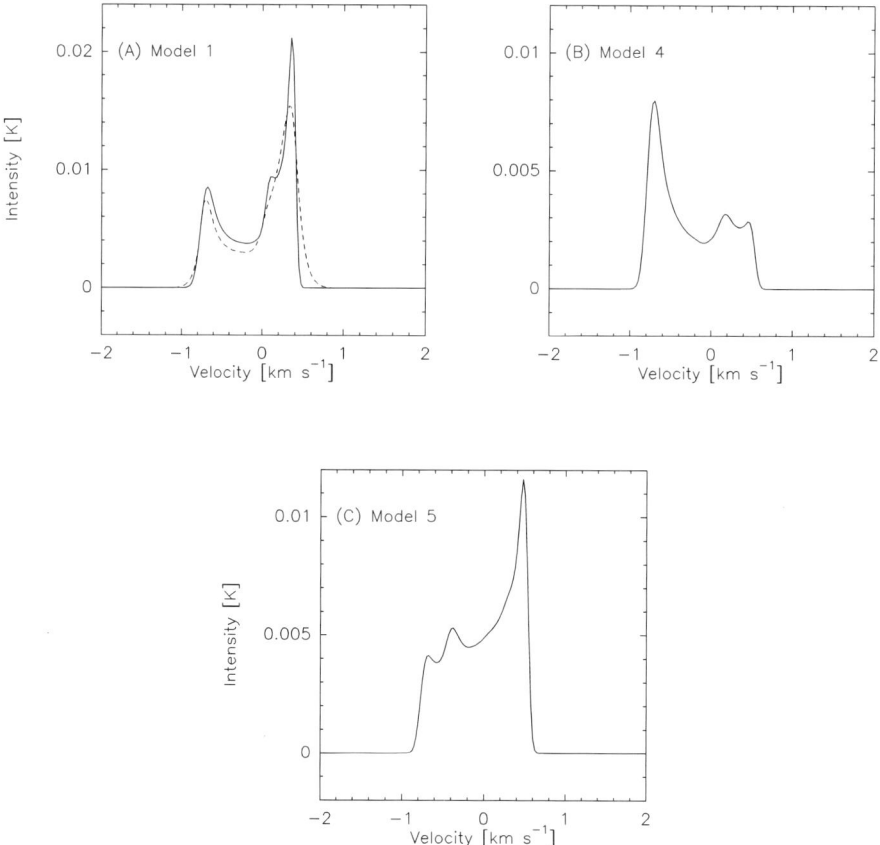

Figure 1. Synthetic CO $J(2–1)$ line profiles for the IRAM observing conditions of 20–24 May 2003. (A) Model 1 (uniform CO production) with $D = 40$ km (plain line) and $D = 80000$ km (dashed line); (B) Model 4 (asymmetric CO production) with $D = 40$ km. (C) Model 5 (starfish nucleus shape with uniform CO production) with $D = 40$ km. For (A) and (B), the phase angle is taken equal to $0°$. For (C), the line of sight is along the smallest nucleus axis and the phase angle is $45°$. The spectral resolution is 0.025 km s^{-1}.

= 0.25–0.50), but with good atmospheric stability. ON–OFF observations were taken during $\sim$ 1 hour per day, and the results are summarized in Table 1.

The data reduction was performed using the NIC softwares. The basic reduction process consists in: i) removing data spikes; ii) computing the ON minus OFF phases; iii) eliminating the sky noise by applying a noise removal algorithm based on beam to beam correlation; iv) correcting the central channel flux from the mean flux observed on adjacent channels, in order to remove possible residual continuum offsets.

We finally conclude that the comet dust continuum is not detected in our bolometric observations and will adopt a 1–σ upper limit of 1 mJy for its 1.2 mm flux.

Upper limit on dust production

Continuum flux radiation in the millimetre spectral range is dominated by thermal emission from cometary solids, mainly dust when the comet is active. (an order of magnitude for the thermal emission from the nucleus of 67P is 0.02 mJy at 1.2 mm, obtained by assuming a 2 km radius sphere and a surface temperature $T = 182$ K). To interpret the observations in terms of dust production, we followed the simple approach described by Jewitt and Luu (1992). As a first approximation, the temperature of the grains is taken to be the blackbody equilibrium temperature for a fast rotator (154 K at 3.15 AU). Indeed, though submicrometric grains may be superheated with respect to the blackbody temperature, they are inefficient emitters of millimetre radiation. From this, we deduce a 1–σ upper limit for the effective blackbody cross-section at 1.2 mm of 7.5×10^8 m^2 within the IRAM beam.

The blackbody cross-section $C(\lambda)$ is related to the grain mass M by $C(\lambda) = \kappa(\lambda)M$, where $\kappa(\lambda)$ (m^2 kg^{-1}) is the opacity at wavelength λ. κ is a function of particle size distribution, composition, shape, and varies with λ. There is a large uncertainty on the value of $\kappa(1$ mm) pertaining to cometary dust. Values adopted in the literature range from 0.05 m^2 kg^{-1} (Jewitt and Matthews 1997) to 7.5 m^2 kg^{-1} (Altenhoff et al. 1999). The former value is that modelled for interstellar (submicrometric) dust, while the latter value characterizes the β Pictoris circumstellar dust dominated by mm-sized particles (see the discussion in Altenhoff et al. 1999). Using this range of $\kappa(1$ mm), we infer an upper limit on M in the range 8.4×10^7 kg to 1.3×10^{10} kg, the smaller value corresponding to the β Pictoris κ value.

In Jewitt and Luu (1992) and following papers, the dust mass production rate Q_d is related to M by the approximate formula $Q_d = M/t$, where $t = p/v_d$ is the residence time of the dust particles in the projected beam of radius p, and v_d is the velocity of the grains emitting efficiently at the observed frequency. For grains that are expanding radially and isotropically, the exact formula (which we use here) is $Q_d = 2\alpha v_d M / \pi p$, where $\alpha = 0.94$ is a coefficient which accounts for the Gaussian shape of the beam ($\alpha = 1$ for a circular aperture) and $p = HPBW/2$.

Multi-wavelength observations in the millimetre-submillimetre range showed that the particles dominating the thermal continuum at 1 mm are typically 1-mm sized. The velocity of such particles in the coma of 67P is uncertain. The calculations performed by Crifo et al. (2004) with the gas production model 1 (see Section 1) give a maximal terminal velocity of 1 m s^{-1} for spherical grains with $a = 9$ mm radius and specific density $\rho_d = 0.1$ g cm^{-3}. As $v_d \propto a^{-0.5}$, $v_d(1 \text{ mm}) \sim 3$ m s^{-1}. For grains with $\rho_d = 1$ g cm^{-3}, $v_d(1 \text{ mm})$ is expected ~ 1 m s^{-1}. Conservatively, we assumed $v_d(1 \text{ mm}) = 3$ m s^{-1}. The inferred (1–σ) upper limit on dust production Q_d is 12.2 kg s^{-1} for $\kappa(1 \text{ mm})$ = 7.5 m^2 kg^{-1}, and 1.9×10^3 kg s^{-1} for $\kappa(1 \text{ mm}) = 0.05$ m^2 kg^{-1}. Better constraints (likely intermediate between these two values) could be obtained using detailed modelling of dust temperature, emissivity, dynamics, and size distribution, but this is beyond the scope of this paper.

3. Discussion

Let us compare the preceding upper limits to other available informations about the gas and dust production, post-perihelion. There exist observations relative to a post-perihelion heliocentric distance $\simeq 2.5$ AU (March 2003). From photometric measurements in the visible performed during this period, Kidger (2003) inferred Q_d =1–2 kg s^{-1}. On the other hand, from studies of the dust tail, Fulle et al. (2004a, b) estimated Q_d to ~ 5–10 kg s^{-1}. The discrepancy between these estimates is not surprising, as visible light photometry is usually insensitive to large dust grains, where most of the ejected mass often lies. Anyhow, these two estimates are well below our present upper limit (at best < 36 kg s^{-1} at 3–σ). Fulle et al. (2004a, b) also determined the dust size distribution, which was found steep, with a maximum size ~ 4 mm. From the well-known balance condition between aerodynamic drag and gravity at the surface, we compute that to eject mm-sized spherical grains from a 2 km radius comet requires a gas production rate (CO and other species) of about 10^{27} s^{-1}, if the nucleus and dust grains specific masses are both equal to 1 g cm^{-3}. Even smaller rates would eject such grains if their specific mass was smaller and/or if their shape was markedly aspherical. Our present Q_{CO} upper limit suggests that this is the case, or that the total (CO+H_2O) production rate reached 10^{27} s^{-1} at 2.5 AU.

Which other constraints can be put on the CO production rate of 67P at 3 AU from the Sun? The water production rate at perihelion ($q = 1.3$ AU) is $\sim 10^{28}$ s^{-1} (see several papers in this book). There are only few constraints on the CO production rate relative to water in Jupiter-family comets near 1

AU from the Sun (see, e.g., Bockelée-Morvan et al. 2004), but reasonnable values are between 1–10%, or even lower. Then, if the CO production is almost constant along the orbit, as predicted by thermal evolution models of cometary nuclei, Q_{CO} is expected to be, at 3 AU, in the range $10^{26} - 10^{27}$ s^{-1}. If, on the contrary, it follows the r_h^{-2} heliocentric dependence observed in comet Hale-Bopp for Q_{CO} (Biver et al. 1999b), we expect Q_{CO} in the range $2 \times 10^{25} - 2 \times 10^{26}$ s^{-1}. Finally, another estimate can be obtained from the total visual magnitude-to-Q_{CO} empirical correlation established by Biver (2001) from observations of comet Hale-Bopp and other comets, which leads here to $Q_{CO} \sim 10^{26}$ s^{-1} at 3 AU post-perihelion.

To summarize, these indirect estimates suggest a low CO production rate at 3 AU on the order 10^{26} s^{-1}, with an uncertainty of an order of magnitude by excess or by default. This uncertainty leaves us with practically the full range of scenarii for the future *Rosetta* landing, e.g. negligible to dominant aerodynamic perturbations on the orbiter. But, in addition, what we are much more interested in knowing, is the production rate of the comet at 3 AU *pre-perihelion* – because the *Rosetta* encounter is scheduled there. There are indications that the comet is much more active pre- than post-perihelion (Fulle et al. 2004a, b). For these reasons, it will be useful to attempt pre-perihelion observations of CO and dust thermal emission at its next passage.

References

Altenhoff, W.J., Bieging J.H., Butler B., Butner H.M., Chini R., Haslam C.G.T., Kreysa E., Martin R.N., Mauersberger R., McMullin J., Muders D., Peters W.L., Schmidt J., Schraml J.B., Sievers A., Stumpff P., Thum C., von Kap–Herr A., Wiesemeyer H., Wink J.E., Zylka R. (1999). Coordinated radio continuum observations of comets Hyakutake and Hale-Bopp from 22 to 860 GHz. *Astron. Astrophys.*, 348:1020–1034.

Biver, N. (1997). Molécules mères cométaires: observations et modélisations. PhD thesis, University of Paris 7.

Biver, N. (2001). Correlation between visual magnitudes and the outgassing rate of CO in comets beyond 3 AU. *International Comet Quarterly*, 23:85–93.

Biver N., Bockelée-Morvan D., Crovisier J., Davies J.K., Matthews H.E., Wink J.E., Rauer H., Colom P., Dent W.R.F., Despois D., Moreno R., Paubert G., Jewitt D., Senay M. (1999a). Spectroscopic monitoring of comet C/1996 B2 (Hyakutake) with the JCMT and IRAM radio telescopes. *Astron. J.*, 118:1850–1872.

Biver, N., Bockelée-Morvan D., Colom P., Crovisier J., Germain B., Lellouch E., Davies J.K., Dent W.R.F., Moreno R., Paubert G., Wink J., Despois D., Lis D.C., Mehringer D., Benford D., Gardner M., Phillips T.G., Gunnarsson M., Rickman H., Winnberg A., Bergman P., Johansson L.E.B., Rauer H. (1999b). Long-term evolution of the outgassing of comet Hale-Bopp from radio observations. *Earth Moon and Planets*, 78:5–11.

Bockelée-Morvan, D., Biver, N., Colom, P., Crovisier, J., Henry, F. Lecacheux, A., Davies, J. K., Dent, W. R. F., and Weaver, H. A. (2004). The outgassing and composition of Comet 19P/Borrelly from radio observations. *Icarus*, 167:113–128.

Crifo, J.F., Lukyanov, G.A., Zakharov, V.V., and Rodionov, A.V. (2004). Physical model of the coma of comet 67P/Churyumov-Gerasimenko. *This volume*.

Crovisier J. (1987). Rotational and vibrational synthetic spectra of linear parent molecules in comets. *Astron. Astrophys. Suppl. Ser.*, 68:223–258.

Fulle, M., Barbieri, C., Cremonese, G., Rauer, H., Weiler, M., Milani, A., and Ligustri, R. (2004a). The dust environment of comet 67P/Churyumov-Gerasimenko. *Astron. Astrophys.*, 422:357–368.

Fulle, M., Barbieri, C., Cremonese, G., Rauer, H., Weiler, M., Milani, A., and Ligustri, R. (2004b). Dust environment of comet 67P/Churyumov-Gerasimenko. *This volume*

Jewitt, D., and Luu, J. (1992). Submillimeter continuum emission from comets. *Icarus*, 100:187–196.

Jewitt, D. and Matthews, H.E. (1997). Submillimeter continuum observations of comet Hyakutake (1996 B2). *Astron. J.*, 113:1145–1151.

Kidger, M.R. (2003). Dust production and coma morphology of 67P/Churyumov-Gerasimenko during the 2002–2003 apparition. *Astron. Astrophys.*, 408: 767–774.

Lamy, Ph., Toth, I., and Weaver, H. (2003). Press release STScI-2003-26.

Schulz, R., Stüve, J.A., and Boehnhardt, H. (2004). Monitoring comet 67P/Churyumov-Gerasimenko from ESO in 2003. *This volume*

THE DUST ACTIVITY OF
COMET 67P/CHURYUMOV–GERASIMENKO

M. Weiler, J. Knollenberg and H. Rauer
Institute of Planetary Research, DLR, Rutherfordstrasse 2, 12489 Berlin, Germany

Abstract   Comet 67P/Churyumov–Gerasimenko is a dusty comet. In a previous work (Weiler et al., 2004) it was found that the mean dust to gas mass ratio could be as high as 4.8. In this work we study the influence of such a high dust mass loading of the gas flow in the coma on the terminal dust velocities used to compute the dust production rates. It was found that the mass loading is not significant for determining the dust production rates of comet 67P/Churyumov–Gerasimenko. Using new dust parameters determined for comet 67P/Churyumov–Gerasimenko by Fulle et al. (2004) we recomputed the dust production rates and dust to gas mass ratios. From the results of our gasdynamical model we can also estimate the dust particle fluxes as a function of dust size and nucleocentric distance.

1.   Introduction

Comet 67P/Churyumov–Gerasimenko (hereafter 67P/C–G) was found to be a dusty comet by a number of authors (Hanner et al., 1985; Osip et al., 1992). A very high mean dust to gas mass ratio was also determined in a previous work (Weiler et al., 2004). The size–dependent dust velocities necessary to compute the dust production rates were calculated in a test particle approach in this work neglecting the influence of the dust particles onto the gas flow in which the grains are accelerated. This approach was used for all observations given in the Lowell Observatory Cometary Database (LOCD, described by A'Hearn et al., 1995) when Afρ and the OH production rate were determined at the same time. The gas activity took only OH into account as an estimate of the water production rate. Since no production rates for CO are available for comet 67P/C–G, this species was neglegted. For the nucleus of comet 67P/C–G a radius of 2 km (Lamy et al., 2003), a density of 1000 kg m^{-3} and an active surface fraction of 10% were assumed. A peak dust production of $2.1 \cdot 10^3$ kg sec^{-1} at 1.36 AU postperihelion was determined, decreasing to only 95 kg sec^{-1} at 1.85 AU postperihelion. These values are higher compared to other publications (e.g., Hanner at al., 1985; Krishna Swamy, 1991) and correspond to a maximum dust to gas mass ratio of 8.4. The average dust

L. Colangeli et al. (eds.), The New ROSETTA Targets, 37–46.

to gas ratio determined from all available data was 4.8.

The high dust mass loading of the gas flow carrying the dust away from the comet nucleus raised the question whether the influence of the dust particles onto the gas flow, neglegted in the computation of the dust production rates by Weiler et al. (2004), has to be taken into account. Furthermore, due to the lack of specific information only a "standard" dust size distribution was used by Weiler et al. (2004). By now detailed analysis of the dust coma and the neck–line observed in comet 67P/C–G in spring 2003 by Fulle et al. (2004) suggests new parameters for the dust size distribution and provides estimates of the cometocentric dust velocity. In this work we study the effect of neglecting the influence of the dust onto the gas flow. In addition, we vary the dust size parameters and study the influence on the dust to gas mass ratio of comet 67P/C–G.

2.   Dust Velocities

In Weiler et al. (2004) the dust particles were treated as test particles in a gas flow resulting from sublimation of ices on or inside the nucleus. The details of this model are described by Weiler et al. (2003). To estimate the error caused by the treatment of dust grains as test particles when computing the dust velocities, we assume the maximum value for the dust to gas mass ratio and compute the dust velocity again in a continuum approach, taking a reaction of the dust onto the gas flow into account. The set of equations to be solved is then (Knollenberg, 1994):

$$\frac{\partial}{\partial t}\mathbf{w} + \frac{1}{r^2}\frac{\partial\, r^2\mathbf{G}}{\partial r} = \mathbf{S}. \tag{1}$$

Here is

$$\mathbf{w} = \begin{pmatrix} \rho_{gas} \\ \rho_{gas}u_{gas} \\ \rho_{gas}\,e \\ \rho_1 \\ \rho_1 v_1 \\ \dots \\ \rho_n \\ \rho_n v_n \end{pmatrix}, \tag{2}$$

$$G = \begin{pmatrix} \rho_{gas} u_{gas} \\ \rho_{gas} u_{gas}^2 + p_{gas} \\ u_{gas}(\rho_{gas} e + p_{gas}) \\ \rho_1 v_1 \\ \rho_1 v_1^2 \\ \cdots \\ \rho_n v_n \\ \rho_n v_n^2 \end{pmatrix}, \tag{3}$$

$$S = \begin{pmatrix} 0 \\ \frac{2}{r} p_{gas} - F_{gd} \\ -Q_{gd} \\ 0 \\ f_{gd,1} \\ \cdots \\ 0 \\ f_{gd,n} \end{pmatrix}. \tag{4}$$

In these equations, u_{gas}, ρ_{gas} and p_{gas} denote the gas velocity, density and pressure, respectively, and e is the total specific energy of the gas. The source terms F_{gd} and Q_{gd} are given by

$$F_{gd} = \sum_{i=1}^{n} \frac{1}{2} C_D \frac{\pi a^2}{m_{dust}} (u_{dust} - u_{gas})^2 \rho_{dust} \rho_{gas} = \sum_{i=1}^{n} f_{gd,i} \tag{5}$$

$$Q_{gd} = 4 \frac{\pi a^2}{m_{dust}} |u_{dust} - u_{gas}| \rho_{dust} S_t \frac{\gamma}{\gamma - 1} \frac{k_B}{m_{gas}} \rho_{gas} (T_r - T_{dust}). \tag{6}$$

C_D is the drag coefficient, S_t the Stanton number and T_r is the recovery temperature. These quantities are given by Probstein (1969) and Kitamura (1986). In the set of equations (1), the first three rows are the continuity equation and the conservation of momentum and energy written for the gas component. The indices 1 to n mark the continuity and conservation of momentum equations of dust particles in the size range i, were i runs from 1 to n and n is the number of discrete intervals of the dust size range $[a_1, a_2]$. The equations of energy conservation for the dust particles are not considered here since the temperature of the dust grains is assumed to be constant in our approach. The temperature values are determined using the equation from Divine (1981):

$$T_{dust} = 310 \, K \cdot \left(\frac{r_h}{1 \, AU} \right)^{-0.58}. \tag{7}$$

The system of differential equations (1) was solved using a Godunov–type scheme of second order for the gas equations and an upwind scheme of second order for the dust equations. The boundary conditions on the nucleus surface for the gas equations were determined as described in Weiler et al. (2003). An active fraction of 10% for the nucleus of comet 67P/C–G is assumed. The dust velocity at the nucleus surface is zero and the dust density in the size interval i is computed using the dust size distribution and the dust to gas mass ratio. For the dust size distribution we use the function given by Newburn & Spinrad (1985):

$$f(a) = \tilde{N} \left(1 - \frac{a_1}{a}\right)^M \left(\frac{a_1}{a}\right)^N. \tag{8}$$

This function has two parameters, N gives the decrease of frequency with increasing particle radius, and the second parameter M then determines the position of the peak of the frequency distribution. $\tilde{N}$ is a normalization factor, chosen in a way that the integral over $f(a)$ is equal to unity. In Fig. 1 the dust velocities computed with the test particle approach are shown for the maximum of activity in the LOCD data on December 14, 1982 at 1.36 AU heliocentric distance. The solid line shows the dust velocity computed with the dust bulk density function of Newburn & Spinrad (1985),

$$\rho = \rho_0 - \rho_1 \cdot \frac{a}{a + \tilde{a}}, \tag{9}$$

with $\rho_0 = 3000$ kg m^{-3}, $\rho_1 = 2200$ kg m^{-3} and $\tilde{a} = 2$ μm. The dashed line gives the dust velocities for a constant density of 1000 kg m^{-3}. The differences are caused by a density larger than 1000 kg m^{-3} for small particles and a lower density for larger particles. The dotted lines show the dust velocities determined by Fulle et al. (2004) at perihelion (upper line) and 50 days after perihelion (lower line) in 2002. The dates correspond to 1.29 AU and 1.47 AU heliocentric distance. These velocities follow a dependency of $v(a) \sim 1/\sqrt{a}$ where the proportional factor was determined from inverse dust tail modeling. The dust velocities from this work are slightly higher than the velocities by Fulle et al. (2004), but they are computed for the maximum of activity. Keeping also in mind that the results from this work and Fulle et al. (2004) correspond to different perihelion passages, the results from the two completely independent models are in fairly good agreement.

The crosses show velocities computed with the continuum model, assuming a dust to gass mass ratio of 8.5 and assuming the parameters $N = 3.5$ and $M = 15.4$, corresponding to a peak in the dust size distribution at $a = 0.54$ μm (Hanner et al., 1985). A dust temperature of 259 K was used. The deviation in velocity between the two methods is about 2.8%.

Assuming the dust size parameters $N = 4.0$ and $M = 17.6$, leading to the same position of the peak in the size distribution, gives a dust to gas mass ratio of 0.89 in the test particle approach. Using these parameters to compute the velocities in the continuum approach results in dust velocities differing also less than 3% from the results of the test particle approach. The dust mass loading can therefore be neglected for computation of the dust production rates of comet 67P/C–G. The parameters $N = 4.5$ and $M = 19.8$ lead to a even lower dust to gas mass ratio and also to a difference less than 3%.

3.   Dust Production Rates

Assuming an isotropic emission of dust from a spherical nucleus, the observational parameter Afρ (A'Hearn et al., 1984) is related to the dust number production rate Q_N by (Jorda, 1995)

$$Q_N = \frac{\mathrm{Af}\rho}{2\pi^2 A_B(\lambda)D(\beta)} \left[\int_{a_1}^{a_2} \frac{f(a)a^2}{v(a)}\, da \right]^{-1}, \tag{10}$$

and the mass production rate Q_M is given by

$$Q_M = Q_N \frac{4\pi}{3} \int_{a_1}^{a_2} \rho(a)\, a^3 f(a)\, da. \tag{11}$$

Here a denotes the radius of the dust particles, A_B their Bond–albedo, $\rho(a)$ the density of a dust particle with radius a, $f(a)$ is the normalized size frequency distribution of dust particles, $v(a)$ is the size–dependent dust velocity with respect to the cometary nucleus and $D(\beta)$ denotes the phase function. The parameters a_1 and a_2 are the minimum and maximum dust grain radii.

We use the Bond–albedo of 0.2 and a phase function according to Divine (1981). A dust bulk density function according to Newburn & Spinrad (1985) and a minimum grain radius a_1 of 0.1 μm is used.

After determining the dust velocities, the dust production rates can be determined using equations (10) and (11). Fulle et al. (2004) found that the dust size parameter N was close to 3.5 and constant from 150 days before perihelion in August 2002 to approximately 50 days after perihelion passage. Between 50 days and 150 days after the perihelion passage, N raises to approximately 4.5. Since the development of activity of comet 67P/C–G with heliocentric distance seems to be very similar from one perihelion passage to another (Weiler et al., 2004), we estimate what effect such different values of N would have based on the data from LOCD, corresponding to the 1982 perihelion passage. In Fig. 2, upper left panel, the dust mass production rates computed with $N = 3.5$, 4.0 and 4.5 and a peak of the size distribution function at 0.54 μm are shown. The constant position of the peak implies the parameter M to be 15.4, 17.6

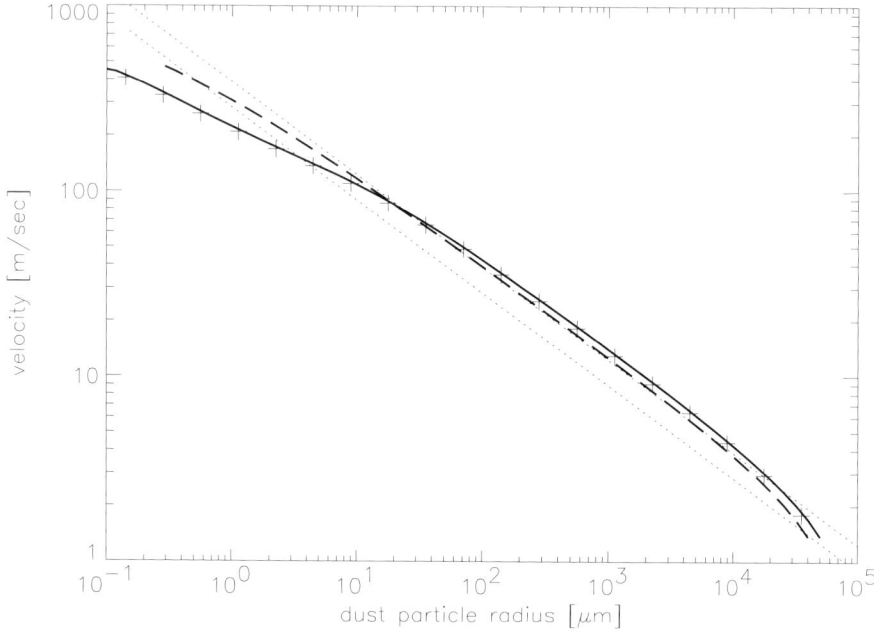

Figure 1. Cometocentric dust velocities as a function of the dust size. The solid line shows the velocities computed with the test particle approach and the dust particle density function according to Newburn & Spinrad (1985). The dashed line is valid for a constant density of 1000 $kg\ m^{-3}$. For comparison, the dotted lines give the dust velocities at the 2002 perihelion (upper line) and 50 days after perihelion (lower line) according to Fulle et al. (2004). The crosses mark dust velocities computed in the continuum approach assuming a dust to gas mass ratio of 8.5 and the dust size parameters $N = 3.5$ and $M = 17.6$ and the density as given by Newburn & Spinrad (1985).

and 19.8. The corresponding number production rates, Q_N, are shown in the upper right panel. A higher value of N, meaning a steeper decrease of the frequency of large particles, causes higher number production rates. Because of the smaller frequency of large particles a higher number of particles in total is required to obtain the observed scattering area of cometary dust.

If a strong increase of the parameter N would have occured in the 1982 apparition of comet 67P/C–G, the dust production rates would have dropped about one order of magnitude compared to the case of a constant N close to 3.5.

In Fig. 2, lower left panel, the dust to gas mass ratios determined with the different parameters are shown. With $N = 3.5$, the dust to gas mass ratio has an increase during the maximum activity of the comet after perihelion and is close to 4 during the other observations. This increase is caused by the strong increase of the maximum radius of grains which can be lifted from the nucleus

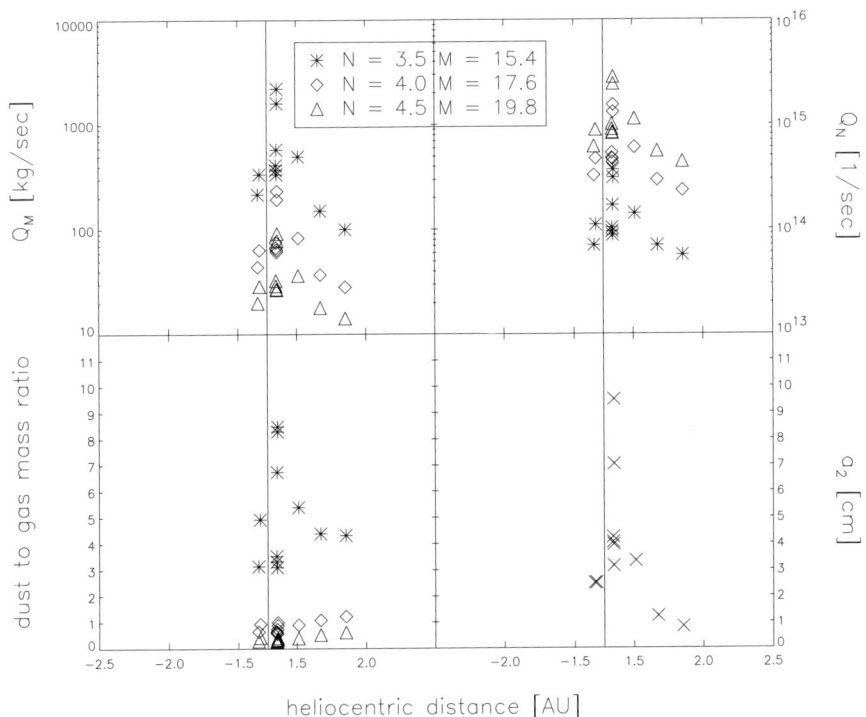

Figure 2. Upper left panel: Dust mass production rates Q_M as a function of heliocentric distance. The different symbols correspond to different dust size distribution parameters N and M. The peak of the size distribution function is at 0.54 μm in all cases. The dust production rates are computed from all LOCD data with simultaneous measurements of Afρ and OH production rates. Upper right panel: Dust number production rates Q_N for the different parameters used. Lower left panel: Dust to gas mass ratios as function of heliocentric distance for the dust production rates shown in the upper panel. Lower right panel: Maximum dust particle radius a_2 as a function of heliocentric distance. A density of the dust particles according to Newburn & Spinrad (1985) is used.

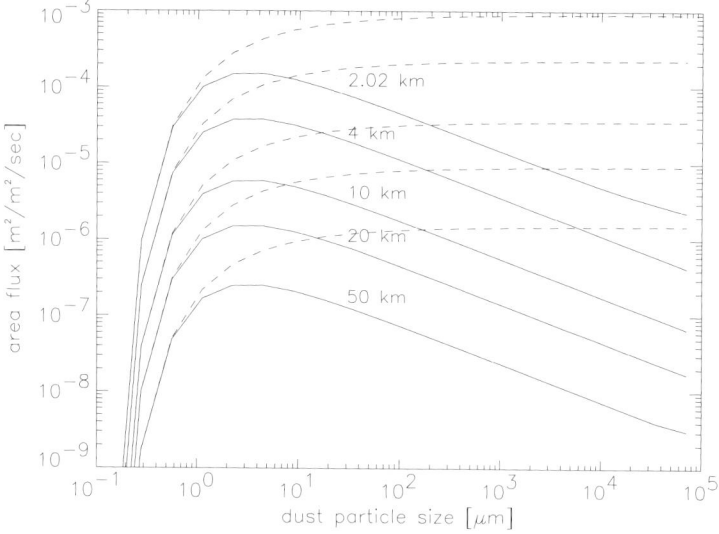

Figure 3.   Fluxes of projected dust area as a fuction of dust size for five different nucleocentric distances. The radius of the nucleus is assumed to be 2 km. The dashed lines give the cumulative area flux. The values are computed at the maximum of cometary activity in the LOCD data.

surface. The values of this maximum radius, a_2, are shown in the lower right panel. In case of a higher value of N, this increase dissapears because the large grains, although they can be lifted from the surface, are very rare and the main contribution to the dust to gas mass ratio comes from the small dust grains. Larger values of N lead to a slight increase in the dust to gas mass ratio and to values about nearly one order of magnitude lower than for $N = 3.5$. An increase of N with time after perihelion would therefore mean a strong decrease of the dust to gas mass ratio.

4.   Dust Flux

From the solution of equations (1) one obtains the density of dust particles and their nucleocentric velocities. It is therefore possible to determine the fluxes of dust particles as a function of dust size and nucleocentric distance. In Fig. 3 the area fluxes, which is the projected area of dust passing through an area unit per time interval, are shown for five different nucleocentric distances. The data shown are computed for $N = 3.5$ and $M = 15.4$. The area fluxes are the highest for dust particles of a few μm in size. For large particles with radii of some cm, the area fluxes correspond to number fluxes in the order of only 10^{-6} sec^{-1} m^{-2} at 20 km distance from the center of the nucleus. The dashed lines give the cumulative area fluxes. The total area fluxes integrated over all dust particle radii are $9.1 \cdot 10^{-4}$ m^2 (m^2 sec)$^{-1}$ at 20 m above the nucleus surface and this decreases to $9.1 \cdot 10^{-6} \text{ m}^2 (\text{m}^2 \text{ sec})^{-1}$ at 20 km distance. These results are computed at the maximum activity of comet 67P/C–G and with an active fraction of 10%. Therefore the results represent the worst case scenario. Since the activity is significantly lower at larger heliocentric distances, the dust fluxes are much smaller there.

5.   Summary and Conclusions

The dust velocities as a function of particle size were computed based on the Afρ value at the maximum OH production rate from LOCD. At this time (December 14, 1982, r_h = 1.36 AU) the dust to gas mass ratio determined by Weiler et al. (2004) was maximal.

The dust velocities computed in a test particle approach by Weiler et al. (2004) for this date are compared with dust velocities determined using a continuum approach taking the dust–gas interaction into account. The new values were found to be only about 3% smaller than the results from the test particle approach and thus in very good agreement. The mass loading of the gas flow can therefore be neglected when computing the dust production rates of comet 67P/C–G.

The exact size distribution of dust particles in comet 67P/C–G is not known. A value of N close to 3.5 is usually assumed for comets, but changes of N along the comets orbit have also been suggested for comet 67P/C–G (Fulle et al., 2004). To estimate the effect on the dust to gas mass ratio, we varied N from 3.5 to 4.5. Using a high value for N leads to a decrease of the dust to gas mass ratio from about 4.8 to values close to 0.6. Thus, if N varies along the orbit as suggested, the change in the dust to gas mass ratio would be very high.

From the results of the gasdynamical model we obtain also the dust particle fluxes in the coma of comet 67P/C–G. As an estimate at the maximum of cometary activity we compute a total area flux of $9.1 \cdot 10^{-6} \text{ m}^2 (\text{m}^2 \text{ sec})^{-1}$ at 20 km nucleocentric distance. However, the dust flux is significantly reduced at larger heliocentric distances, and the dust production rate is reduced by an order of magnitude compared to the peak value at 1.86 AU heliocentric distance.

References

A'Hearn, M. F., Schleicher, D. G., Millis, R. L., et al. (1984). *AJ*, 89:579–591.
A'Hearn, M. F., Millis, R. L., Schleicher, D. G., et al. (1995). *Icarus*, 118:223–270.
Divine, N. (1981). ESA–SP 174.
Fulle, M., Barbieri, G., Cremonese, G., et al. (2004). *A&A*, submitted.
Hanner, M. S., Tedesco, E., Tokunaga, A. T., et al. (1985). *Icarus*, 64:11–19.
Jorda, L. (1995). PhD thesis, Observatoire de Paris-Meudon.
Kitamura, Y. (1986). *Icarus*, 66:241–257.
Knollenberg, J. (1994). PhD thesis, Georg-August-Universitaet zu Goettingen.

Krishna Swamy, K. S. (1991). *A&A*, 241:260–266.

Lamy, P. L., Toth , I., Weaver, H., et al. (2003). AAS/Division for Planet. Sci. Meeting, 35.

Newburn, R. L., Spinrad, H. (1985). *AJ*, 90:2591–2608.

Osip, D. J., Schleicher, D. G., Millis, R. L. (1992). *Icarus*, 98:115–124.

Probstein, R. F. (1969). in "Problems of Hydrodynamics and Continuum Mechanics", 568.

Weiler, M., Rauer, H., Knollenberg, J., et al. (2003). *A&A*, 403:313–322.

Weiler, M., Rauer, H., Helbert, J. (2004). *A&A*, in press.

OBSERVATIONS OF COMET 67P/CHURYUMOV-GERASIMENKO WITH THE INTERNATIONAL ULTRAVIOLET EXPLORER AT PERIHELION IN 1982

Paul D. Feldman
Department of Physics and Astronomy, The Johns Hopkins University, Baltimore, Maryland 21218, USA
pdf@pha.jhu.edu

Michael F. A'Hearn
Astronomy Department, University of Maryland, College Park, Maryland 20742, USA

Michel C. Festou
Observatoire Midi-Pyrénées, 14, avenue E. Belin, F-31400 Toulouse, France

Abstract

Comet 67P/Churyumov-Gerasimenko was observed with the *International Ultraviolet Explorer* (*IUE*) satellite observatory on 7 November 1982, five days before perihelion. At the time of the observation the heliocentric distance was 1.307 AU and the geocentric distance 0.414 AU. Only two relatively short exposures were obtained, one each with the short (SWP) and long wavelength (LWR) cameras. In the LWR exposure, OH, continuum, and weak CS emission are detected. Only H I Lyman-α is detected in the SWP exposure. From the LWR data we derive a water production rate of 6.6×10^{27} molecules s^{-1} and a value of $Af\rho$ at 2960 Å of 180 cm. These results are consistent with those derived from ground-based photometry and radio observations. Compared with comet 46P/Wirtanen, whose ultraviolet spectrum was observed with the Faint Object Spectrograph on *HST* in January 1997, also near perihelion, the water production rate of comet 67P is comparable but the dust environment is richer by a factor of ten. An attempt to observe the comet again in January 1996 was unsuccessful due to scattered light in the tracking camera of *IUE*.

L. Colangeli et al. (eds.), The New ROSETTA Targets, 47–52.

1.   The International Ultraviolet Explorer

The *International Ultraviolet Explorer* (*IUE*) satellite was launched in January 1978 and operated jointly by NASA (GSFC) and ESA (Vilspa) until September 1996 (the last comet observed was C/1995 O1 Hale-Bopp). The instruments consisted of two spectrographs covering the spectral range from 1180 to 3400 Å at ∼10 Å resolution in low dispersion mode. Over 50 comets were observed (some on several apparitions) during this period (see Festou and Feldman, 1987 for a review of early results). Some of the brightest comets were also observed in high dispersion mode giving a spectral resolving power of 3,000. Comet 67P/Churyumov-Gerasimenko was observed briefly in 1982 five days before its perihelion. The observing conditions are listed in Table 1. An attempt to observe the comet again in January 1996 was unsuccessful due to scattered light in the tracking camera (FES) of *IUE*.

Table 1.   Observational Parameters

Comet	67P/Churyumov-Gerasimenko	6P/D'Arrest	46P/Wirtanen
Date	7 November 1982	2 October 1982	15 January 1997
r (AU)	1.307	1.309	1.308
Δ (AU)	0.414	0.808	1.690
$\dot{r}$ (km s^{-1})	−0.83	3.39	−12.50
Phase angle (degrees)	33.9	49.8	35.3
Telescope	IUE		HST/FOS
Aperture	$9'' \times 15''$		$1.''29 \times 3.''66$
Image ID/	SWP 18498	SWP 18177	—
Exposure Time (s)	450	1200	—
	LWR 14579	LWR 14317	y3dj1302t
	2400	5400	1280

2.   Analysis

For our analysis we used reprocessed (NEWSIPS, Garhart et al., 1997) data taken from the *IUE* Final Archive (available from the Multimission Archive at STScI – http://archive.stsci.edu/mast.html). The spectra are shown in Figure 1 and the extracted brightnesses of the principal emission features are given in Table 2. The top panel of Figure 1 also shows H I Lyman-α emission in a short wavelength spectrum (SWP 18497) recorded 2 arc-minutes (36,000 km) east of the comet, which was intended to provide a measure of the geocoronal background but includes a small cometary component. For comparison the table also includes the same emissions in comet 6P/D'Arrest observed by *IUE*

Table 2. Observational Results

Comet	67P/Churyumov-Gerasimenko	6P/D'Arrest	46P/Wirtanen[a]
Brightness (rayleighs)			
OH(0,0) 3085 Å	910	2330	1180
CS(0,0) 2576 Å	40	33	11
Lyman-α 1216 Å	3600	7800	—
$Af\rho$ (at 2950 Å) (cm)	180	< 110	15 ± 3
$Q(H_2O)$ (molecules s^{-1})	6.6×10^{27}	1.4×10^{28}	5.1×10^{27}
$Q(CS_2)$ (molecules s^{-1})	6.6×10^{24}	9.2×10^{24}	1.4×10^{24}
$Q(CS_2)/Q(H_2O)$	1.0×10^{-3}	6.5×10^{-4}	2.8×10^{-4}
Active area (%)	4^b	—	—

[a] Stern et al. (1998); $\tau(CS_2)$ at 1 AU taken to be 345 s.
[b] Nucleus radius from Lamy et al. (2003).

during the same epoch at the same heliocentric distance (Festou et al., 1992) and with 46P/Wirtanen observed by *HST*/FOS in January 1997, also at same heliocentric distance (Stern et al., 1998). The water production rate, also given in Table 2, was evaluated from the observed brightness of the OH $A\,^2\Sigma^+$ – $X\,^2\Pi$ (0,0) band at 3085 Å using the vectorial model of Festou (1981) together with heliocentric velocity dependent fluorescence efficiencies from Schleicher and A'Hearn (1988) and solar activity dependent photodissociation lifetimes from Budzien et al. (1994). Note that at this heliocentric distance the activity levels of comets Churyumov-Gerasimenko and Wirtanen, the present and former *Rosetta* mission targets, are comparable, while comet d'Arrest has a gas production rate about a factor of two higher.

A similar analysis was carried out for CS. In this case the photodissociation lifetime of CS_2 (assumed to be the primary parent of CS) at 1 AU was taken to be 1000 s based on recent *HST*/STIS observations of comets C/1999 H1 (Lee) (Feldman et al., 1999) and 153P/Ikeya-Zhang. The literature gives a range of lifetimes of 300–500 s (eg Prisant and Jackson, 1987) and a value of 345 s was used in the analysis of the comet Wirtanen spectra by Stern et al. (1998). The value of $Q(CS_2)$ in comet Wirtanen, when a CS_2 lifetime of 1000 s is used becomes 4.2×10^{24} molecules s^{-1} and the ratio $Q(CS_2)/Q(H_2O)$ becomes 8.4×10^{-4}, so that all three comets exhibit comparable relative abundance of the CS parent.

We also derive a value of $Af\rho$ at 2950 Å, a quantity that is a measure of the dust production rate, which appears to be a factor of ten higher in comet Churyumov-Gerasimenko than in comet Wirtanen. However, this enhance-

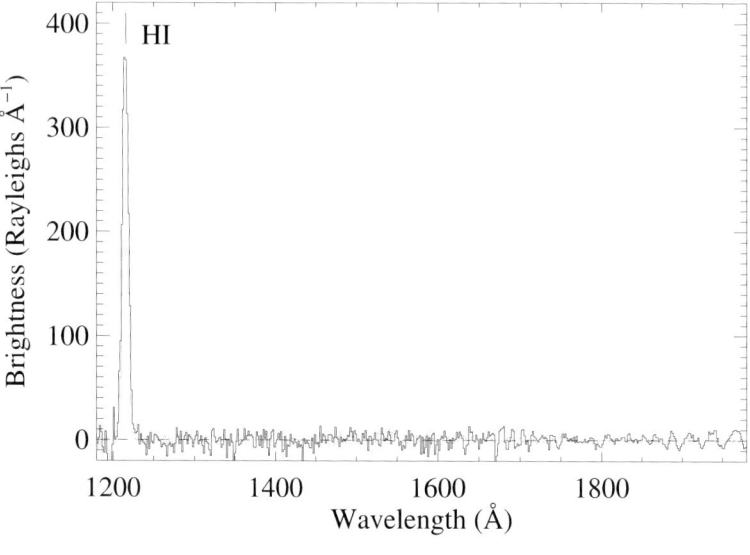

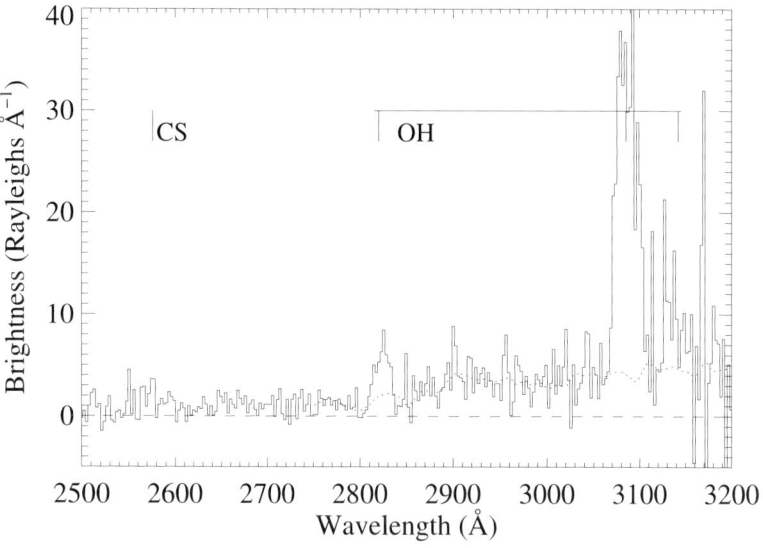

Figure 1. *IUE* spectra of comet 67P/Churyumov-Gerasimenko observed on 7 November 1982. Top: SWP 18498, 450 second exposure. The gray line is the previous exposure, SWP 18497, 900 seconds, recorded at a position 2 arc minutes east of the comet. Bottom: LWR 14579, 2400 second exposure. The emissions are identified in the figure. The dotted line is a scaled solar spectrum convolved to the instrument resolution.

ment does not seem to be as pronounced in the visible and the discrepancy between the ultraviolet and visible values of $Af\rho$ in comet Wirtanen (Schulz et al., 1998) remains to be accounted for.

3.   Comparison with Ground-based Observations

The 1982 apparition was very favorable for ground-based telescopes and there are many reports in the literature with which our results may be compared. Limiting ourselves to pre-perihelion observations, from photometry on October 16–17, 1982 ($r = 1.34$ AU), Osip et al. (1992) derived $Q(OH) = 3.6 \times 10^{27}$ molecules s^{-1} and $Af\rho$ (visible) = 170–180 cm. From 18 cm radio observations at Nançay October 9–25, 1982 ($r = 1.32 - 1.37$ AU), Crovisier et al. (2002) report a maximum derived $Q(OH) \approx 9 \times 10^{27}$ molecules s^{-1}. These are both consistent with our derived values.

There appear to be less data from the pre-perihelion period in 2002, presumably because comet Churyumov-Gerasimenko was not selected as the new *Rosetta* mission target until spring 2003 when a campaign of observations was begun. Kidger (2003) reports $Af\rho$ derived from R-band photometry just prior to perihelion of $\sim$140 cm. This result, together with preliminary results from the post-perihelion campaign presented in various papers in this volume, suggest that the activity level of Churyumov-Gerasimenko has not changed appreciably since the 1982 apparition.

4.   Summary

Observations of comet 67P/Churyumov-Gerasimenko were made with the *IUE* on 7 November 1982, five days before perihelion. Ultraviolet spectra showed the emissions of H, OH and CS, together with dust scattered continuum. From the OH emission we derive a water production rate of 6.6×10^{27} molecules s^{-1} and the continuum gives a value of $Af\rho$ at 2960 Å of 180 cm. Compared with comet 46P/Wirtanen, observed by *HST* at the same heliocentric distance in January 1997, the water production rate of comet 67P is comparable but the dust environment is richer by a factor of ten. Emissions of O, C, S and CO, which are amongst the primary objectives of the *ALICE* ultraviolet spectrometer investigation on *Rosetta*, were not detected in the SWP spectrograph because of the very long exposure times required by *IUE* for comets at comparable levels of activity at 1.3 AU.

References

Budzien, S. A., Festou, M. C., and Feldman, P. D. (1994). Solar Flux Variability and the Lifetimes of Cometary H$_2$O and OH. *Icarus*, 107:164–188.

Crovisier, J., Colom, P., Gérard, E., Bockelée-Morvan, D., and Bourgois, G. (2002). Observations at Nançay of the OH 18-cm lines in comets. The data base. Observations made from 1982 to 1999. *Astron. Astrophys.*, 393:1053–1064.

Feldman, P. D., Weaver, H. A., A'Hearn, M. F., Festou, M. C., McPhate, J. B., and Tozzi, G.-P. (1999). Ultraviolet Imaging Spectroscopy of Comet Lee (C/1999 H1) with HST/STIS. *BAAS*, 31:1127.

Festou, M. C. (1981). The density distribution of neutral compounds in cometary atmospheres. I - Models and equations. *Astron. Astrophys.*, 95:69–79.

Festou, M. C. and Feldman, P. D. (1987). Comets. In Kondo, Y., editor, *ASSL Vol. 129: Exploring the Universe with the IUE Satellite*, pages 101–118.

Festou, M. C., Feldman, P. D., and A'Hearn, M. F. (1992). The gas production rate of periodic comet d'Arrest. In Harris, A. W. and Bowell, E., editors, *Asteroids, Comets, Meteors 1991*, pages 177–182. Houston:LPI.

Garhart, M. P., Smith, M. A., Turnrose, B. E., Levay, K. L., and Thompson, R. W. (1997). International Ultraviolet Explorer New Spectral Image Processing System Information Manual: Version 2.0. *IUE NASA Newsletter*, 57:1–267.

Kidger, M. R. (2003). Dust production and coma morphology of 67P/Churyumov-Gerasimenko during the 2002-2003 apparition. *Astron. Astrophys.*, 408:767–774.

Lamy, P. L., Toth, I., Weaver, H., Jorda, L., and Kaasalainen, M. (2003). The Nucleus of Comet 67P/Churyumov-Gerasimenko, the New Target of the Rosetta Mission. *BAAS*, 35:30.04.

Osip, D. J., Schleicher, D. G., and Millis, R. L. (1992). Comets - Groundbased observations of spacecraft mission candidates. *Icarus*, 98:115–124.

Prisant, M. G. and Jackson, W. M. (1987). A rotational-state population analysis of the high-resolution IUE observation of CS emission in comet P/Halley. *Astron. Astrophys.*, 187:489–496.

Schleicher, D. G. and A'Hearn, M. F. (1988). The fluorescence of cometary OH. *Astrophys. J.*, 331:1058–1077.

Schulz, R., Arpigny, C., Manfroid, J., Stuewe, J. A., Tozzi, G. P., Cremonese, G., Rembor, K., and Peschke, S. (1998). Spectral evolution of Rosetta target comet 46P/Wirtanen. *Astron. Astrophys.*, 335:L46–L49.

Stern, S. A., Parker, J. W., Festou, M. C., A'Hearn, M. F., Feldman, P. D., Schwehm, G., Schulz, R., Bertaux, J., and Slater, D. C. (1998). HST mid-ultraviolet spectroscopy of comet 46P/Wirtanen during its approach to perihelion in 1996-1997. *Astron. Astrophys.*, 335:L30–L36.

OBSERVATIONS OF THE
NEW ROSETTA TARGETS

C. Barbieri, S. Fornasier
Dipartimento di Astronomia, Università di Padova

G. Cremonese
INAF–Osservatorio Astronomico di Padova

I. Bertini
Dipartimento di Astronomia, Università di Padova; Physikalisches Institut, Bern

M. Fulle
INAF–Osservatorio di Trieste

A. Magazzú
INAF–Centro Galileo Galilei, La Palma

Abstract   At the beginning of January 2003 the Rosetta launch was delayed and the primary target was changed. As soon as a new short period comet, Churyumov–Gerasimenko, was defined as the new target, we asked for observing time at the 3.5m Galileo Telescope (TNG) on La Palma. On February 20, 2003, we obtained images in the B, V, and R broad band filters, and a low resolution spectrum. Since the images showed interesting dust features, we asked for other deeper images in the R band, which were obtained on March 27. These images show a very pronounced dust tail structure, called Neck Line. During the February run, the heliocentric and geocentric distances were 2.36 AU and 1.42 AU, respectively, and 2.62 AU and 1.70 AU in the March run. The spectrum obtained on Feb. 20 with the low resolution spectrograph in the visible range doesn't show any particular emission, most likely because the comet appears very dusty and it was far away from the Sun. Only a weak CN emission around 388 nm could be clearly identified in the spectrum.

In addition to the comet, we have observed two asteroids that could be selected as target in the new scenario mission, 437 Rhodia and 21 Lutetia. We have obtained one low resolution spectrum of 437 Rhodia, covering the 550–900 nm

L. Colangeli et al. (eds.), The New ROSETTA Targets, 53–60.

spectral range, with the AFOSC instrument at the 182cm telescope at Asiago on March 8, 2003. This asteroid, having one of the highest albedos, seems to reveal a spectroscopic behaviour consistent with the S type, and in particular with the Sl type in the Bus taxonomy, even if this is in contrast with the high albedo typical of an E class asteroid.

For 21 Lutetia, previously observed with B&C spectrograph at 1.22m Asiago telescope on August 24, 2000, we obtained a featureless visible spectrum that confirms the M–class classification of this object.

1.   Introduction

The Rosetta scenario mission was completely revisited after the launch postponement due to problems with the Ariane 5 launcher. The main consequence was the change of the mission targets. The main target of Rosetta, whose first launch opportunity will be on 26 February 2004, has been changed from comet 46P/Wirtanen to comet 67P/Churyumov–Gerasimenko. After a journey of nearly 10 years the spacecraft will rendez-vous with the comet in mid 2014 at an heliocentric distance of 4 AU.

Comet 67P is a Jupiter family comet as the previous target, with an orbital period of 6.57 year. This comet is much larger than comet Wirtanen: the nucleus has an estimated diameter of 3×5 km.

If the primary target is fixed in the new Rosetta scenario mission, the asteroids targets are still unknown and their choice will be performed only after the mission launch, depending on the residual Δv. If the available Δv will be low, then only a single asteroid fly-by might be performed by Rosetta, otherwise a double fly-by with two different asteroids might be planned. On December 2003 the possible asteroid targets have been discussed at ESTEC on the basis of the Δv required during the fly-bys, their physical properties and the best scientific return for the mission. A list of candidates have been proposed both for the single and double asteroid(s) fly-by scenario.

At the present time the principal target for a single asteroid baseline is 2513 Baetsle, according to the minimum Δv (16 m/s) required during its fly-by. For the double fly-by scenario, a Δv of about 130–150 m/s is needed and the couples of asteroids 5538 Luichewoo & 21 Lutetia and 437 Rhodia & 21 Lutetia have been selected as primary and secondary choice. The scientific community asked for selecting 21 Lutetia as main asteroid target because, despite the high Δv required also for a single fly-by (about 130 m/s), it is the only asteroid in the candidates list large enough (96 km) to allow the mass determination (and consequently the density determination) by the radio science experiment. Most of the asteroid candidates are quite small and, except Lutetia, their physical properties are unknown or poorly studied.

In order to improve the knowledge of the new Rosetta targets we observed comet 67P and the two possible asteroid targets 437 Rhodia and 21 Lutetia.

2.   Comet 67P

Comet 67P Churyumov-Gerasimenko was observed with the 3.5m Galileo Telescope (TNG) on La Palma. Spectrophotometric data were obtained on February 20, 2003 with DOLORES instrument using B, V, and R broad band filters and the low resolution LR-B grism to cover the wavelength range 370-800 nm. Images in R filter were taken also on March 27, 2003 (Table 1). Images were bias and flat field corrected before their analysis. No photometric standard stars were observed, so we cannot give the magnitude of the comet. The image obtained on March 2003 has been used, together with other images taken by other observers, to provide important data on the dust production and to interpret the dust tail structure. Two models have been applied (Fulle, 1989; Fulle and Sedmak, 1988) allowing to identify the dust tail structure as a Neck-Line providing a dust mass loss rate of 300 kg s^{-1} before perihelion and 10 kg s^{-1} after perihelion. In Figure 1 we show the best model fit referred to the isotropic dust ejection.

Further details and results on the image are reported in a separate work by Fulle et al. (2004).

Table 1.   Comet 67P: observational circumstances, where r and Δ are respectively the heliocentric and geocentric distances, α the phase angle and m$_v$ the predicted magnitude from JPL ephemeris service.

date	UT start	T_{exp}	Mode	m_v	r (AU)	$\Delta(AU)$	$\alpha(^o)$
21 Feb 03	00:29	120	B imaging	14.0	2.361	1.416	9.1
21 Feb 03	00:33	60	V imaging	14.0	2.361	1.416	9.1
21 Feb 03	00:35	60	R imaging	14.0	2.361	1.416	9.1
21 Feb 03	00:53	1800	spectrum	14.0	2.361	1.416	9.1
27 Mar 03	00:25	600	R imaging	14.8	2.621	1.695	10.0

During 20 February night we obtained also a spectrum of comet 67P. The spectrum was bias and flat field corrected, then cosmic rays were removed and the 2 dimensional spectrum collapsed to one dimension, considering the lines within 2 FWHM of the assumed Gaussian comet profile. Extinction correction was performed using the La Palma coefficients extinction table. The spectrum was wavelength calibrated using the reference spectrum of an helium lamp, acquired during the same night.

Finally the spectrum was flux calibrated using the reference spectrum of the spectrophotometric standard star HD 93521, observed just after the comet spectrum acquisition.

In Figure 2 we report the comet reflectance spectrum without the sky background removal: in fact a sky spectrum was not acquired and the comet was quite dusty making impossible to properly estimate the sky background nearby the comet spectrum. In Figure 2 the three emission lines around 5577, 5890

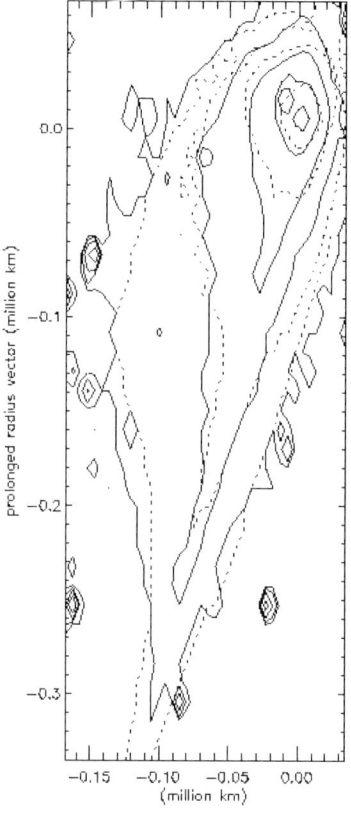

Figure 1. Image of comet 67P obtained on 27 March 2003 in the R band. The isophote steps differ by a factor 2 in intensity. Continuous line: isophotes of the observed image. Dashed lines: isophotes computed by the Montecarlo Tail model.

and 6300 Å are due to the oxygen and sodium sky emission. So at the time of observations, corresponding to an heliocentric distance of 2.36 AU, the comet had a very low cometary gas emission, as only a weak CN emission at 3880 Å could be clearly seen in our spectrum (Fig. 2).

The flux calibrated comet spectrum is shown in Figure 3. Only the oxygen and sodium sky lines, clearly identified by the comparison with the standard star background, were removed on the comet spectrum.

3. Asteroids

437 Rhodia, one of the possible new asteroidal targets was observed on March 2003 at the 1.8m Asiago telescope with the low resolution grism #2 in the spectral range 520-900 nm. We recover also the spectrum of 21 Lutetia,

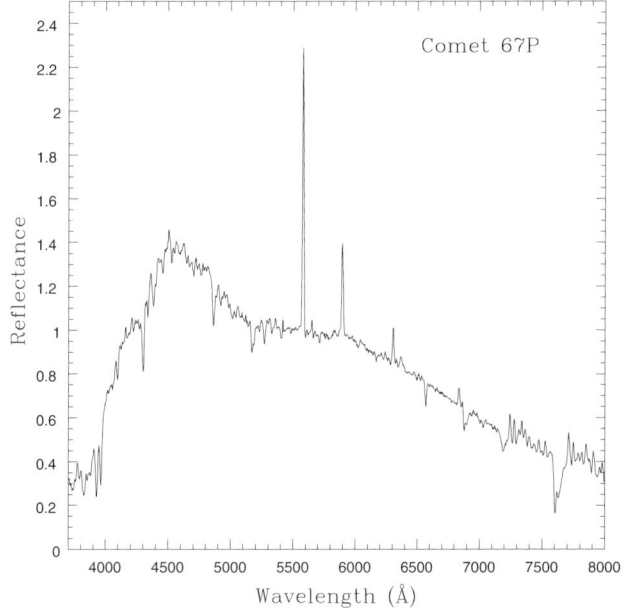

Figure 2. Spectrum of comet 67P, not corrected for the sky background.

another possible asteroidal new target, obtained on August 2000 with the 1.2m Asiago telescope equipped with the B&C spectrograph. Here we used the 150 gr/mm grating to cover the spectral range from 400 to 850 nm. The spectra obtained were reduced using ordinary procedures of data reduction with the software package Midas, including:

- subtraction of the bias from the raw data

- flattening of the data in order to remove large scale structures

- cosmic ray removal

- background subtraction

- collapsing the two dimensional spectra

- wavelength calibration

- atmospheric extinction correction

The reflectivity of the asteroids was then obtained by dividing the spectra of the objects by that of a solar analog spectrum acquired at similar airmass. Spectra have been normalized at 5500 Å and smoothed with a median filter technique.

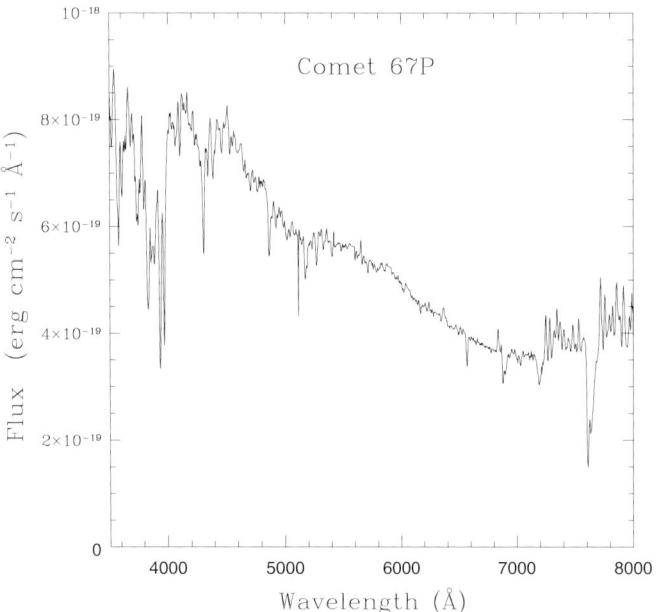

Figure 3. Flux calibrated spectrum of comet 67P, not corrected for the sky background; only the oxygen and sodium sky lines were removed.

437 Rhodia

437 Rhodia was observed on 8 March 2003 at a phase angle of 20° and an heliocentric distances of 2.82 AU. From IRAS data it has a diameter of 14 km and the highest albedo in the asteroid population (0.56 ± 0.03, Tedesco and Veeder, 1992). This high albedo is associated to igneous high thermally evolved E type asteroids.

However our visible spectrum (Fig. 4) is not similar to that of E type asteroids but it shows a good match with the Sl type (a subtype of the S class) following Bus classification scheme (1999). This is quite surprising as S type albedo is typically less than 0.3.

Observations in the near infrared region are necessary to clearly identify the spectral class of 437 Rhodia: if the 1 and 2 micron bands associated to olivine and pyroxene will be revealed in the infrared spectrum, than 437 Rhodia might really be an S-type.

Also polarimetric observations are welcome to obtain an independent measurement of the albedo and to verify the high value derived from IRAS observations.

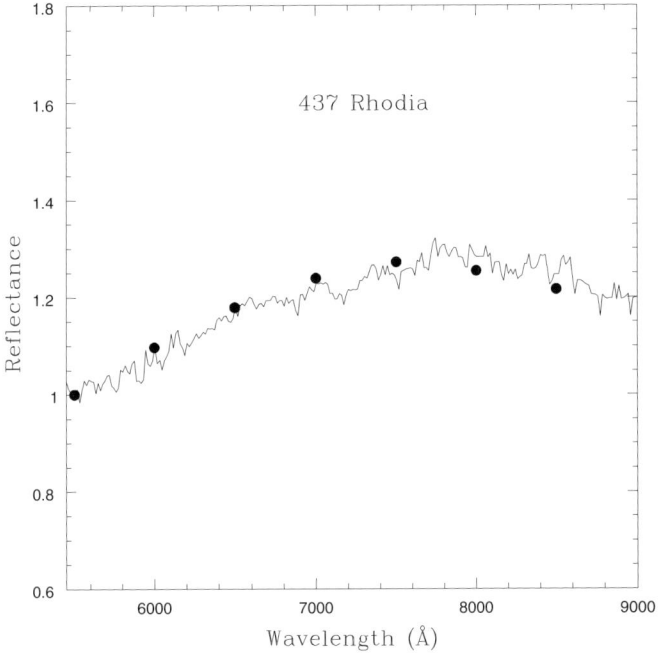

Figure 4. Visible spectrum of 347 Rhodia. The points represent the mean Sl type reflectance (from Bus, 1999).

21 Lutetia

21 Lutetia was observed on 24 August 2000 at a phase angle of 18.4^o and an heliocentric distances of 2.06 AU. From IRAS data it has a diameter of about 96 km and an albedo of 0.22; it is a well studied asteroid. Our spectrum (Fig. 5) confirms the spectral data previously published in literature; from our visible spectrum and according to its albedo (0.22±0.02, Tedesco and Veeder, 1992), 21 Lutetia has a featureless reddish spectrum typical of M-type asteroids (Gregnanin et al, 2003); it is probably composed of metals such iron or nickel. M type asteroids are believed to be the progenitors both of differentiated meteorites such as iron meteorites and of undifferentiated meteorites like the enstatite chondrites.

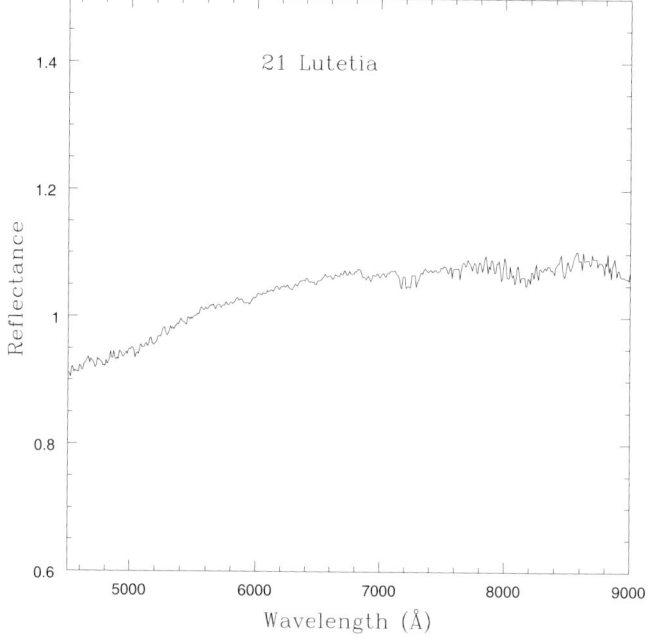

Figure 5.   Visible spectrum of 21 Lutetia

Acknowledgments

The authors thank Dr. Dino Fugazza for his help in performing observations at the TNG telescope.

References

Bus, S.H., 1999. Compositional structure in the asteroid belt: Results of a spectroscopic survey. *MIT PhD thesis.* DAI-B, 61/01, 311.

Fulle, M. & Sedmak, G. (1988). Photometrical analysis of the Neck-Line structure of Comet Bennet 1970II. *Icarus*, 74:383.

Fulle, M. (1989). Evaluation of cometary dust parameters from numerical simulations - Comparison with an analytical approach and the role of anisotropic emissions. *A&A*, 217:283.

Fulle, M., Barbieri, C., Cremonese, G., Rauer, H., Weiler, M., Milani, G., Ligustri, R. (2004). The dust environment of comet 67P/Churyumov-Gerasimenko: a challenge for the Rosetta probe? *A&A*, submitted.

Gregnanin, A., Fornasier, S., Barbieri, C. (2003). Visible spectroscopy of minor bodies from the 1.22m Asiago telescope. *V Convegno di Scienze Planetarie*, Gallipoli, September 2003.

Tedesco, E. F., and Veeder, G. F. (1992). IMPS albedos and diameter catalog. In Tedesco, E. F., Veeder, G. F., Fowler, J. W., and Chillemi, J. R., editors, *The IRAS Minor Planet Survey*, Tech. Rep. PL–TR–92–2049. Phillips Laboratory, Hanscom AF Base, MA.

WATER PRODUCTION RATE OF COMET 67P/CHURYUMOV–GERASIMENKO

J. Teemu T. Mäkinen

ESA/RSSD, Keplerlaan 1, SCI–SB, 2201 AZ Noordwijk ZH, The Netherlands

tmakinen@rssd.esa.int

Abstract   The Solar Wind Anisotropies (SWAN) instrument is a scanning Lyman–alpha imager on board the Solar and Heliospheric Observatory (SOHO) spacecraft. Since becoming operational in January 1996 SWAN has been producing full sky Lyman–alpha maps which are primarily used to study the interaction between solar wind and the interplanetary neutral hydrogen. In addition to that SWAN images can be used to study the hydrogen coma of comets down to about a visual magnitude of 12. After the retargeting decision of the Rosetta mission the SWAN archive was checked for possible occurrences of 67P/Churyumov–Gerasimenko. Five values were obtained for the 1996 apparition but none for the 2002 apparition because of degraded instrument sensitivity and larger observing distance. The observations suggest a perihelion water production rate of about $8 \cdot 10^{27}\,\mathrm{s}^{-1}$ and possible postperihelion increase of activity.

Keywords: comets, water production, ultraviolet observations, Rosetta

1.   Introduction

The Solar Wind Anisotropies (SWAN) instrument (Bertaux et al., 1995) is a scanning Lyman–alpha imager on board the Solar and Heliospheric Observatory (SOHO) spacecraft. It consists of two identical sensors each capable of accessing one ecliptic hemisphere with $5° \times 5°$ instantaneous field–of–view and $1° \times 1°$ resolution. In the normal operation mode SWAN is able to cover the entire sky in about one day. The main scientific objective of the instrument is to study the properties of the solar wind through its ionizing effect on the interplanetary medium (IPM) of neutral hydrogen atoms. In addition to that several comets have been detected based on reflected solar Lyman–alpha light from their hydrogen coma (Mäkinen et al., 2001). Because cometary neutral hydrogen comes predominantly from photodissociation of water, SWAN observations can be used to estimate the water production rate of a comet. High spatial and temporal coverage combined with the location of SOHO in the first Lagrangian point between the Earth and the Sun, far away from the

L. Colangeli et al. (eds.), The New ROSETTA Targets, 61–67.
© 2004 *Kluwer Academic Publishers. Printed in the Netherlands.*

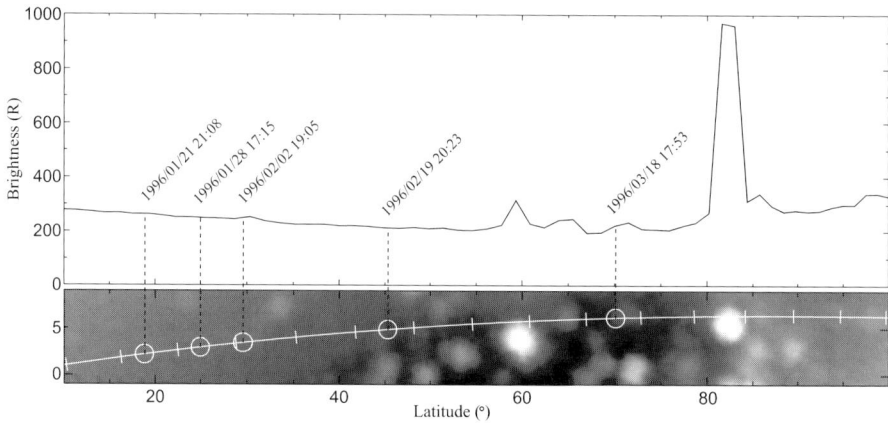

Figure 1. The apparent path of 67P on the sky with tick marks at seven day intervals in ecliptic coordinates as seen by SWAN, and the intensity of the background Lyman–alpha radiation from interplanetary neutral hydrogen and UV bright stars along the path. Comet locations during the SWAN observations are depicted by circles.

Lyman–alpha interference by the geocorona, makes the SWAN cometary data set unique both in extent and internal consistency.

Because the instrument was designed for observing the large–scale structure of IPM there are, however, several shortcomings for cometary observations. First of all, the relatively high background contribution from the IPM restricts the sensitivity of the instrument to comets of approximately visual magnitude 12 or brighter. Because of the fairly wide spectral window of the instrument (115 – 180 nm) the images contain a large number of UV bright stars whose interference is made even worse by the noticeable chromatic aberration and poor line–of–sight retrieval capability (around 1°) of the instrument. Therefore a long–term programming effort has been taken to develop suitable tools for SWAN cometary data analysis, with the operational stage of the software just recently reached.

SOHO was launched in December 1995 and the first SWAN full sky map was completed on Jan 21, 1996, right after the perihelion passage of 67P/Churyumov-Gerasimenko which was chosen as the target comet for the Rosetta mission after the January 2003 launch window for the original target 46P/Wirtanen was abandoned because of launcher difficulties. Furthermore, the apparent path of 67P went through an area with relatively low star density in the downwind direction of the IPM where background intensity is at its lowest (Fig. 1). With these nearly optimal conditions it was possible to make several unambiguous detections of 67P from those SWAN images that were complete enough to contain the area in question.

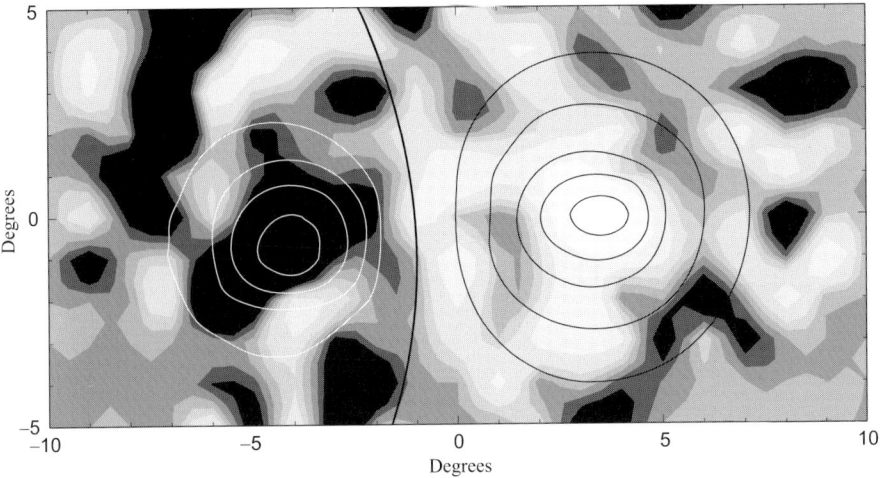

Figure 2. Difference image of January 21 (positive) and 28 (negative), 1996, SWAN full sky maps depicting two instances of the coma of 67P overlaid by best–fit model contours. Sun is on the right hand side.

2. Data analysis

Because of the intensity of the background (200 – 300 Rayleigh) was much higher than peak intensity of the coma (around 20 R), it was necessary to analyse the images by pair differencing. A pair of observations was subtracted from each other, leaving some residual background caused by varying solar illumination between observing dates, and one negative and one positive instance of the comet. A coma model for each instance and a simple polynomial for the residual background was least squares fitted to estimate the water production rate (Fig. 2). Because of the line–of–sight uncertainty, the exact location of the coma had to be found as well. Therefore a Powell–type optimizing method was used.

The model used to produce the coma profiles was a new hybrid simulation–analytical synchrone model (paper submission pending) which takes into account all relevant physical phenomena except multiple scattering, and driven by a realistic velocity distribution for hydrogen escaping the collision zone gives profiles practically identical to those obtained by full scale Monte Carlo simulations. It can be seen as an extension of the syndyname model of Keller and Meier, 1976 with varying conditions between different parts of the hydrogen cloud also properly taken into account by limited particle simulation and propagation of corrective coefficients to an analytical formula of column inten-

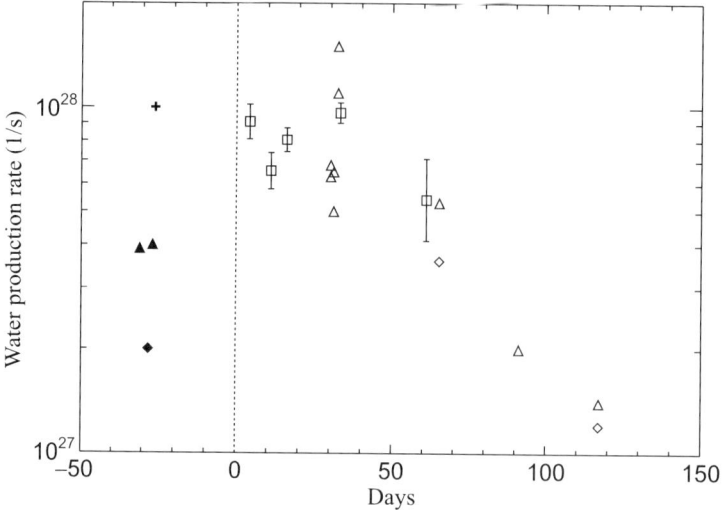

Figure 3. SWAN water production estimates (squares) for the 1996 apparition with one sigma error bars as a function of time. Other water production rate estimates from the 1982 apparition include Hanner et al., 1985 (diamonds), Osip et al., 1992 (triangles) and Crovisier et al., 2002 (cross). Preperihelion estimates are depicted as solid.

sity for hollow shells. The relevant model parameter was a hydrogen lifetime of $1.5 \cdot 10^6$ s.

3. Results

The comet could be detected in five different full sky observations between Jan 21 and Mar 18, 1996. The water production rates for these observations are given in Table 1. Because of the large extent of the hydrogen coma and a static model fit, each value represents several days worth of averaged activity.

Table 1. SWAN data analysis results giving the date as days after perihelion in 1996, heliocentric and SOHO–comet distance, applied g–factor and water production rates and one sigma error margins.

T (Days)	r_H (AU)	Δ (AU)	g_0 ($\times 10^{-3} \mathrm{s}^{-1}$)	Q_{H_2O} ($\times 10^{27} \mathrm{s}^{-1}$)	σ_Q ($\times 10^{27} \mathrm{s}^{-1}$)
4.22	1.301	1.100	1.368	9.10	1.11
11.06	1.307	1.124	1.366	6.55	0.84
16.14	1.314	1.146	1.366	8.06	0.68
33.19	1.359	1.236	1.367	9.65	0.68
61.09	1.485	1.460	1.376	5.41	1.69

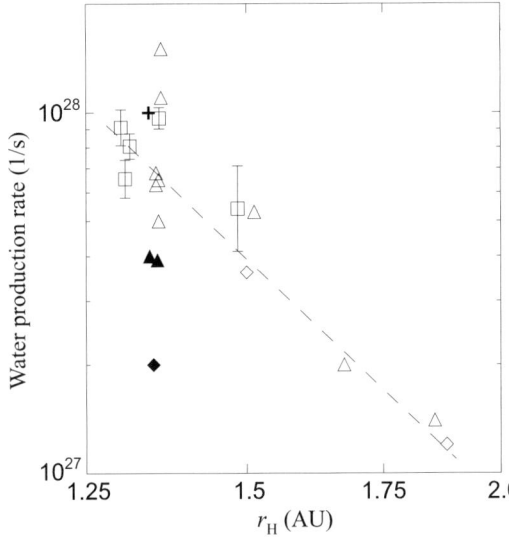

Figure 4. Water production estimates as a function of heliocentric distance. Symbols are the same as in Fig. 3. The linear heliocentric distance dependence of water production rate is depicted with a slashed line.

The SWAN values are plotted with other estimates from the 1982 apparition in Fig. 3 as a function of time. Considering the scattering of estimates the agreement between SWAN and other values is rather good especially for profiles from different apparitions. In Fig. 4 the estimates are plotted as a function of heliocentric distance. The linear fit to all values suggests a $r^{-5.4}$ heliocentric distance dependence for the water production rate but the deviation from a straight slope is quite notable.

In addition to water production estimates from ground–based observations the results can be compared with visual magnitude estimates given in the International Comet Quarterly. The total visual magnitudes m_1 must first be converted to absolute magnitudes H_0 by

$$H_0 = m_1 - 5 \log_{10} r_H - 5 \log_{10} \Delta \qquad (1)$$

where r_H is the heliocentric and Δ the geocentric distance of the comet. Jorda et al., 1992 suggest a simple relation between the absolute magnitude and water production rate Q_{H_2O} in the form of

$$\log_{10} Q_{H_2O} = a - bH_0 \qquad (2)$$

with two adjustable parameters. The available visual magnitude estimates for the 1996 apparition are predominantly preperihelion ones. Therefore a comet–specific value of $a = 30.45$ can only be given to the first fitting parameter with the second one taken to have the default value of $b = 0.24$. With these values the water production rate estimates from visual magnitudes agree with SWAN and other observations as plotted in Fig. 5.

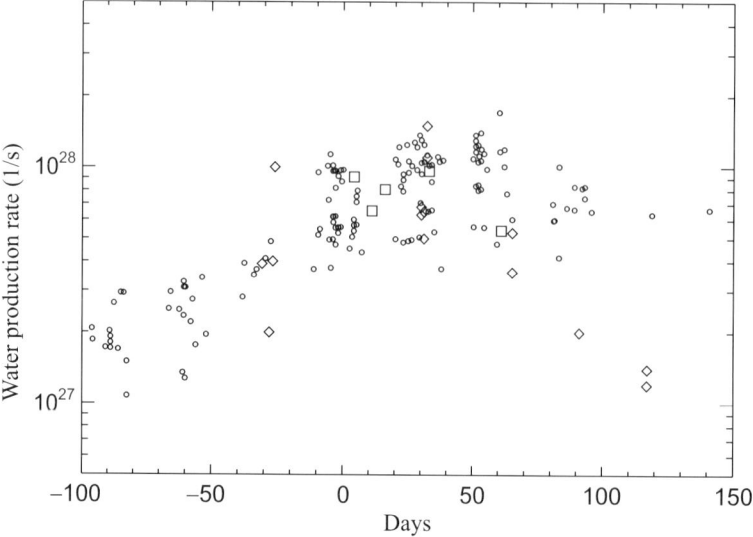

Figure 5. Water production rates from SWAN (squares) and other observations (diamonds) are compared with water production estimates derived from visual magnitudes of the 1996 apparition as given in ICQs 96 – 101 and 108 (issues 17/4 through 19/1 and 20/4).

4. Conclusions

The positive detection of 67P during the 1996 apparition depended on many conditions that happened to be favourable. In contrast the 2002 apparition was clearly beyond the sensitivity limit of the instrument because of larger geocentric distance of 1.7 AU and instrument degradation. Based on the 1982 and 1996 apparitions the comet seems to have a fairly permanent water production profile from one apparition to another. A persistent feature of the profile appears to be a postperihelion increase of activity by a factor of around 2 which can be seen both in SWAN and visual magnitude data, and in other observations to the extent that their consistency allows. This type of behaviour strongly suggests a seasonal effect.

References

Bertaux, J. L., Kyrola, E., Quemerais, E., Pellinen, R., Lallement, R., Schmidt, W., Berthe, M., Dimarellis, E., Goutail, J. P., Taulemesse, C., Bernard, C., Leppelmeier, G., Summanen, T., Hannula, H., Huomo, H., Kehla, V., Korpela, S., Leppala, K., Strommer, E., Torsti, J., Viherkanto, K., Hochedez, J. F., Chretiennot, G., Peyroux, R., and Holzer, T. (1995). SWAN: A Study of Solar Wind Anisotropies on SOHO with Lyman Alpha Sky Mapping. *Solar Phys.*, 162:403–439.

Crovisier, J., Colom, P., Gérard, E., Bockelée-Morvan, D., and Bourgois, G. (2002). Observations at Nançay of the OH 18-cm lines in comets. The data base. Observations made from 1982 to 1999. *Astron. Astrophys.*, 393:1053–1064.

Hanner, M. S., Tedesco, E., Tokunaga, A. T., Veeder, G. J., Lester, D. F., Witteborn, F. C., Bregman, J. D., Gradie, J., and Lebofsky, L. (1985). The Dust Coma of Periodic Comet Churyumov-Gerasimenko (1982 VIII). *Icarus*, 64:11–19.

Jorda, L., Crovisier, J., and Green, D. W. E. (1992). The correlation between water production rates and visual magnitudes in comets. In *Asteroids, Comets, Meteors 1991*, pages 285–288.

Keller, H. U. and Meier, R. R. (1976). A cometary hydrogen model for arbitrary observational geometry. *Astron. Astrophys.*, 52:273–281.

Mäkinen, J. T. T., Bertaux, J.-L., Pulkkinen, T. I., Schmidt, W., Kyrölä, E., Summanen, T., Quémerais, E., and Lallement, R. (2001). Comets in full sky L_α maps of the SWAN instrument. I. Survey from 1996 to 1998. *Astron. Astrophys.*, 368:292–297.

Osip, D. J., Schleicher, D. G., and Millis, R. L. (1992). Comets - Groundbased observations of spacecraft mission candidates. *Icarus*, 98:115–124.

ROSETTA ASTEROID CANDIDATES

M.A. Barucci
LESIA, Observatoire de Paris, France

M. Fulchignoni
LESIA, Observatoire de Paris, France

I. Belskaya
Astronomical Observatory of Kharkiv National University, Ukraine

P. Vernazza
LESIA, Observatoire de Paris, France

E. Dotto
INAF-Osservatorio Astronomico di Roma, Monte Porzio Catone, Italy

M. Birlan
IMCCE, Observatoire de Paris, France

Abstract   The new scenario of the Rosetta mission to comet 67/P Churyumov-Gerasimenko (launch on February 2004), includes as baseline the fly-by of the asteroid 2513 Baetsle. Several other asteroids are possible fly-by candidates (single or double) within the available resources. Other candidates whose fly-bys require a larger Δv can be considered if the execution of the Rosetta interplanetary orbit insertion maneuver will allow the Rosetta Project to make available, for the asteroid fly-by, a fraction or the totality of the contingency Δv.

This paper present the history of Rosetta asteroid target selection as well as the situation concerning the choice of the Rosetta asteroid targets as it is at January 2003. A particular attention is devoted to the asteroid 21 Lutetia which represents the most interesting candidate.

L. Colangeli et al. (eds.), The New ROSETTA Targets, 69–78.

1.   Introduction

In late 1993 the European Space Agency (ESA) selected the mission Rosetta as the Planetary Cornerstone of its "Horizon 2000" program. The mission baseline included a rendez-vous with in situ investigations of a comet and fly-bys with two asteroids. The mission was named from the Rosetta Stone, due to the importance it had for the archeologists, allowing them to decipher the Egyptian hieroglyphics. In fact the knowledge of the nature of the primordial material (composing comets and asteroids) is considered by the planetologists of fundamental importance for the understanding of the origin and evolution of our solar system. The aim of the mission is to investigate the formation, the composition of planetesimals and their evolution over the last 4.5 billion years. The mission was scheduled to be launched by ESA using European technology and infrastructure. Considering a dedicated Ariane V launch followed by two or three gravity-assist swing-bys with the Earth, Venus and/or Mars, the ESA science team identified a number of mission scenarios which include different possible targets.

2.   History of the asteroid target selection

On the Rosetta phase A Report (ESA SCI(93)7), ESOC listed many possible asteroid fly-bys both for rendezvous with the comet P/ Wirtanen and for some others comets rendezvous opportunities.

To select the most primitive targets and to complete the scenario of the asteroid already visited by space missions, an international workshop was organized at Max-Planck Institute (Katlenbur-Lindau) in May 1994. The scientific community underlined the necessity to observe all the asteroid candidates suggested by ESA. Asteroids represent a vast heterogeneous population of small bodies with a wide range of orbital, physical and compositional characteristics. Although some asteroids can be differentiated and/or have experienced a collisional evolution, most of them have undergone relatively little thermal and geological evolution since their formation. A considerable amount of information regarding some of the primordial processes which governed the evolution of the whole solar system, immediately after the collapse of the protoplanetary nebula and before the formation of the planets is "frozen" in the asteroid population. The asteroids belonging to the taxonomic classes of C (carbonaceous chondrite-like material) and D (volatile-rich ultracarbonaceous material) are considered quasi-unaltered, volatile-rich objects and for these reasons the scientific community recommended to include in the mission the fly-by of these primitive objects.

Among various targets, the comet P/Wirtanen was selected as baseline with two fly-bys of the asteroids 3840 Mimistrobel and 2530 Shipka on the basis of the minimum Δv expenses criterium. The announcement of Opportunity published by ESA on March 1995 established the launch date in January 22, 2003.

Barucci and Lazzarin (1995) observed the two targets spectroscopically at CFHT (Mauna Kea observatory) deducing that Mimistrobel and Shipka revealed to belong to S and B classes respectively. As for each selected comet, several asteroid fly-bys were possible, Barucci and Lazzarin suggested to find other less evolved targets.

In 1996, refining the spacecraft trajectory, ESA redefined the mission baseline, changing the second target (Shipka) to 2703 Rodari, selected by the Rosetta Project always on the basis of the minimum Δv cost criterium. Barucci et al. (1998) observed the new target together with all the other possible candidates. The conclusion of their work was that Rodari was again another S type asteroid, as 951 Gaspra and 243 Ida already visited by the Galileo mission and 433 Eros target of the NEAR mission. On the basis of their spectral analysis they concluded that 140 Siwa was the best target. The obtained data indicated that Siwa is a more primitive object, belonging to the C taxonomic class. Due to the spectral type and its large diameter (110 km), 140 Siwa was strongly pushed to be the primary asteroid target of the mission. After Barucci et al. (1998) study and suggestion, ESA selected in early 1999, 140 Siwa and 4979 Otawara as the asteroid targets in the new baseline for the Rosetta mission. The asteroid Otawara was added for the small increase of the total Δv. During its cruise to 46P/Wirtanen (rendez-vous), the Rosetta spacecraft was supposed to encounter Otawara on 11 July 2006 (heliocentric distance of 1.86 AU, minimum encounter distance of 2200 km and a relative velocity of 10.63 km s^{-1}) and Siwa on 24 July 2008 (at 2.75 AU from the Sun, at a minimum distance of 3500 km, and a relative velocity of 17.04 km s^{-1}).

3.   140 Siwa and 4979 Otawara

Many international observational campaigns followed to characterize the nature of Otawara and Siwa in order to optimize both the mission trajectory and the science operations. Doressoundiram et al. (1999) determined the synodic rotational period of 2.707 $\pm$ 0.005 hr for Otawara. On the basis of visual spectrum a possible taxonomic class S or V was associated at the object. Le Bras et al. (2001) observed the two candidates obtaining a precise determination of the synodic rotational period of Siwa (18.495 $\pm$ 0.005 hr) and confirming the previous determination of that of Otawara. The phase functions allowed them to determine the H and G parameters for both asteroids. The near-IR spectrum of Siwa confirmed the C/P type nature of Siwa. Just few months before the programmed Rosetta launch, Fornasier et al. (2003) gave a complete portrait of Otawara. The spin vector is presented with a retrograde sense of rotation and the axial ratio a/b=1.21 $\pm$ 0.05. The visible and the near-IR spectra allow them to classify Otawara as S type asteroid and more specifically, on the basis of the analysis of band depths and slopes, in the S(IV) subgroup, suggesting a similarity to ordinary chondrite meteorites.

4.   New baseline

In January 2003 the European Space Agency decided to postpone the launch of the spacecraft due to problems with Arianne V launcher, the new launch date has been fixed at the end of February 2004. A new baseline of the Rosetta mission including a long orbital rendez-vous with the 67/P Churyumov-Gerasimenko comet nucleus (in 2014) and one or two asteroids flybys (in the time span 2008-2010). The selection of the asteroid target(s) depends on the Δv available after the Rosetta probe interplanetary orbit insertion manoever. A few meter/sec Δv are available for the asteroid science in the pre-launch resource budget; but there is the possibility to allocate to the asteroid some of the remnant Δv, now reserved as contingency for the insertion manoeuvre, as soon as the Rosetta probe will be on its way toward the comet.

On December 2003 at ESTEC, the asteroid 2513 Baetsle was selected as target of the baseline mission. In fact a minimum extra Δv of 19 m/s is necessary to allow the spacecraft to flyby this asteroid on August 8, 2010 with a relative velocity of 8.6 km/s. 2513 Baetsle is a very small asteroid with an IRAS albedo of 0.028 and an estimated diameter of 16.7 km. The asteroid, on the basis of its orbital parameters, has been assigned to the Flora family, which members have spectra characteristic of S type. This fact is in contrast with its low albedo, typical of a C or D type objects, and for this reason it could be an interloper of the Flora family.

Many other possible targets have been individuated for the mission. In table I the single fly by and in table II the double fly-bys opportunities are listed. On December 2003 at ESTEC a priority was given also for the double fly-bys within a Δv range 30–160 m/s. The first priority was given to 437 Rhodia and 21 Lutetia and as the second one to 2867 Steins and 21 Lutetia. The scientific community strongly pushed to include Lutetia as asteroid target.

The idea is to wait the insertion maneuver and to consider the new available budget (in terms of Δv) to perform the asteroid fly-by(s). For this reason an international observational campaign is started with the aim to increase at the maximum level the characterization of the possible asteroid targets of the mission (Birlan et al. 2004).

Following the scientific objective of the mission, *21 Lutetia* represents one of the most interesting candidate.

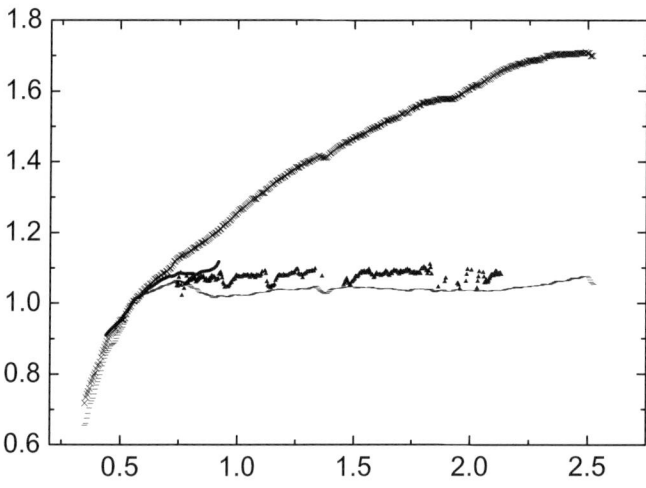

Figure 1. Spectra of Lutetia obtained by Birlan et al. (2004). The comparison with iron meteorite and carbonaceous chondrite reveals the similarity of the asteroid to C class objects.

Table 1. Single fly-by asteroid with the main physical characteristics and encounter data

Asteroid	Diameter (km)	IRAS albedo	Tax. type	Extra Δv m/s	Fly-by date dd.mm.yy	Rel. Vel. (km/s)
21 Lutetia	96	0.221±0.020	C	125	10.07.10	14.9
437 Rhodia	13	0.703±0.084	E?	97	17.09.08	11.2
1393 Sofala	–	–	?	111	11.09.08	6.9
2181 Fogelin	–	–	S	18	25.05.10	13.6
2513 Baetsle	17	0.028±0.007	C-D?	19	10.05.08	8.6
2867 Steins	–	–	?	57	07.09.08	8.6
3050 Carrera	–	–	?	102	30.07.08	11.3
3418 Izvekov	27	0.066±0.013	?	15	05.12.10	11.2
5538 Luichewoo	–	–	S?	32	08.04.09	5.6

5. 21 Lutetia: the best possible choice

21 Lutetia is the largest asteroid available in the list of the possible candidates. On the basis of IRAS observations, the estimated diameter is 95.8 ± 4.1 km with an albedo of 0.221 ± 0.020 (Tedesco and Veeder, 1992). Its synodic period is of 8.17 ± 0.01 hr (Zappala et al. 1984). Previously classified as M type (Barucci et al., 1987 and Tholen 1989), Lutetia was supposed to

74

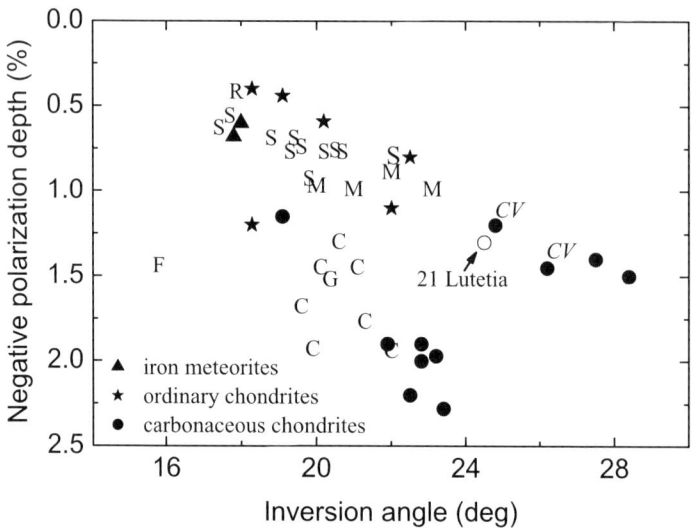

Figure 2. Diagram of the negative polarization depth versus the inversion angle of asteroids and meteorites.

be a parent body of iron meteorites. Further observations have shown that 21 Lutetia is atypical for the M taxonomic class. Its infrared spectrum is unusually flat as compared to other M-asteroids (Howell et al., 1994; Burbine and Binzel, 2002). New observations in near-infrared (Birlan et al. 2004) suggest a similarity with the carbonaceous chondrite spectra classifying this objects as a C-type asteroid (Fig 1). Its polarimetric properties are also better interpreted by a carbonaceous chondrite composition (Belskaya and Lagerkvist, 1996). The asteroid 21 Lutetia presents the largest inversion angle ever observed for asteroids and very similar to those found in laboratory for the carbonaceous chondrites of the CV type (Fig. 2). According to radar observations (Magri et al., 1999) Lutetia has the lowest radar albedo of any other M-asteroids which excludes metallic surface composition. Rivkin et al.(2000) reported the detection of the water-of-hydration absorption feature at 3 micron in its spectrum. All mentioned properties of 21 Lutetia are consistent with a carbonaceous chondrite composition of this asteroid though the high IRAS albedo leads some controversy. A lower albedo of 0.09 has been reported by Zellner et al. (1977) from both polarimetry and radiometry measurements. New measurements of Lutetia's albedo is very important for solving the controversy.

Table 2.   Double asteroid fly-bys with relative extra Δv

1st encounter between Earth and Earth	2nd encounter between Earth and comet	extra Δv m/s
437 Rhodia	21 Lutetia	159
5538 Luichewoo	21 Lutetia	129
2867 Steins	21 Lutetia	139
437 Rhodia	2181 Fogelin	113
2513 Baetsle	2181 Fogelin	79
2867 Steins	2181 Fogelin	83
5538 Luichewoo	2181 Fogelin	35
437 Rhodia	3419 Izvekov	112
1393 Sofala	3419 Izvekov	146
2513 Baetsle	3419 Izvekov	73
2867 Steins	3419 Izvekov	77
5538 Luichewoo	3418 Izvekov	32

21 Lutetia represents one of the most interesting candidates, infact it's the only one which will allow us to obtain mass determination by radio science experiments, and consequently it will be possible to determine its density. Moreover, if the chondritic character of this object will be confirmed, it will cope with the scientific objectives of the mission: the exploration of the primitive bodies of the planetary system.

6.   Possible double fly-by

After the Rosetta probe interplanetary orbit insertion maneuver, the final Δv available will be known and only at that time we will know if a double fly-bys will be possible. In table 2, a list with double asteroid fly-bys is presented. The first three represent the order of scientific interest priority. The asteroid *437 Rhodia* is an intriguing object due to its very high IRAS albedo 0.70 ± 0.08 (Tedesco and Veeder, 1992) which is the largest one ever observed for asteroids. The synodical period of ≥ 56 hours (Binzel, 1987) allows us to consider it within the slow rotator asteroid group which can imply a binary possibility for the object. There is any available spectral observations of the asteroid. However the measured B-V color of Rhodia (Binzel, 1987) together with high albedo are consistent with the E-type composition (Fig. 3).

All the other candidates are in general very small with an unknown taxonomic type. If we consider the family members having spectra quite homogeneous (Florczak et al, 1998), we can assume that the asteroid *5538 Luichewoo* belong to the S type which is characteristic of the members of the Flora family.

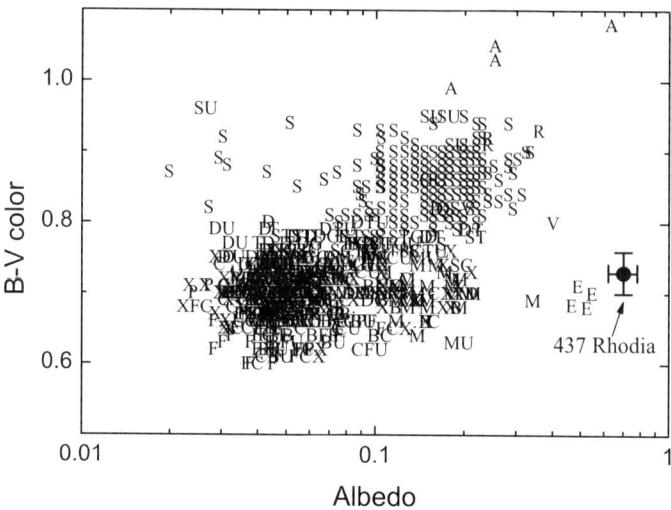

Figure 3. The diagram of the B-V colors versus albedo for asteroids of different types.

7. Conclusion

The asteroid fly-bys were part of the science definition of the Rosetta mission. The scientific community requires to have one or two asteroid visited by Rosetta during the cruise to the comet.

The final decision on the asteroid candidates will be taken after the launch when the remnant Δv will be known. Information on all the possible target asteroids is important to be able to contribute to the best choice of the targets and to optimise the mission science return.

21 Lutetia represents one of the most interesting candidates. Infact due to its large dimention, the radio science experiment will be able to obtain mass determination, and consequently it will be possible to determine its density. Moreover, if the chondritic character of this object will be confirmed, it will cope with the scientific objectives of the mission: the exploration of primordial material of the planetary system.

If Lutetia cannot be selected due to the lack of available Δv, the asteroid candidate choice has to be done favouring the objects characterized by: 1) the more primitive compositional types (C, P, D); 2) the slower fly-by relative velocity; and 3) the larger diameter.

The Mission has been successfully launched on March 2nd, 2004. Due to optimal launch conditions and the precision of the interplanetary orbital insertion, the available delta-V resulted enough to fly by two asteroids. 2867 Steins and 21 Lutetia are the asteroid targets of the Rosetta mission.

8. Acknowledgement

I.Belskaya is grateful to the European Space Agency who granted her fellowship to spend a year at LESIA, Observatoire de Paris, as visiting scientist.

References

Barucci, M. A., Capria, M. T., Coradini, A., and Fulchignoni, M. (1987). Classification of asteroids using G-mode analysis. *Icarus*, 72:304–324.

Barucci, M.A., and Lazzarin, M. (1995). Visible Spectroscopy of the Rosetta Asteroid Targets: 3840 Mimistrobel and 2530 Shipka. *Icarus* 118: 216–218.

Barucci, M.A., Doressoundiram, A., Fulchignoni, M., Florzac, M., Lazzarin, M., and Angeli C. (1998) Composition type characterization of Rosetta asteroid targets. *Planet Space Sci.* 46: 75–82.

Belskaya, I. N., and Lagerkvist, C.-I. (1996). Physical properties of M-class asteroids. *Planet. Space Sci.*, 44:783–794.

Binzel, R. P. (1987). A photoelectric survey of 130 asteroids. *Icarus*, 72:135–208.

Birlan, M., Barucci, M. A., Vernazza, P., Fulchignoni, M., Binzel, R. P., Bus, S. J., Belskaya, I., and, Fornasier, S. (2004). Near-IR spectroscopy of asteroids 21 Lutetia, 89 Julia, 140 Siwa, 2181 Fogelin and 5480 (1989 YK8), potential targets for the Rosetta mission; remote observations campaign. *New Astronomy*, in press.

Burbine, T. H., and Binzel, R. P. (2002). Small main-belt asteroid spectroscopic survey in the near-infrared. *Icarus*, 159:468–499.

Doressoundiram, A., Weissman, P.R., Fulchignoni, M. Barucci, M.A., Le Bras, A., Colas,F., Lecacheux, J., Birlan, M., Lazzarin, M., Fornasier, S., Dotto, E., Barbieri, C., Sykes, M.V., Larson, S., Hergenrother, C.(1999) 4979 Otawara: fly-by target of the Rosetta mission. *Astron. Astrophy.* 352: 697-702.

Florczak, M., Barucci M.A., Doressoundiram A., Lazzaro D., Angeli C.A., and Dotto, E. (1998). A visible spectroscopic survey of the Flora clan. *Icarus* 133: 233–246.

Fornasier, S., Barucci M.A., Binzel R.P., Birlan M., Fulchignoni M., Barbieri, C., Bus, S.J., Harris, A.W., Rivkin, A.S., Lazzarin,M., Dotto,E., Michalowski, T., Doressoundiram, A., Bertini, I., and Peixinho, N (2003). A portrait of 4979 Otawara, target of the Rosetta space mission. *Astron. Astroph.* 398: 327–333.

Howell, E. S., Merenyi, E., and Lebovsky, L. A. (1994). Classification of asteroid spectra using a neural network. *J. Geophys. Res.*, 99:10848–10865.

Le Bras, A., Dotto, E., Fulchignoni, M., Doressoundiram, A., Barucci, M.A., Le Mouelic, S., Forni, O., and Quirico, E. (2001). The 2000 Rosetta asteroid targets observational campaign: 140 Siwa and 4979 Otawara. *Astron. Astroph* 379: 660–663.

Magri, C., Ostro, S. J., Rosema, K. D., Thomas, M. L., Mitchell, D. L., Campbell, D. B., Chandler, J. F., Shapiro, I. I., Giorgini, J. D., and Yeomans, D. K. (1999). Mainbelt asteroids: results of Arecibo and Goldstone radar observations of 37 objects during 1980–1995. *Icarus*, 140:379–407.

78

Rivkin, A. S., Howell, E. S., Lebofsky, L. A., Clark, B. E., and Britt, D. T. (2000). The nature of M–class asteroids from 3 μm observations. *Icarus*, 145:351–368.

Tedesco, E. F., and Veeder, G. F. (1992). IMPS albedos and diameter catalog. In Tedesco, E. F., Veeder, G. F., Fowler, J. W., and Chillemi, J. R., editors, *The IRAS Minor Planet Survey*, Tech. Rep. PL–TR–92–2049. Phillips Laboratory, Hanscom AF Base, MA.

Tholen, D. (1989). Asteroid taxonomic classifications. In: Binzel, R.P., Gehrels T., and Matthews, M.S., editors, *Asteroids II*, Univ. of Arizona Press, Tucson:1139–1150.

Zappala, V., Di Martino, M., Knezevic, Z., and Djurasevic, G. (1984). New evidence for the effect of phase angle on asteroid lightcurve shape - 21 Lutetia. *Astron. Astroph* 130: 208–210.

Zellner, B., Leake, M., LeBerte, T., Duseaux, M., and Dollfus, A. (1977). The asteroid albedo scale. II. Laboratory polarimetry of meteorites. In *Proc. Lunar Sci. Conf.*, 8th:1091–1110. Pergamon Press, Oxford.

CHARACTERIZING 21 LUTETIA WITH ITS REFLECTANCE SPECTRA

V. V. Busarev
Lunar and Planetary Department, Sternberg State Astronomical Institute, Moscow University, 119992 Moscow, Universitetskij pr., 13, Russian Federation

V. V. Bochkov
Crimean Astrophysical Observatory, Nauchny, Crimea, Ukraine

V. V. Prokof'eva
Crimean Astrophysical Observatory, Nauchny, Crimea, Ukraine

M. N. Taran
Institute of Geochemistry, Mineralogy and Ore Formation, 03142 Kiev, pr. Palladina 34, Ukraine

Abstract   As follows from the visible-range spectrophotometric observations of 21 Lutetia and laboratory reflectance spectra of suitable analog samples, the asteroid is probably an M-type body covered with uneven layer of hydrated silicates

Keywords: 21 Lutetia, asteroid spectrophotometric observations, M-type asteroids, terrestrial analog samples, hydrated silicates

Introduction

Asteroid 21 Lutetia is selected by ESA as a possible target for Rosetta Mission. Lutetia belongs to the newly discovered sub-class of hydrated M-asteroids (Busarev and Krugly, 1995; Rivkin et al., 1995; Busarev, 1998). According to results of the IR observations in the region of 3-μm (Rivkin et al., 2000), about 25% of the known M-class asteroids may enter into the sub-class. Nature of the bodies remains quite controversial so far. We have obtained previously reflectance spectra of Lutetia and laboratory reflectance spectra of suitable analog samples which could be used for the asteroid characterization.

L. Colangeli et al. (eds.), The New ROSETTA Targets, 79–83.

Observations and measurements

Spectrophotometric observations of Lutetia had been carrying out for three months (end of August through November) in the year 2000 in Crimean Observatory at 0.5-m meniscus telescope of Maksutov (MTM-500) with a television detection facility and 0.6-m reflector with a CCD-spectrograph in the visible spectral range (0.38-0.72 μm). The method of differential spectrophotometry was used for the observations. Reduction of the observational data was performed according to corresponding techniques for the television detection facility (Abramenko et al., 2000) or the CCD-spectrograph (Hanisch, 1992). Standard star, HD10307 (G2V) being also a solar analog (Mironov and Kharitonov, 2001), was taken for the observations and calculations of approximated asteroid reflectance spectra. The spectra were obtained mainly at good atmospheric conditions and air masses no more than 1.5. The relative median square deviation in the reflectance spectra does not exceed 2% in the whole. We have registered the visible and near-IR reflectance spectra of terrestrial analog samples as powders of <0.20-0.30 mm grain size. The spectra were scanned in the 0.4-1.0 μm range with a single-beam microspectrophotometer based on a SpectraPro-275 triple-grating monochromator and controlled by an IBM 486 PC. The incident and reflected beam angles and the light beam diameter were 45°, 0° and 5 mm, respectively. Compressed powder of MgO was used as the reflectance standard. The root mean square relative error (RMSRE) for the reflectance spectra does not exceed $0.5-1.0\%$ in the visible region and increases gradually to 2% the red end of the operational spectral region.

Results and conclusions

Overall analysis of Lutetia reflectance spectra and their comparison with data of laboratory spectrophotometric measurements led us to the following.

A general shape of the asteroid averaged over three months reflectance spectrum corresponds to that of an M-type body (Fig. 1, curves 1 and 2). It agrees with the asteroid classification (Tolen and Barucci, 1989). However, in the reflectance spectra (e.g., curves 3 and 4 in Fig. 1) obtained at some asteroid rotational phases we have found absorption bands at 0.44 and 0.67 μm (with relative intensity up to 10%) which are non-typical for a solid body composed by only high-temperature silicates (pyroxenes, olivines and so on) and/or metals. An absorption feature at 0.44 μm in the asteroid spectra is similar to one detected in our laboratory reflectance spectra of powdered terrestrial serpentine samples (Fig. 2). From the laboratory reflectance spectra of serpentine samples we suppose that the absorption band could be considered as an indicator of the ferric iron presence in silicate matter and, hence, a sign of hydrated or oxidized state of the substance. However, to solve the problem accurately, we plan to perform additional investigations of the samples (Mossbauer, X-ray

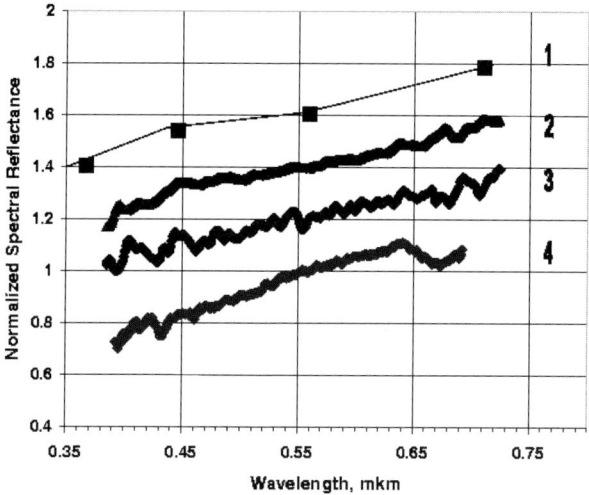

Figure 1.   (1) Averaged and normalized (at 0.55 μm) reflectance spectrum of an M- class asteroid (Tolen and Barucci, 1989); (2) averaged and normalized reflectance spectrum of 21 Lutetia over three months (Sept.-Nov. 2000; phase angles 2.7-23.2 degrees); (3) normalized reflectance spectrum of 21 Lutetia obtained 25/10/00 (rotational phase 0.16) at MTM-500 with a television detection facility; (d) normalized reflectance spectrum of 21 Lutetia obtained 31/08-01/09/00 (rotational phase 0.16) at 0.6-m reflector with a CCD-spectrograph (offsets for the curves are 0.2).

and so on). The very mechanism of the spectral feature is poorly understood because of its relatively high intensity in reflectance spectra of some serpentine samples (up to about 20%, see Fig. 2, curves 1 and 2). If the band was a crystal-field one, it should have a lower intensity (Burns, 1970). An absorption band at 0.67 μm in the asteroid spectra resembles to ones which we see in our laboratory reflectance spectra of some carbonaceous chondrites (e. g., Murchison) (Busarev and Taran, 2002) and terrestrial serpentines and chlorites (Fig. 2). Probably, it arises due to $Fe^{2+} \longrightarrow Fe^{3+}$ intervalence charge transfer (Khomenko and Platonov, 1987; Bakhtin and Gorobets, 1992). As follows from the mentioned and other investigations, these spectral features are typical for phyllosilicates (serpentines, cholorites, etc.) and may be considered as confirmations of their presence on Lutetia's surface. Furthermore, an absorption band at 3.0 μm being characteristic of bound H_2O detected on the asteroid is an additional evidence for such interpretation.

It should be also underlined that variations of considered absorption bands and overall slopes of Lutetia's reflectance spectra were registered with asteroid rotation (see Fig. 3). This may be indicative of some irregular distribution of hydrated silicates and other mineral species along the asteroid surface.

82

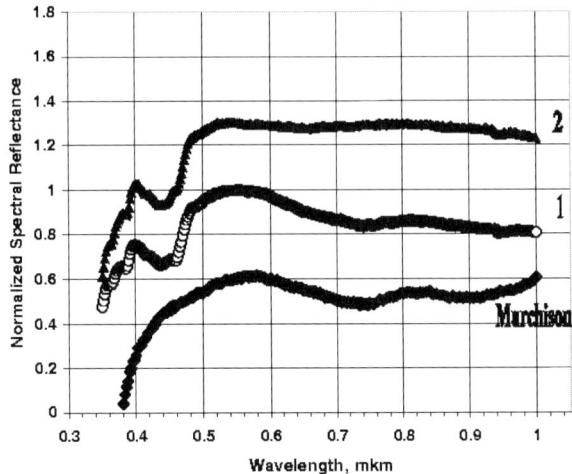

Figure 2.   Laboratory normalized (at 0.55 μm) reflectance spectra of two powdered samples of different terrestrial serpentines (1 and 2) having a relatively intense absorption band at 0.44 μm and carbonaceous chondrite Murchison (CM2) (particle sizes $\leq$0.25 mm) (offsets for curves 2 and Murchison's are 0.3 and 0.4, respectively)

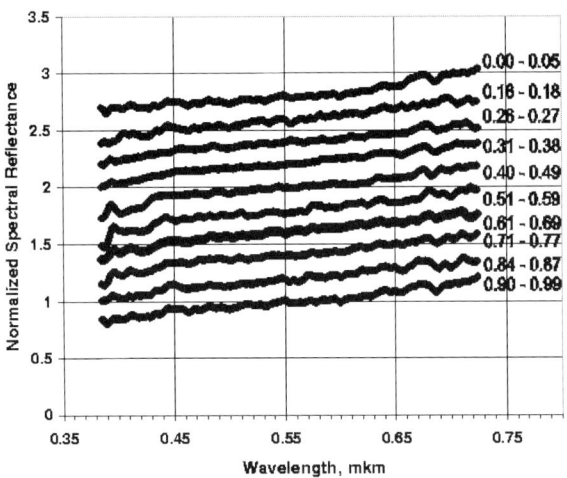

Figure 3.   Averaged and normalized (at 0.55 μm) reflectance spectra of 21 Lutetia at different intervals of asteroid rotational phases counted from the beginning of observations (31/08/00) (offsets for the curves are 0.2)

Thus 21 Lutetia is probably a M-type asteroid covered with uneven layer of hydrated silicates. As far as we observe their combined spectral characteristics of high-temperature silicates (and/or metals) and chondrite-like material, the latter may present as an admixture to Lutetia's regolith or as separated units (craters or spots). The uneven surface layer or units enriched with hydrated silicates could arise on the asteroid due to its past low-velocity collision with some primitive body. Further possibility may be gravitational capture and accumulation of dust particles with high content of hydrosilicates on Lutetia's surface. The particles could arise and move in the vicinity of the asteroid orbit during initial collisional evolution of the asteroid belt. Some scenarios of such events we have considered before (Busarev, 2002).

References

Abramenko, A.N., Bochkov, V.V., and Prokof'eva, V.V. (2000). Facilities of small telescopes, Proc. Soc. of Photo-Optical Instrument. Engineers, 4008: 866–874.

Bakhtin, A.I., and Gorobets, B.S. (1992). Optical spectroscopy of minerals and ores and its application in geological prospecting work, Kazan' Univ. Press, Kazan'. (In Russian).

Burns, R.G. (1970). Mineralogical applications of crystal field theory, Cambridge Univ. Press, Cambridge.

Busarev, V.V. (1998). Spectral Features of M Asteroids: 75 Eurydike and 201 Penelope. *Icarus*, 131: 32–40.

Busarev, V.V. (2002). Hydrated silicates on M–, S–, and E–type asteroids as possible traces of collisions with bodies from the Jupiter growth zone. *Solar Sys. Res.*, 36: 39–47.

Busarev, V.V. and Krugly, Yu.N. (1995). A spot of hydrated silicates on the M- asteroid 201 Penelope? Lunar Planet. Sci. Conf., XXVI, 197–198.

Busarev, V.V., and Taran, M.N. (2002). On the spectral similarity of carbonaseous chondrites and some hydrated and oxidized asteroids. Proc. of Asteroids, Comets, Meteors (ACM 2002), Technical Univ. Berlin, ESA–SP–500, Berlin, 933-936.

Hanisch, R.J. (1992). Image processing, data analysis software, and computer systems for CCD data reduction and analysis, in *Astronomical CCD observing and reduction techniques*, ASP Conference Series 23, Steve B. Howell. Ed, San Francisco, 285–318.

Khomenko, V.M. and Platonov, A.N. (1987). Rock–forming pyroxenes: optical spectra, colouring and pleochroism, Naukova Dumka Press, Kiev. (In Russian).

Mironov, A.V. and Kharitonov, A.V. (2001). Selections of solar analogs on a base of different colour indices, in *Transactions of Sternberg Astron. Inst. 71*, G.V. Yakunina. Ed, Moscow, 94–101 (In Russian)

Rivkin, A.S., Howell, E.S., Britt, D.T., Lebofsky, L.A., Nolan, M.C. and Branston, D.D. (1995). 3-μm spectrophotometric survey of M– and E-class asteroids. *Icarus*, 117: 90–100.

Rivkin, A.S., Howell, E.S., Lebofsky, L.A., Clark, B.E., and Britt, D.T. (2000). The nature of M–class asteroids from 3-μm observations. *Icarus*, 145: 351–368.

Tolen, D.J., and Barucci, M.A. (1989). Asteroid taxonomy, in *Asteroids II*, R.P. Binzel, T. Gehrels, and M.S. Matthews. Eds, Univ. of Arizona Press, Tucson, 298–315.

DETECTION OF PARENT MOLECULES IN COMETS USING UV AND VISIBLE SPECTROSCOPY

William M. Jackson and Alessandra Scodinu
Department of Chemistry, University of California, One Shields Ave. Davis, California 95616

Abstract   A new method for detecting and characterizing comets is presented. Theoretical calculations using CS_2 as an example are presented to support the possibility of using this method to identify parent molecules in comets. Laboratory experiments are suggested that can be used to provide the kind of data that is needed to make this proposal successful.

1.   Introduction

One of the principle goals of Planetary Science is to determine the parent molecules that are responsible for the short-lived species in comets. Infrared emission spectroscopy and radio telescopes have been particularly effective in accomplishing this goal. Mass spectroscopy measurements in cometary flyby experiments have also been used to identify and characterize the molecules in comets. It is expected that direct sampling in cometary rendezvous missions will provide us information about new unidentified molecules. There are many unidentified spectral lines in cometary spectra that in principle could provide additional information about comets. If information were available that could be used to identify the unidentified lines we would gain even more information about composition of molecules in comets. In this paper laboratory experiments are described that in principle will provide new information about emission lines from molecules that can be compared to the unidentified lines in comets. If new molecules are discovered, a variety of new cometary observations can be envisioned. These include telescopic observations from earth and space, as well as new and cheaper space missions to comets and planets. Cameras with specific filters could be used to isolate particular molecular emissions so that the distributions of these molecules in comets can be determined. Since these are parent molecules the radial distribution could be compared with the radial distribution of the daughter fragments to confirm the mechanism of production of these daughter molecules. These distributions will also give us in-

85

L. Colangeli et al. (eds.), The New ROSETTA Targets, 85–95.

formation about the flow characteristics in the coma which will help to confirm detailed theoretical models of the coma. If direct determination of a variety of cometary parent molecules can be determined from ground observations they will help to pin down compositional variations in comets which in turn reveal where in the preplanetary nebula these objects were formed. Finally, because we expect most of these lines to be in the visible regions of the spectrum, the sensitivity for detection will be higher so that more objects can be studied in any type of investigation that one chooses to do.

A specific molecule, CS_2, that is expected to be in comets (Jackson et al., 1986) will be analyzed to show how it could be emitting and contributing to the unidentified lines in the visible cometary spectra. The principles that are involved can be applied to other molecules such as C_2N_2 and C_6H_6. None of these molecules has been identified in comets. Both CS_2 and C_2N_2 are linear molecules in their ground states and are therefore not active in the radio region of the spectrum. Their symmetry makes them weak emitters in IR region because only few fundamental vibrations are infrared active and these are in low frequency region where the detectors are not as sensitive. We will show in this proposal that CS_2 and molecules with similar characteristics can be detected in the visible and UV spectra of comets. Detailed calculations have been done on CS_2, which will illustrate the characteristics that are needed for parent molecules in comets to be detected by their visible, and UV emission spectrum.

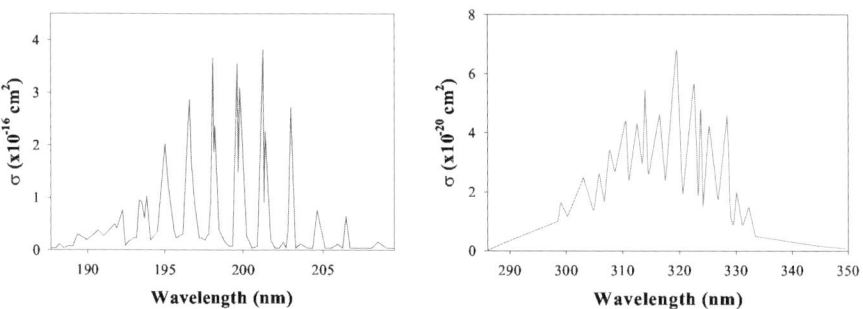

Figure 1. Absorption cross-section for CS_2 reproduced from Rabalais et al., 1971; Hemley et al., 1983.

In Fig. 1, the absorption spectrum of CS_2 is shown in two wavelength regions (Rabalais et al., 1971; Hemley et al., 1983). In the first region between 185 nm to 210 nm, the absorption of a photon will lead to dissociation because the energy of the photon is greater than the bond dissociation energy (NIST, 2003) and the excited state is mixed with a dissociation continuum. Various laboratory experiments have been performed in this wavelength region (Xu et al., 2004; Barry et al., 1986; Black and Jusinski, 1986; Frey and Felder, 1996; McCrary et al., 1985; McGivern et al., 2000; Tzeng et al., 1988; Waller and

Hepburn, 1987; Yang et al., 1980; Kitsopoulos et al., 2001). They show that the dissociation produces CS and S. The CS is both vibrationally and rotationally excited and the S atoms are produced in the ^{3}P and ^{1}D state (McGivern et al., 2000; Tzeng et al., 1988; Waller and Hepburn, 1987; Kitsopoulos et al., 2001; Dornhoefer et al., 1984; Mank et al., 1996). Other experiments have been reported that have shown that the quantum yield for fluorescence is $\sim 10^{-3}$ (Hara and Phillips, 1978). Excitation spectra of CS_2 taken by monitoring the S or CS fragments produced during photodissociation have shown that even though some rotational features can be identified they are all substantially broadened which indicates that predissociation is fast (Mank et al., 1996). This and the evidence for anisotropy in the dissociation agree with the low quantum yield for fluorescence.

The wavelength region between 285 nm and 350 nm is the fluorescent region because the photon does not have enough energy to break the S-CS bond so that in the absence of collisions the excited molecule has to fluoresce. When the comet is one AU from the sun, the equations given by Jackson and Donn (1966) can be used to show that at the surface of a cometary nucleus, the mean free path is equal to 71 cm. This means that the time between collisions is 1.4 ms, which is long, compared to the radiative lifetime of the excited CS_2, molecule, which is only 1 to 10 μs (Douglas, 1966; Heicklen, 1963; Orita et al., 1981; Brus, 1971; Liou et al., 1991). Quenching collisions are even less important than these numbers imply because, as will be shown below, the lifetime for excitation of an excited CS_2 molecule is 167 s, so that most of the molecules will absorb radiation when the molecules are a few hundred kilometers away from the nucleus.

The absorption rate constants in the two spectral regions can be used to determine the relative importance of fluorescence versus photodissociation in these two spectral regions, since collisions are not important in quenching the excited state of the CS_2. The rate constant for absorption at a given wavelength i, is given by the product absorption coefficient, $\sigma(\lambda_i)$, times the solar flux, $\Phi(\lambda_i)$, at that wavelength, as shown in Eq. 1 (Huebner et al., 1992).

$$k(\lambda_i) = \sigma(\lambda_i) \cdot \Phi(\lambda_i) \tag{1}$$

Then the total rate in a given wavelength region is the sum of the individual rate constants in that region as given by Eq. 2.

$$k_T = \sum_i g(\lambda_i) \tag{2}$$

Using these equations, the calculated rate constants for the 185-210 nm and the 285-350 nm regions are 7.3 x 10^{-3} s^{-1} and 6.0 x 10^{-3} s^{-1}, respectively. This means that absorption rates for the two regions are nearly identical and as a result if hundred molecules absorb solar radiation at a given distance approximately, half of them will dissociate and half of them will return to the ground state. The half of them that return to the ground state will absorb light again in about equal proportions in the dissociation and the fluorescence region of the spectrum. Thus, the total fluorescence rate in the coma will be about equal to the total evaporation rate of CS_2 molecules from the nucleus. Various satellite measurements of the amount of CS observed in comets suggests that at one AU a typical comet is evaporating about 8 x 10^{24} s^{-1} (A' Hearn, 2003). Using this value and assuming that comet is 1AU away from the earth will yield a signal of 400 R for a one arc second viewing angle on the earth. This is detectable from earth if all of this radiation is in one line but in general, we would expect it to be disbursed among many lines, which will make it more difficult to detect from earth. Some comets get closer to the earth than one AU so that they provide an opportunity to obtain measurements with a higher probability for success in these comets because you will be able to collect more light from the inner region of the coma of the comet.

The theoretical calculations suggests that there should be UV and visible radiation from excited CS_2 molecules in comets but the immediate question that comes to mind is why have we not been able to identify such emissions in comets? In reality there are many unidentified spectral features observed in comets that could be due CS_2 emission. If they are due to this molecule, they will be difficult to theoretically identify because of the complexity of the emission spectra of this molecules.

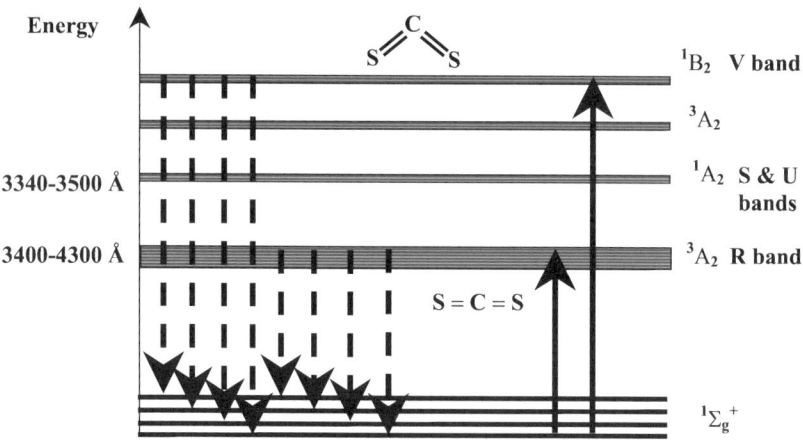

Figure 2.   Schematic energy level diagram of CS_2

The energy level diagram (Mulliken, 1958) in Fig. 2 helps to explain why the emission spectrum is so complicated. The ground state of CS_2 is linear and all of the excited states are bent which immediately complicates the rotational structure associated with the emission spectrum (Jungen et al., 1973). Further complications arise because there are many low-lying triplet states that interact with the singlet states through spin-orbit coupling (Zhang and Vaccaro, 1995; Loge et al., 1986). Spin-orbit coupling is particularly strong, since the molecule has two sulfur atoms. The emission spectrum (Douglas, 1966; Lambert and Kimbell, 1973; Mills and Zare, 1970) will be shifted to the red of the absorption spectrum (Jungen et al., 1973; Kleman, 1963; Jungen et al., 1972) because of the Stokes shift. This arises because usually the bond strengths in the excited states of molecules are weaker than they are in the ground state. This in turn shifts the upper potential curves to longer equilibrium distances, which in turn shifts the emission spectrum. The final states of the CS_2 molecules that are accessed after the molecule is excited is the difference between the exciting photon and the emitted photon. This might complicate the emission spectra but at 1 AU, the time between absorption of a UV photon is of the order of hundreds of seconds thus any vibrationally or electronically excited molecule will have ample time to emit before they absorb another photon.

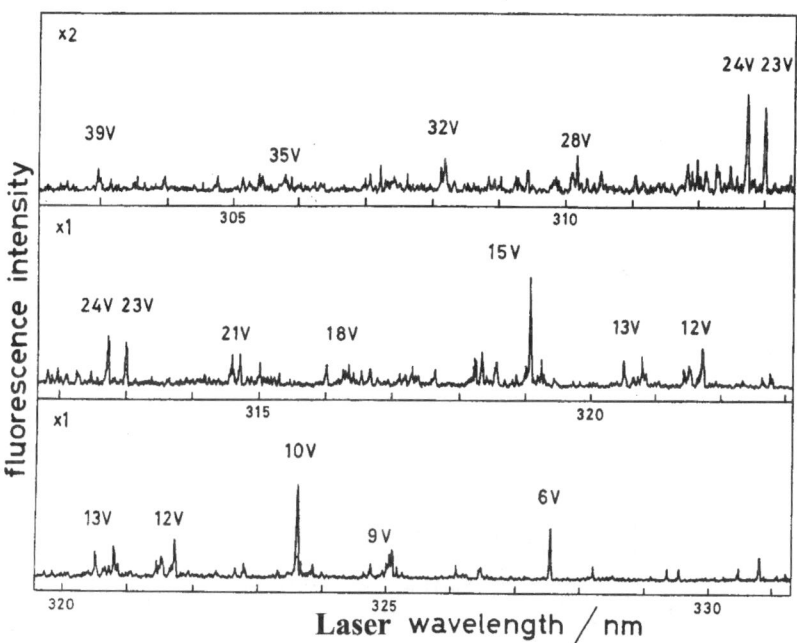

Figure 3. Laser fluorescent excitation spectra from Ochi (1987).

How complicated can we really expect the emission spectrum of CS_2 to be? To answer the question we only need to consider some of the laser excited spectra that have been reported in the literature (Silvers et al., 1976; Pique et al., 1992; Mikami et al., 1981; Kasahara et al., 1984; Ochi et al., 1987).

A laser excitation spectrum of CS_2, which is equivalent to an absorption spectrum, is shown in Fig. 3. This spectrum covers the spectral region between 302 to 332 nm, which covers most of fluorescent region in Fig. 1. One can easily identify 63 absorption features that will all be excited in comets by the broad band light source which is the sun. To obtain an idea of how many emission lines one could expect to obtain from excitation of just one of these vibrational bands consider the emission spectra shown in Figs. 4 and 5, which result from the excitation of the 10 V band shown in Fig. 3 (Ochi et al., 1987). The experimental spectrum in Fig. 4 (Kasahara et al., 1984) shows that $\sim$ 32 emission lines are observed between 480 nm and 320 nm. However, the theoretical spectrum in Fig. 5 (Mikami et al., 1981) shows that one should have seen 63 lines over a spectral range from 666 nm to 320 nm. In the spectral range that was observed the theory predicts that one should have seen 39 lines with an intensity distribution completely unlike what was observed. This clearly indicates that theory is not yet able to adequately predict the number and intensity of the lines that should be observed.

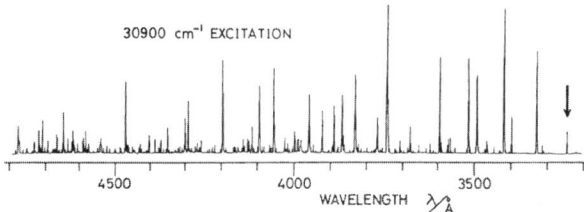

Figure 4.   Experimental emission spectrum of CS_2 after laser excitation of the 10 V (Kasahara et al., 1984.)

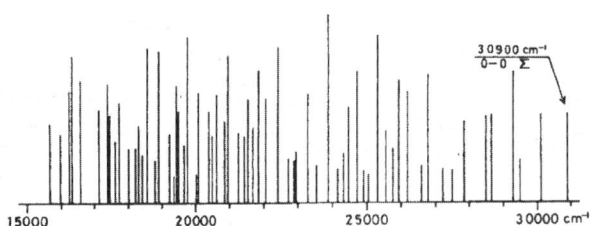

Figure 5.   Theoretical emission spectrum of CS_2 after laser excitation of the 10 V band (Mikami et al., 1981.)

We need to determine the number and relative intensity of lines that may be excited by the solar excitation of CS_2, if we are going to really know what the limits of delectability of visible and UV emissions in comets are. To do this we need to prepare a sample of CS_2 that mimics the conditions of this molecule in comets. The conditions are not completely known but it is expected that the rotational and vibrational distribution will be very cold because of the expansion and the time available for emission of infrared radiation of the vibrational and rotational energy in the molecule to space. We can accomplish this by adjusting the conditions for the expansion of the pulsed molecular beam. Laser excitation spectra of the 13 V band in CS_2 are shown in Fig. 6 (Ochi et al., 1987).

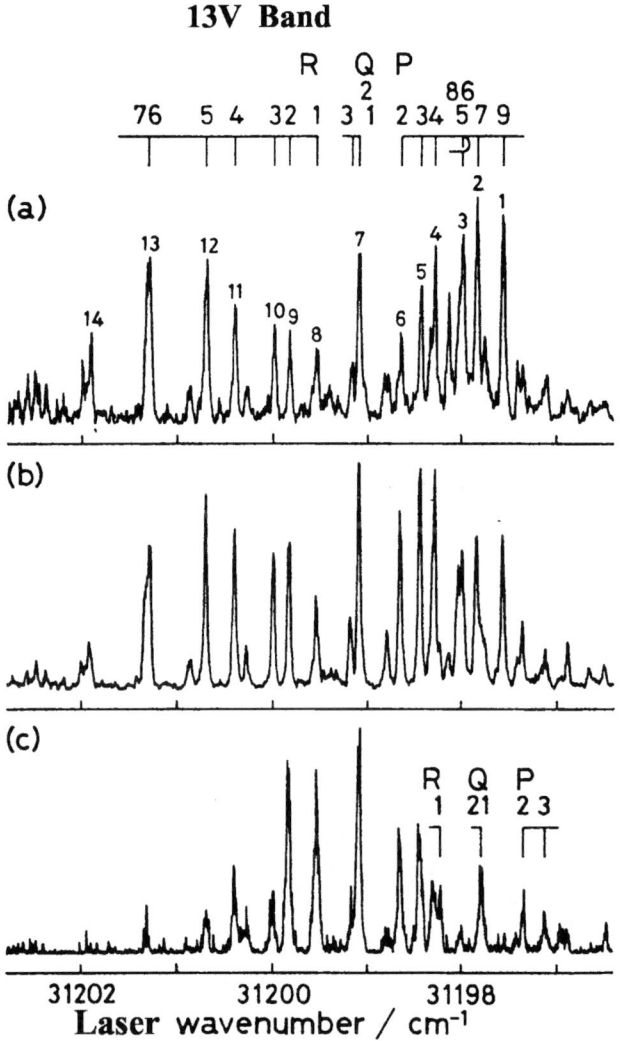

13V Band

Figure 6. Reproduced from Ochi (1987). (a) p_0 = 120 Torr, T_R = 20 K; (b) p_0 = 390 Torr, T_R = 7 K; (c) p_0 = 2 Atm, T_R = 1.5 K

This figure shows that the rotational temperature of this molecule can be varied from 20 K to 1.5 K by varying the backing pressure of the gas in the pulsed valve. This figure also shows that laser excitation can be used to characterize the rotational temperature of the gas in the beam. This will allow us to use the catalog spectra to determine the rotational temperature of the CS_2 molecule in comets because the observed number and intensity of lines should be a function of this temperature.

2. Proposed Experiment

A pulsed molecular beam apparatus that is shown in Fig. 7 will be used to obtain a catalog of emission lines from CS_2 excited by solar radiation.

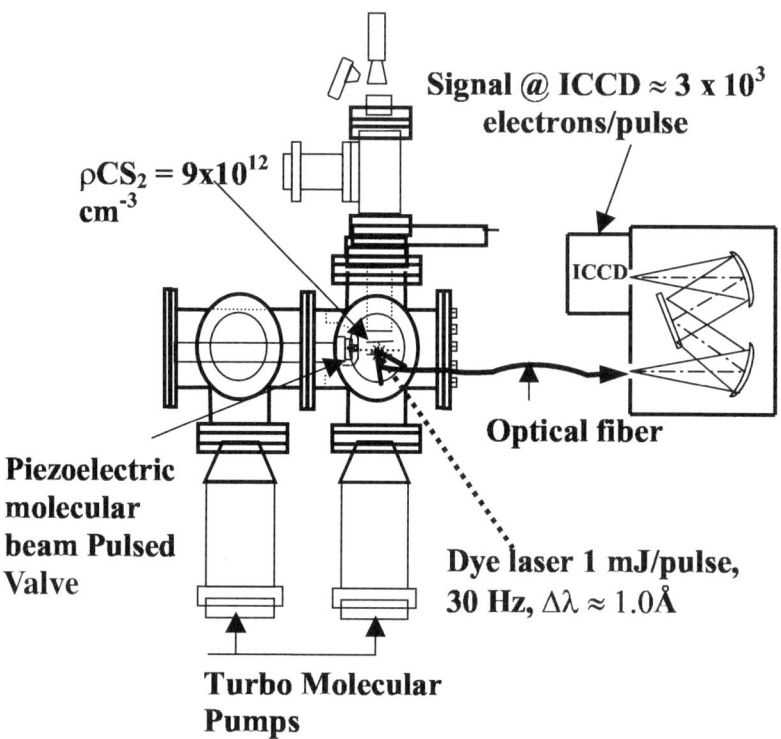

Figure 7. Proposed experimental apparatus.

The pulsed molecular beam already exists in the laboratory (Jackson and Xu, 2000; Huang et al., 2001; Xu et al., 2001). The only modifications that are required are the installation of a fiber optic probe to collect the emission of the radiation from the interaction zone where the laser beam and the molecular beam cross each other. This radiation is then transported to 0.303 m imaging spectrograph with an intensified CCD detector, where the spectrum is recorded and transferred to the computer memory. The intensity of the laser used to excite the emission spectrum also needs to be recorded with the spectrum so that it can be used to properly normalize the spectra to mimic the distribution of the expected solar radiation. The vibrational bands in this molecule are about 0.4 to 0.6 Å wide and because there are about 32 bands it means that we will have to accumulate about 32 complete spectra to simulate what one would expect to be excited by solar radiation. Figure 4b suggests that complete emission spectra have to cover the wavelength region from about 300 nm to 700 nm. A 0.3 m spectrograph with a 25 mm ICCD detector and 1200 l/mm covers 65 nm at a time so that it will take seven runs to obtain one complete spectra from one vibrational band. This should be done as a function of the expansion conditions so that it will be known how the internal temperature of the CS_2 affects the results. Thus, at least 224 runs will have to be done to cover all of the bands that can be excited in a CS_2 molecule exposed to solar radiation.

3.   Conclusions

Theoretical calculations of the emission probability for UV and visible line of CS_2 in comets suggest that it is possible to detect the emission spectrum of CS_2 in a suitable comet. Laboratory experiments are described that should provide the kind of information that will allow modeling of the relative intensities of any observed molecular emission lines in comets. The kind of calculations that have been done for this molecule also point out how molecules with chromophores that allow them to absorb radiation below their dissociation threshold should also emit in comets and may be the source of some of the unidentified lines in comets.

Acknowledgments

This work was supported by NASA planetary atmospheres program under grant number NAG5471. We also wish to thank Professor Mike A'Hearn for his helpful discussions.

References

A' Hearn, M. F. (2003). *Private Communication.*

Barry, M. D., Johnson, N. P., and Gorry, P. A. (1986). A fast (30 μs) pulsed supersonic nozzle beam source: application to the photodissociation of carbon disulfide at 193 nm. *Journal of Physics E: Scientific Instruments*, 19(10): 815–819.

Black, G. and Jusinski, L. E. (1986). Branching ratio for $S(^3P)$ and $S(^1D_2)$ atom production in the photodissociation of CS_2. *Chemical Physics Letters*, 124(1):90–92.

Brus, L. E. (1971). Two exponential decay of 3,371 Å laser excited CS_2 fluorescence. *Chemical Physics Letters*, 12(1): 116–119.

Dornhoefer, G., Hack, W., and Langel, W. (1984). Electronic excitation and quenching of CS formed in the ArF laser photolysis of CS_2. *Journal of Physical Chemistry*, 88:3060–3069.

Douglas, A. E. (1966). Anomalously long radiative lifetimes of molecular excited states. *Journal of Chemical Physics*, 45(3):1007–1015.

Frey, J. G. and Felder, P. (1996). Photodissociation of CS_2 at 193 nm investigated by polarised photofragment translational spectroscopy. *Chemical Physics*, 202(2-3):397–406.

Hara, K. and Phillips, D. (1978). Fluorescence of CS_2 excited to the third excited singlet state. *Journal of the Chemical Society Faraday Transactions II*, 74(8):1441–1445.

Heicklen, J. (1963). The fluorescence of carbon disulfide vapor. *Journal of the American Chemical Society*, 85(22):3562–3565.

Hemley, R. J., Leopold, D. G., Roebber, J. L., and Vaida, V. (1983). The direct ultraviolet absorption spectrum of the $^1\Sigma_g^+ \rightarrow {}^1B_2({}^1\Sigma_u^+)$ transition of jet-cooled carbon disulfide. *Journal of Chemical Physics*, 79(11):5219–5227.

Huang, J. H., Xu, D. D., Fink, W. H., and Jackson, W. M. (2001). Photodissociation of the dibromomethane cation at 355 nm by means of ion velocity imaging. *Journal of Chemical Physics*, 115(13):6012–6017.

Huebner, W. F., Keady, J. J., and Lyon, S. P. (1992). Solar photo rates for planetary atmospheres and atmospheric pollutants. *Astrophysics and Space Science*, 195(1):1–294.

Jackson, W. M., Butterworth, P. S., and Ballard, D. (1986). The orogin of CS in Comet IRAS-Araki-Alcock 1983d. *Astrophysical Journal*, 304(1):515–518.

Jackson, W. M. and Xu, D. D. (2000). Photodissociation of the acetone cation at 355 nm using the velocity imaging technique. *Journal of Chemical Physics*, 113(9):3651–3657.

Jackson, W. M. and Donn, B. (1966). Collisional processes in the inner coma. *Memories de la Societe Royale des Sciences de Liege, Collection in VIII*, 12(1):133–140.

Jungen, Ch., Malm, D. N., and Merer, A. J. (1973). Analysis of a $^1\Delta_u \rightarrow {}^1\Sigma_g^+$ transition of CS_2 in the near ultraviolet. *Canadian Journal of Physics*, 51:1471–1490.

Jungen, Ch., Malm, D. N., and Merer, A. J. (1972). Ultraviolet absorption of CS_2 near the N_2 laser wavelengths (3,371 Å). *Chemical Physics Letters*, 16(2):302–305.

Kasahara, H., Mikami, N., Ito, M., Iwata, S., and Suzuki, I. (1984). Excitation and dispersed fluorescence spectra of the $^1B_2(V) \rightarrow {}^1\Sigma_g^+(X)$ transition of jet-cooled CS_2. *Chemical Physics*, 86(1-2):173–188.

Kitsopoulos, T. N., Gebhardt, C. R., and Rakitzis, T. P. (2001). Photodissociation study of CS_2 at 193 nm using slice imaging. *Journal of Chemical Physics*, 115(21):9727–9732.

Kleman, B. (1963). The near-ultraviolet absorption spectrum of CS_2. *Canadian Journal of Physics*, 41:2034–2036.

Lambert, C. and Kimbell, G. H. (1973). The fluorescence of CS_2 vapor (collisional quenching). *Canadian Journal of Chemistry*, 51(16):2601–2608.

Liou, H. T., Yang, H., Wang, N. C., and Joy, R. W. (1991) Successive single rotational level radiative decay lifetime measurements of CS_2: evidence of state mixing caused by rotational coupling. *Chemical Physics Letters*, 178(1):80–88.

Loge, G. W., Tiee, J. J., and Wampler, F. B. (1986). Fluorescence lifetimes and Zeeman quantum beats of single rotational levels in $^3B_2$ carbon disulfide. *Chemical Physics*, 84(7):3624–3629.

Mank, A., Starrs, C., Jego, M. N., and Hepburn, J. W. (1996). A detailed study of the predissociation dynamics of the $^1B_2({}^1\Sigma_u^+)$ state of CS_2. *Journal of Chemical Physics*, 104(10):3609–3619.

McCrary, V. R., Lu, R., Zakheim, D., Russell, J. A., Halpern, J. B., and Jackson, W. M. (1985). Coaxial measurement of the translational energy distribution of carbon monosulfide (CS) produced in the laser photolysis of carbon disulfide at 193 nm. *Journal of Chemical Physics*, 83(7):3481–3490.

McGivern, W. S., Sorkhabi, O., Rizvi, A. H., Suits, A. G., and North, S. W. (2000). Photofragment traslational spectroscopy with state-selective "universal detection": The ultraviolet photodissociation of CS_2. *Journal of Chemical Physics*, 112(12):5301–5307.

Mikami, N., Kasahara, H., and Ito, M. (1981). SVL fluorescence spectra from the $^1B_2$ state of CS_2 cooled in a supersonic free jet. *Chemical Physics Letters*, 83(3):488–492.

Mills, J. W. and Zare, R. H. (1970). Magnetic depolarization of CS_2 vapour fluorescence. *Chemical Physics Letters*, 5(1):37–41.

Mulliken, R. S. (1958). The lower excited states of some simple molecules. *Canadian Journal of Chemistry*, 36:10–23.

NIST Chemistry WebBook (2003). *http://webbook.nist.gov/chemistry*.

Ochi, N., Watanabe, H., Tsuchiya, S., and Koda, S. (1987). Rottationally resolved laser-induced fluorescence and Zeeman quantum beat spectroscopy of the V $^1B_2$ of jet-cooled CS_2. *Chemical Physics*, 113(2):271–285.

Orita, H., Morita, H., and Nagakura, S. (1981). Collisional quenching constants and collision-free lifetimes of fluorescence of gaseous carbon disulfide. *Chemical Physics Letters*, 81(1):33–36.

Pique, J. P., Manners, J., Sitja, G., and Joyeux, M. (1992). Intra-inter polyad mixing and breaking of symmetric-antisymmetric selection rule in the vibrational spectra of CS_2 molecule. *Journal of Chemical Physics*, 96(9):6495–6508.

Rabalais, J. W., McDonald, J. M., Scherr, V., and McGlynn, S. P. (1971). Electronic spectroscopy of isoelectronic molecules. II. Linear triatomic groupings containig sixteen valence electrons. *Chemical Reviews*, 71:73–108.

Silvers, S. J. and McKeever, M. R. (1976). Time and frequency resolution of CS_2 fluorescence excited by a nitrogen laser. *Chemical Physics*, 18(3-4):333–339.

Tzeng, W. B., Yin, H. M., Leung, W. Y., Luo, J. Y., Nourbakhsh, S., Flesch, G. D., and Ng, C. Y. (1988). A 193 nm laser photofragmentation time-of-flight mass spectrometric study of CS_2 and CS_2 clusters. *Journal of Chemical Physics*, 88(3):1658–1669.

Waller, I. M. and Hepburn, J. W. (1987). Photofragment spectroscopy of CS_2 at 193 nm: direct resolution of singlet and triplet channels. *Journal of Chemical Physics*, 87(6):3261–3268.

Xu, D. D., Huang, J. H., and Jackson, W. M. (2004). Reinvestigation of CS_2 dissociation at 193 nm by means of product state-selective VUV laser ionization and velocity imaging. *Journal of Chemical Physics*, in press.

Xu, D. D., Price, R. J., Huang, J. H., and Jackson, W. M. (2001). Photodissociation of the ethyl bromide cation at 355 nm by means of TOF-MS and ion velocity imaging techniques. *Zeitschrift fur Physikalische Chemie*, 215(2):253–271.

Yang, S. C., Freedman, A., Kawasaki, M., and Bersohn, R. (1980). Energy distribution of the fragments produced by photodissociation of CS_2 at 193 nm. *Journal of Chemical Physics*, 72(7):4058–4062.

Zhang, Q. and Vaccaro, P. H. (1995). *Ab Initio* studies of electronically excited carbon disulfide. *Journal of Physical Chemistry*, 99:1799–1813.

GRAIN SIZES OF EJECTED COMET DUST. CONDENSED DUST ANALOGS, INTERPLANETARY DUST PARTICLES AND METEORS

Frans J.M. Rietmeijer
Department of Earth and Planetary Science, MSC03–2040, 1 University of New Mexico, Albuquerque, NM 87131–0001, U.S.A.

Joseph A. Nuth III
Laboratory for Extraterrestrial Physics, Code 691, NASA Goddard Space Flight Center, Greenbelt, MD 20771, U.S.A.

Abstract

Dust ejected from comet nuclei includes up to millimeter–sized highly porous and collapsed aggregates plus small pebbles with mass/size distributions and compositions defined by the hypothesis of hierarchical dust accretion based on collected interplanetary dust particles and meteor properties.

Keywords: Interplanetary dust particles, cluster IDPs, cosmic dust, comet dust, meteors, silicate dust analogs, vapor condensation experiments, hierarchical dust accretion, principal components, comet dust detection, comet Halley, comet 67P/Churyumov–Gerasimenko, Leonids, Perseids, Rosetta mission

1.   Introduction

The idea that all comets are homogeneous, dirty snowballs has evolved to recognize that most comet nuclei are rubble piles wherein self–gravitation and dirty–ice/icy–dirt "glue" are holding the building blocks together (Gombosi & Houpis, 1986), viz.

- km–size to hundreds of meters, probably ice–free boulders similar to hydrated CI, possibly even CM, carbonaceous meteorite materials (Rietmeijer, 2000) or anhydrous, Si–rich proto–CI material (Rietmeijer & Nuth, 2000);

L. Colangeli et al. (eds.), The New ROSETTA Targets, 97–110.
© 2004 *Kluwer Academic Publishers. Printed in the Netherlands.*

- meter– and cm–sized pebbles of compacted, massive dust–ice mixtures with variable degrees of hydration of embedded dust, and

- dust in the so–called "Whipple glue" (Gombosi & Houpis, 1986, Rietmeijer, 2002a).

Most of this dust is ejected in collimated jets as was first seen at comet Halley. Its dust was analyzed at considerable distance from the nucleus but still included dust clusters (or agglomerates), 10^{-13}g to 10^{-10}g, that disintegrated once unspecified glue had eroded (Simpson et al., 1989). Very small dust, 10^{-20}g $<$mass$<10^{-17}$g, near the cometopause was probably "graphite", metal–oxides or a mixture (Sagdeev et al., 1989). Aggregate interplanetary dust particles (IDPs) collected in the Earth's stratosphere (Rietmeijer, 2000, Rietmeijer, 2002b, Zolensky et al., 1994) are the best natural analogs for comet dust (Fig. 1). Combining the compositions (Jessberger et al., 1988) and mass (Fomenkova et al., 1992) of comet Halley's dust, Rietmeijer (1998, 2002b) concluded that it were aggregates of principal components in the matrix of aggregate IDPs (Fig. 1) wherein embedded sulfide and silicate grains are scarce.

The ROSETTA mission will analyze dust near and at the nucleus of the target comet and will thus encounter larger particles that probably include large cluster IDP, aggregates, and even larger, up to cm–sized, debris such as detected among the annual Perseid meteors and the recent Leonid storm meteors (Rietmeijer, 2002a, Rietmeijer, 2003, Jenniskens et al., 2000). Large particles near to a nucleus probably include fluffy aggregates with ice filling pore spaces and compact dust that are collapsed anhydrous aggregates and massive, partially or fully hydrated aggregates with porosity ranging from 0.01 g/cm^3 to $\approx$3 g/cm^3 (Rietmeijer, 1998). The collected IDPs support the co–existence of anhydrous and hydrated dusts formed in comet nuclei (Rietmeijer, 2002b, Mackinnon & Rietmeijer, 1987, Sandford, 1987). Surprisingly collected IDPs do not appear to include fragments of the black refractory organic mantle.

The hypothesis of *hierarchical dust accretion*, based on the analyses of laboratory–condensed dust analogs, aggregate and cluster IDPs plus meteor size, structure and compositions, states that a limited number of circumstellar (or molecular cloud) dust types will grow in size as accretion proceeds in an evolving circumstellar dust cloud or disk (Rietmeijer, 2002b, Rietmeijer, 1998). Increasing dust size will be is accompanied by chemical and mineralogical complexity of newly– formed, larger dust during dust recycling through (pre)–protoplanet processing and re–introduction of coarser–grained in the evolving circum–stellar dust cloud or disk. Dust coarsening is driven by solid–state growth of larger grains to reduce the surface free energy, or Ostwald ripening.

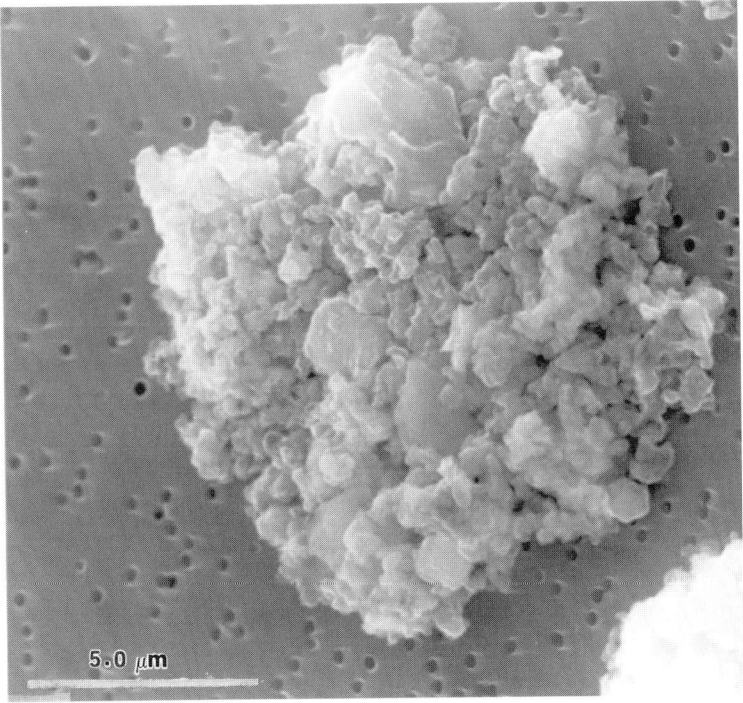

Figure 1. Scanning electron microscope image of carbon–rich, porous (fluffy) chondritic aggregate IDP W7029B13 collected in the lower stratosphere. It is placed on a nucleopore filter (background) for scanning electron microscope analysis. The image shows several large platy silicates embedded in the matrix of (partially fused) principal components (PCs) that include (1) carbonaceous PCs, (2) PCs with silicates and sulfides embedded in a carbonaceous matrix and (3) (C–free) ferromagnesiosilica PCs. These three distinct PC types resemble the CHON (carbon–hydrogen–oxygen–nitrogen), "mixed" and "silicate" (or "rock") particles, respectively, in the coma of comet Halley (Rietmeijer, 2002a, Rietmeijer, 2002b); Courtesy the National Aeronautics and Space Administration (NASA number S-82-27575) (scale bar: 5 microns)

As an interesting corollary of this hypothesis the comet dust ejected in collimated jets on a nucleus represents the least–processed dust in a parent body environment but that spent the longest time in a circumstellar nebula where it was exposed to energetic nuclei and radiation from the central star.

In this paper we will highlight changes in size and mass of dust entities in increasingly larger aggregate particles. Judiciously combining size, density and composition will aid the identifications of detected dusts. We will rely heavily on reviews containing original references. We fully acknowledge the work of the original authors.

2.   Hierarchical dust accretion

Laboratory condensation experiments

Dust initially formed by condensation from a circumstellar gas phase. This process was originally considered to proceed at thermodynamic equilibrium conditions that led to the formation of stoichiometric silicate minerals such as olivine and pyroxene. Not only is it unlikely but also the meteorite record does not confirm this type of condensation. Recent, laboratory experiments showed that non–equilibrium condensation might be a viable alternative to the ultimate formation of common silicate minerals (Nuth et al., 1999, Nuth et al., 2002). The dust formed by condensation of an Mg-Fe-SiO-H_2-O_2 vapor included amorphous, mixed 'MgSiO' and 'FeSiO' dusts plus SiO_2, MgO and Fe–oxide nanograins (Fig. 2). The condensed mixed dusts had metastable eutectic serpentine dehydroxylate and smectite dehydroxylate compositions (Rietmeijer, 2002b) that successfully constrained the pure–Mg silicates around many young stellar objects (Bouwman et al., 2001). These nanometer conden-sates also resemble the small amorphous hydrogenated pure–Mg and Mg–rich silicate grains that could be responsible for the 217.5 nm absorption feature (Steel & Duley, 1987).

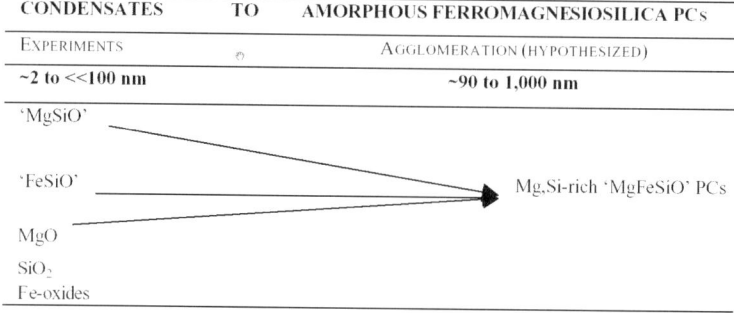

Figure 2.   Condensed circumstellar dust analogs agglomerating to form pure ferromagne-siosilica dust with a Si– and Mg–rich smectite dehydroxylate composition

The agglomeration of these metastable mixed dusts yields amorphous Mg–rich 'MgFeSiO' PCs (Rietmeijer et al., 1999) (Fig. 2) that when thermally an-nealed become compact clusters of coarse–grained Mg–rich (Mg,Fe)–olivine and similar pyroxene plus amorphous, Ca and/or Al–bearing, Si–rich material.

Other abundant but distinctly Fe–rich 'MgFeSiO' PCs (Rietmeijer, 2002b) formed when earlier–agglomerated 'Mg[Fe]SiO' PCs reacted with condensed Fe–oxides (Fig. 2) to form these amorphous Fe–rich PCs, which required an activation energy source that is (partially) supplied by the metastable nature of the reagents The GEMS (glass with embedded metals and sulfides) are a unique subset of the Fe–rich PCs that necessitated amorphization and fusion of pure–Mg pyroxene (enstatite) plus Fe,Ni–sulfide by high energy nuclei (Bradley, 1994).

Mixed–dust analog condensation experiments are complex but progress was reported (Rotundi et al., 2002). Pure carbon condensation has unique challenges caused by graphite stability with the resultant formation of metastable carbons, e.g. C_{60} fullerene and fullerenic nanostructures (Rotundi et al., 1998, Rietmeijer et al., 2004).

Principal components forming matrix aggregates

Condensed dusts are no longer recognizable in aggregate IDPs wherein PCs are the smallest, ≈ 100 to ≈ 1000 nm in diameter, surviving entities (Fig. 3) that agglomerated into ≈ 5 μm–size matrix aggregates. They include, (Rietmeijer, 2002b)

- Carbonaceous PCs (mostly refractory hydrocarbons; amorphous carbons),

- C–bearing ferromagnesiosilica PCs (Figs. 4,5), and

- Ferromagnesiosilica PCs: (I) coarse–grained (up to ≈ 400 nm) Mg–rich smectite dehydroxylate PCs, Fe/(Fe+Mg) *(fe)* = 0 – 0.33, (II) Fe–rich serpentine dehydroxylate PCs, *fe* = 0.3 – 0.83, and (III) GEMS.

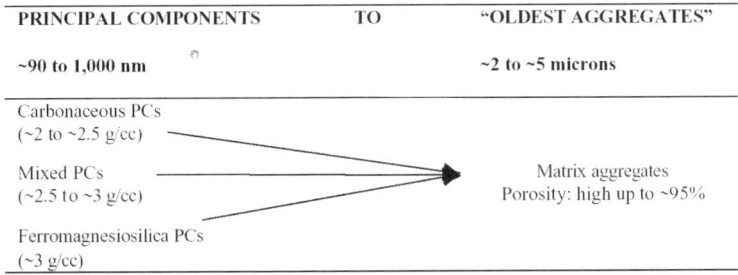

Figure 3.   Matrix aggregates with variable proportions of PCs reflect the availability of these PCs in an evolving nebula at the time and location of accretion. The PC densities are best estimates based on composition; the fractal matrix aggregate density is from Rietmeijer, 2002b.

Dust cluster fragmentation in comet Halley's coma is attributed to the erosion of "glue" believed to be a part of Greenberg's core–mantle grains. The carbonaceous matrix of mixed PCs (Fig. 4) is the only evidence so far for this hypothesized "glue". The MIDAS atomic force microscope onboard ROSETTA

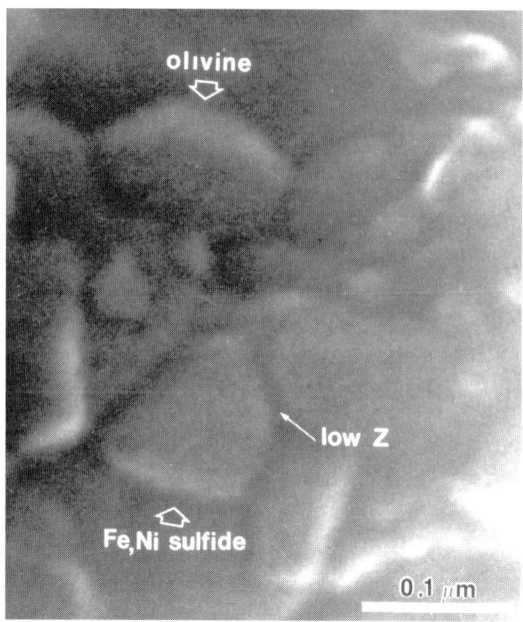

Figure 4. Scanning transmission electron micrograph showing of a mixed PC with 2nm to 50nm, platy Fe,Mg–olivine, –pyroxene and Fe,Ni–sulfide grains in low–Z (atomic number) carbonaceous matrix of refractory hydrocarbons and amorphous carbons (modified after Rietmeijer, 1998)

(Riedler et al., 1998) could recognize this particular mixed PC texture. Its volatile "glue" (Fig. 5) would be conducive to space erosion and would yield platy grains of 10^{-20}g $<$mass$<10^{-17}$g, a perfect match for grains near Halley's cometopause (Sagdeev et al., 1989) but much smaller than the analyzed dust grains (Jessberger et al., 1988, Fomenkova et al., 1992).

Matrix aggregates to aggregate IDPs

With continued hierarchical dust co–accretion matrix aggregates plus silicate and Fe,Ni–sulfide dusts, $\approx$5 μm in size, produced aggregate IDPs (Fig. 6). The silicate dusts were mostly Mg–rich (Mg,Fe)–olivine and pyroxenes ranging from Ca–free (enstatite) to high–Ca (diopside) (Rietmeijer, 2002b, Rietmeijer, 1998). There are also small amounts of amorphous aluminosilica grains with variable SiO_2/Al_2O_3 ratios that formed after coagulation of amorphous Si– rich materials in Mg–rich smectite–dehydroxylate ferromagnesiosilica PCs (Rietmeijer, 2002b). At each level of the accretion hierarchy the relative dust proportions were highly variable to reflect their abundances as a function of time and location in (pre–)protoplanet accretion regions. Compact clusters of individual silicate grains, sulfide grains, or a mixture of both grains,

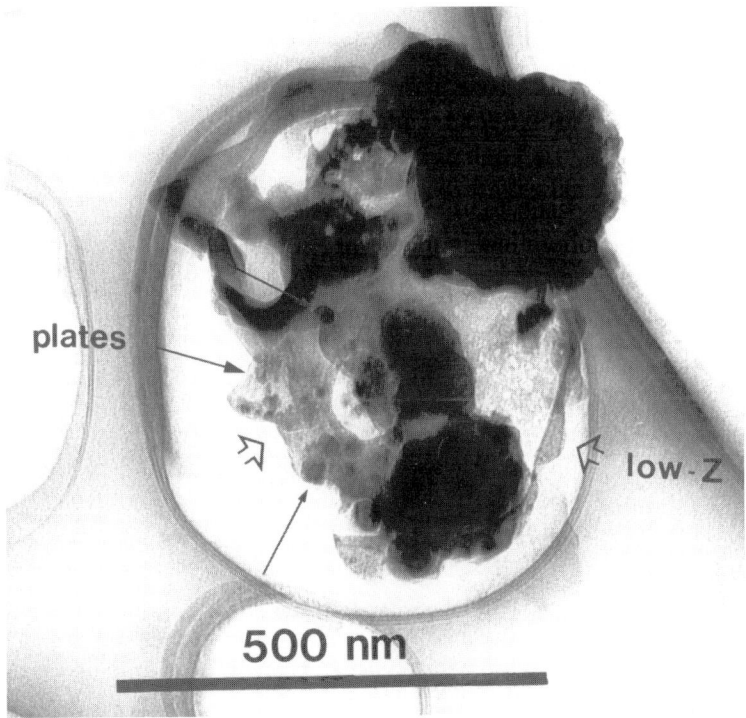

Figure 5. Transmission electron micrograph of another mixed PC with tiny platy (arrows) and larger grains (black areas) are embedded in a volatile carbonaceous (low–Z; open arrow) matrix bubbles that formed during exposure to the incident 200 keV electron beam

with minor co–accreted aggregate matrix material dominated some collected aggregate IDPs.

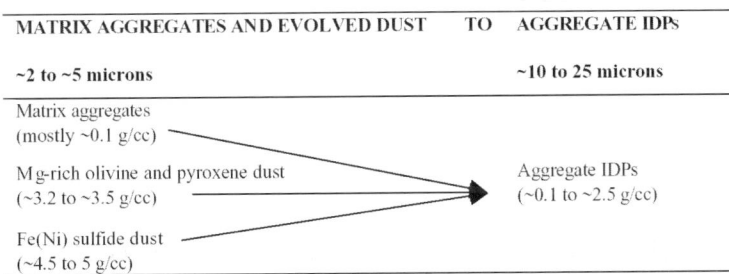

Figure 6. Matrix aggregates and evolved non–chondritic dusts accreting into chondritic aggregate IDPs that can be highly porous anhydrous or collapsed, (partially) hydrated. Silicate and sulfide density from standard mineral textbooks; measured aggregate IDP densities (Rietmeijer, 1998)

Aggregate IDPs to Cluster IDPs

Cluster–IDPs are the next level of dust accretion whereby previously formed IDP– aggregate co–accreted with large silicate fragments that now also include aluminosilicate (i.e. plagioclase) and sulfide particles (Fig. 7). The silicate and sulfide fragments are compact polycrystalline grain clusters wherein coarsening of originally agglomerated grains could ultimately lead to 10–25 μm single–crystal silicate and sulfide fragments. Rare refractory, Ca,Al,Ti– rich clusters ($\approx$3 to $\approx$4 g/cm^3) are probably carried by the aggregate IDP (matrix) component of the clusters.

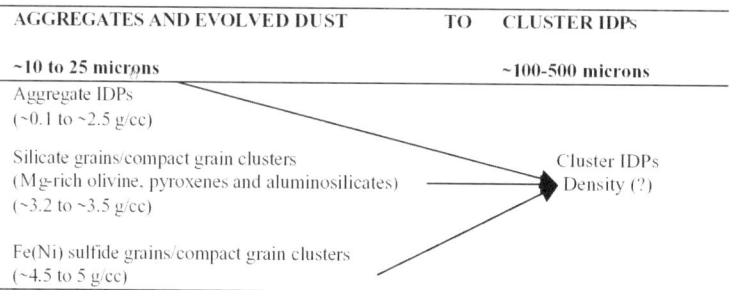

Figure 7. Formation of cluster IDP aggregates from aggregate IDPs and evolved non–chondritic dusts. The clusters are porous aggregates to judge from their appearance on the collectors (Rietmeijer, 2002b, Rietmeijer, 1998). Silicate and sulfide densities are from mineral textbooks.

Cluster IDPs are the most advanced stage of hierarchical dust accretion available for laboratory analyses.

Giant, mega clusters, etc. to meteors

Giant and mega clusters (Fig. 8) do not survive atmospheric entry. Meteors are currently the only way to study dust accretion beyond the largest collected aggregates and before the CI and CM chondrite meteorite matrix.

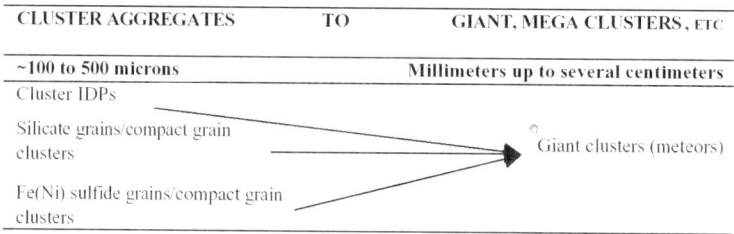

Figure 8. Hypothesized aggregate meteors, and mega and giant cluster agglomeration

Prior to the recent Leonid storms (Jenniskens et al., 2000) meteor research was disconnected from laboratory analyses of collected extraterrestrial materials but this situation has changed (Rietmeijer, 2002a, Rietmeijer, 2003). For example, "humped" light curves are intriguing. They indicate that decelerating meteoroids are a heterogeneous composite object with a dustball (aggregate) component and a massive grain. For one Leonid meteor with two $(2.4 - 2.5) \cdot 10^{-4}$g mass components (Murray et al., 2000), Rietmeijer (2002b) calculated a dustball of 575 μm (1g/cm^3 density) plus either a massive silicate (525 μm) or a 460 μm Fe–sulfide fragment. This meteor was probably a ≈2500 to ≈5000 μm "giant cluster" (Fig. 8).

In another development, (semi) quantitative meteor chemistry is slowly being integrated in studies of collected dust and meteorite compositions (Rietmeijer, 2000, Rietmeijer, 2003). There is a lot of potential. For example, the normalized Mg versus Fe data (Fig. 9) confirm a comet Halley bulk composition of matrix aggregates, i.e. compositions closer to unprocessed molecular cloud dust than to evolved nebular dust bulk compositions. The dotted line through porous aggregate IDPs could represent extremely porous, cm–sized anhydrous aggregates that might survive when not incorporated in a parent body. The other dotted line through the collapsed hydrated aggregate IDPs might support the presence of very small collapsed pebbles in both comets (Fig 9). Both lines delineate the potential extent of structural and chemical variation among comet debris.

These exercises (Figs. 8,9) become speculative but as hierarchical dust accretion continues to produce increasingly larger aggregates wherein most of the mass will be in the silicate and sulfide dusts, we need to consider that

- when silicate or sulfide dusts reach cm–size aggregate density approaches that of these dusts unless aggregate porosity dramatically increases (Rietmeijer & Nuth, 2000, Rietmeijer, 2002b).

- such large grains are not found in CI meteorites (density: 1.6 g/cm^3; 35% porosity) wherein silicates and sulfides are sub–millimeter grains (Rietmeijer, 2002a),

- highly porous aggregates larger than "giant clusters" are rare, and

- Leonid and Perseid meteor data (Rietmeijer, 2002a) suggest a transition from porous aggregates to denser aggregates for dust between ≈1000 and ≈5000 μm.

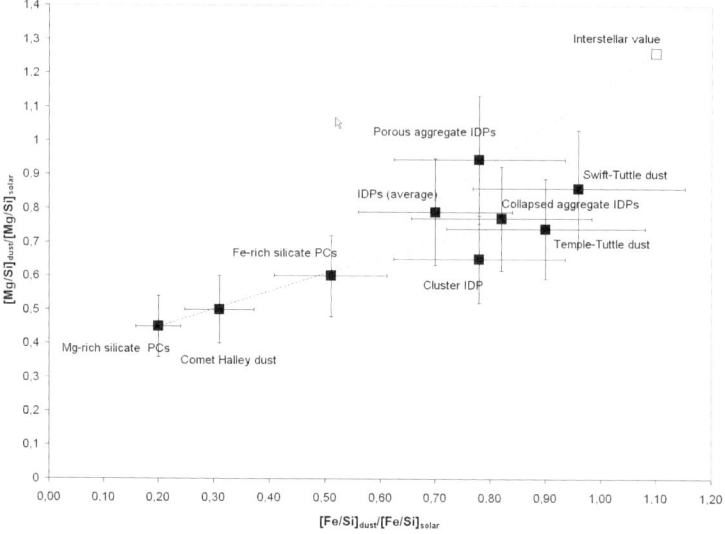

Figure 9. Mg and Fe ratios (Si and solar normalized) in ferromagnesiosilica PCs, aggregate and cluster IDPs (Rietmeijer, 2002a, Rietmeijer, 2002b) (collapsed aggregate IDPs refer to chondritic smooth IDPs), comet Halley dust (Jessberger et al., 1988) and two meteors (1) 250g Leonid meteor (comet Tempel–Tuttle) and (2) 0.2g Perseid meteor (comet Swift–Tuttle) (Trigo-Rodriguez et al., 2003). The dashed line connects Halley's dust composition and the matrix aggregate PCs. Dotted lines are connecting aggregate IDPs and the interstellar value (Snow & Witt, 1996) and the collapsed aggregate IDPs. Error bars set randomly at 20% relative

3. Dust size analyses

Is it possible to reduce these data in a manner that would allow dust characterization when lacking mineralogical and chemical data for dust near and on a comet nucleus? Dust instruments on–board the ROSETTA mission will measure size, mass, texture, morphology and chemical composition but the ideal combination of all these properties for each particle will not be available. The diameters, mass and density of accreting dusts (Table 1) are chosen to cover a wide, but observed, range for dust diameters and aggregate densities. Aggregate particles from a preceding stage become the "matrix component" at the next level. Dust mass (grams) and diameter (microns) (Table 1) define a function described by the equation

$$Mass = 6 \cdot 10^{-13} \cdot (diameter)^{2.959} \qquad (1)$$

This correlation ($r^2 = 0.93$) is a guideline to the stepped changes of these dusts that defined hierarchical dust accretion. It can be used to assess density "giant and mage clusters" and meteors. Equation 1 predicts that a "giant cluster", 2500 μm in diameter, will have a density of 0.8 g/cm^3, which is quite

Table 1. Physical properties of dusts that accrete into continuously larger porous aggregates as described by the hypothesis of hierarchical dust accretion (Rietmeijer, 2002a, Rietmeijer, 2002b, Rietmeijer, 1998)

Diameter (microns)	Mass (g)	Dust type	Density (g/cm^3)
PRINCIPAL COMPONENTS (PCS)			
0.09	$8.59 \cdot 10^{-16}$	CHON	2.25
0.09	$1.05 \cdot 10^{-15}$	Mixed	2.75
0.09	$1.15 \cdot 10^{-15}$	Silicate	3.00
1.00	$1.18 \cdot 10^{-12}$	CHON	2.25
1.00	$1.44 \cdot 10^{-12}$	Mixed	2.75
1.00	$1.57 \cdot 10^{-12}$	Silicate	3.00
MATRIX AGGREGATES; SILICATES, FE–SULFIDES (IDPS)			
2.00	$4.19 \cdot 10^{-14}$	Matrix aggregate	0.01
2.00	$4.19 \cdot 10^{-13}$	Matrix aggregate	0.1
2.00	$4.19 \cdot 10^{-12}$	Matrix aggregate	1.0
5.00	$6.54 \cdot 10^{-13}$	Matrix aggregate	0.01
5.00	$6.54 \cdot 10^{-12}$	Matrix aggregate	0.1
5.00	$6.54 \cdot 10^{-11}$	Matrix aggregate	1.0
2.00	$1.40 \cdot 10^{-11}$	Mg–rich olivine/pyroxene	3.35
2.00	$1.99 \cdot 10^{-11}$	Fe–sulfide	4.75
5.00	$2.19 \cdot 10^{-10}$	Mg–rich olivine/pyroxene	3.35
5.00	$3.11 \cdot 10^{-10}$	Fe–sulfide	4.75
AGGREGATE IDPS; SILICATES, FE–SULFIDES (CLUSTER IDPS)			
10.00	$5.24 \cdot 10^{-11}$	Aggregate IDP	0.1
10.00	$1.31 \cdot 10^{-9}$	Aggregate IDP	2.5
25.00	$8.18 \cdot 10^{-10}$	Aggregate IDP	0.1
25.00	$2.05 \cdot 10^{-8}$	Aggregate IDP	2.5
10.00	$1.75 \cdot 10^{-9}$	Mg–rich olivine/pyroxene	3.35
10.00	$2.49 \cdot 10^{-9}$	Fe–sulfide	4.75
25.00	$2.74 \cdot 10^{-8}$	Mg–rich olivine/pyroxene	3.35
25.00	$3.89 \cdot 10^{-8}$	Fe–sulfide	4.75
CLUSTER IDPS; SILICATES, FE–SULFIDES ("GIANT CLUSTERS")			
100.00	$5.24 \cdot 10^{-8}$	Cluster IDP	0.1
100.00	$1.31 \cdot 10^{-6}$	Cluster IDP	2.5
500.00	$6.54 \cdot 10^{-6}$	Cluster IDP	0.1
500.00	$1.64 \cdot 10^{-4}$	Cluster IDP	2.5
100.00	$1.75 \cdot 10^{-6}$	Mg–rich olivine/pyroxene	3.35
100.00	$2.49 \cdot 10^{-6}$	Fe–sulfide	4.75
500.00	$2.19 \cdot 10^{-4}$	Mg–rich olivine/pyroxene	3.35
500.00	$3.11 \cdot 10^{-4}$	Fe–sulfide	4.75

reasonable. The Perseid meteor, mass = 0.2 g (Fig. 9) has a 3.9 cm diameter (Eq. 1), corresponding to ≈ 0.01g/cm^3 density and implying >90% porosity. Conceivably such highly porous, cm–sized aggregates might exist in "free space" but it is unlikely they are preserved in comet nuclei, which revisits the hypothesized "glue". Except mixed PCs (Figs. 4,5), the collected aggregate IDPs show no evidence for "glue" but they show that mobility in the surface layer of accreted dusts will cause "necking" between individual particles (Rietmeijer, 2002b, Rietmeijer, 1998) resulting in increased material strength of aggregates. Using the CI meteorite density (1.6 g/cm^3), this Perseid meteor was a very small, massive pebble, 0.6 cm in diameter. The shape of its light curve, "humped" or not, would allow making a choice between these extreme textures. Significant structural discontinuities in the collected dusts exist when earlier–formed aggregates had structurally strengthened by surface mobility processes. Such discontinuities were probably most noticeable for aggregate particles <100 microns (Fig. 10).

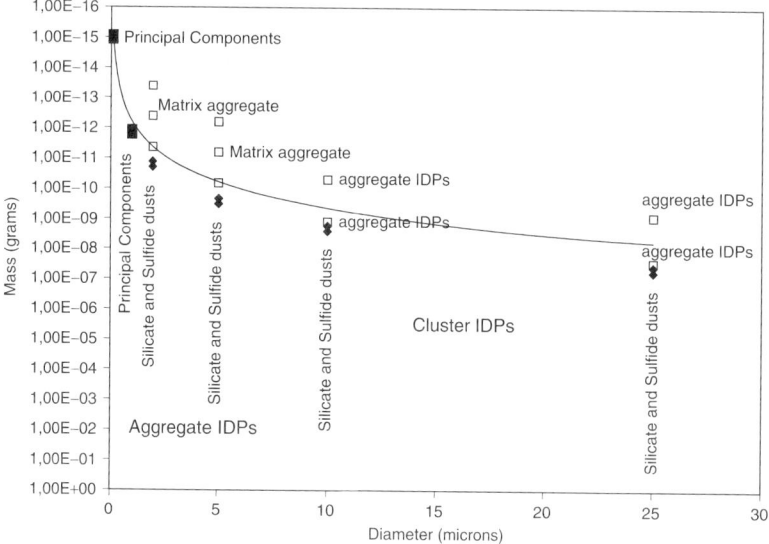

Figure 10. Principal components (black squares) accreting into 10–25 micron aggregate IDPs of matrix aggregates of different density (i.e. porosity) (open squares), sulfide and silicate dusts 10–25 micron, in size. Continued accretion leads to of cluster (Table 1)

The data in Figure 10 are limited to dusts that accreted into the collected aggregate IDPs. The extremely rare refractory aggregates, which are in fact collected as individual IDPs, are not shown here. Unfortunately, the mass and size of the sulfide and silicates dusts and these unique Ca,Al,Ti–rich aggregates are considerably overlapping and chemical data for these particular accreting dusts will be needed to make a reliable identification. Finally, although Figure 10 appears to ignore carbon dust, it is noted that most of the carbon will accrete very early as carbonaceous (CHON) PCs, and to a lesser extent in mixed PCs, when matrix aggregates are formed. The bulk carbon content will thus depend on the relative proportions of these PCs and silicate PCs (Riet-meijer, 2002b). For comparison, the measured Halley dust was 10^{-16}g$<$mass $<10^{-12}$g (Fomenkova et al., 1992) corresponding to PCs and porous, 0.01 and 0.1 g/cm^3, matrix aggregates and perhaps sub–micron silicate and/or Fe–sulfide dust.

4. Summary

Dust ejected from comet 67P/Churyumov–Gerasimenko and at its surface will include a range of sizes and mass ranging from porous aggregates to collapsed aggregates and (very) small pebbles of compacted aggregates. The size/mass ratio of this dust will be stepped in accordance with the hypothesis of hierarchical dust accretion. This accretion will lead to increasingly larger, porous dust aggregates that, when reaching cm–size dimensions will collapse into the less– porous pebbles wherein the constituent dusts will be 0.5 to $\approx$1 millimeter in size. A combination of size/mass ratio and dust chemical composition will define the nature of the collected dust particles and might allow making testable predictions on dust morphology and internal texture.

Acknowledgments

This work was supported by a grant from the National Aeronautics and Space Administration, NAG5-11762 (FJMR).

References

Bouwman, J., Meeus, G., de Koter, A., Hony, S., Dominik, C. and Waters, L.B.F.M. (2001). Processing of silicate dust grains in Herbig Ae/Be systems. *Astron. Astrophys.*, 375, 950-962.

Bradley, J.P. (1994). Chemically anomalous, pre-accretionally irradiated grains in interplanetary dust from comets. *Science*, 265, 925-929.

Fomenkova, M.N., Kerridge, J.F., Marti, K. and McFadden, L.-A. (1992). Compositional trends in rock-forming elements of comet Halley dust. *Science*, 258, 266-269.

Gombosi, T.I. and Houpis, H.L.F. (1986). An icy-glue model of cometary nuclei. *Nature*, 324, 43-44.

Jenniskens, P., Rietmeijer, F.J.M., Brosch, N. and Fonda, M. (Eds) (2000). Leonid Storm Research, 606p. *Kluwer Academic Publishers*, Dordrecht, the Netherlands.

Jessberger, E.K., Christoforidis, A. and Kissel, J. (1988). Aspects of major element composition of Halley's dust. *Nature*, 332, 691-695.

Mackinnon, I.D.R. and Rietmeijer, F.J.M. (1987). Mineralogy of chondritic interplanetary dust particles. *Revs. Geophys.*, 25, 1527-1553.

Murray, I.S., Beech, M., Taylor, M.J. and Hawkes, R.L. (2000). Comparison of 1998 and 1999 Leonid light curve morphology and meteoroid structure. *Earth, Moon, Planets*, 82-83, 351-367.

Nuth III, J.A., Hallenbeck, S.L. and Rietmeijer, F.J.M. (1999). Interstellar and interplanetary grains, Recent developments and new opportunities for experimental chemistry. *In Labo-*

110

ratory Astrophysics and Space Research, P. Ehrenfreund, K. Krafft, H. Kochan and V. Pir-ronello, Eds., 143-182, Kluwer Acad. Publ., Dordrecht.

Nuth III, J.A., Rietmeijer, F.J.M. and Hill, H.G.M. (2002). Condensation processes in astrophys-ical environments: The composition and structure of cometary grains. *Meteoritics Planet. Sci.*, 37, 1579-1590.

Riedler, W. et al. (1998). The MIDAS experiment for the ROSETTA mission. *Adv. Space Sci.*, 21, 1547-1556.

Rietmeijer, F.J.M. (1998). Interplanetary Dust Particles. *In Planetary Materials, Revs. Mineral.*, vol. 36 (J.J. Papike, ed.), 2-1 – 2-95. The Mineralogical Society of America, Washington, DC.

Rietmeijer, F.J.M., Nuth III, J.A. and Karner, J.M. (1999). Metastable eutectic condensation in a Mg-Fe-SiO-H2-O2 vapor: Analogs to circumstellar dust. *Astrophys. J.*, 527, 395-404.

Rietmeijer, F.J.M. (2000). Interrelationships among meteoric metals, meteors, inter-planetary dust, micrometeorites, and meteorites. *Meteoritics Planet. Sci.*, 35, 1025-1041.

Rietmeijer, F.J.M. and Nuth III, J.A. (2000). Collected extraterrestrial materials: Constraints on meteor and fireball compositions. *Earth, Moon, Planets*, 82/83, 325-350.

Rietmeijer, F.J.M. (2002a). Shower Meteoroids: Constraints from interplanetary dust particles and Leonid meteors. *Earth, Moon, Planets*, 88, 35-58.

Rietmeijer, F.J.M. (2002b). The earliest chemical dust evolution in the solar nebula. *Chemie der Erde*, 62, 1-45.

Rietmeijer, F.J.M. (2003). Meteors: The other interplanetary dust particles. *In Proc. 2002 In-ternl. Sci. Symp. Leonid Meteor Storms*, H. Yano, A. Abe and M. Yoshikawa, Eds., ISAS Rept SP-16, 139-147, Inst. Space and Astronautical Science, Sagamihara, Japan.

Rietmeijer, F.J.M., Rotundi, A. and Heymann, D. (2004). C60 and giant fullerenes in soot con-densed in vapors with variable C/H2 ratio. *Fullerenes, Nanotubes, Carbon Nanostructures*, 12, 659–680.

Rotundi, A., Rietmeijer, F.J.M., Colangeli, L., Mennella, V., Palumbo, P. and Bussoletti, E. (1998). Identification of carbon forms in soot materials of astrophysical interest. *Astron. Astrophys.*, 329, 1087-1096.

Rotundi, A., Brucato, J.R., Colangeli, L., Ferrini, G., Menella, V., Palomba, E. and Palumbo, P. (2002). Production, processing and characterization techniques for cosmic dust analogues. *Meteoritics Planet. Sci.*, 37, 1623-1635.

Sagdeev, R.Z., Evlanov, E.N., Fomenkova, M.N., Prilutskii, O.F. and Zubkov, B.V. (1989). Small-size dust particles near Halley's comet. *Adv. Space Sci.*, 9, 263-267.

Sandford, S.A. (1987). The collection and analysis of extraterrestrial dust particles. *Fund. Cos-mic Physics*, 12, 1-73.

Simpson, J.A., Tuzzolino, A.J., Ksanfomality, L.V., Sagdeev, R.Z. and Vaisberg, O.L. (1989). Confirmation of dust clusters in the coma of comet Halley. *Adv. Space Sci.*, 9, 259-262.

Steel, T.M. and Duley, W.W. (1987). A 217.5 nanometer absorption feature in the spectrum of small silicate particles. *Astrophys. J.*, 315, 337-339.

Snow, T.P. and Witt, A.N. (1996). Interstellar depletions updated: Where all the atoms went. *Astrophys. J.*, 468, L65-L68.

Trigo-Rodriguez, J.M., Llorca, J., Borovička, J. and Fabregat, J. (2003). Chemical abundances determined from meteor spectra: I. Ratios of the main chemical elements. *Meteoritics Planet. Sci.*, 38, 1283-1294.

Zolensky, M.E., Wilson, T.L, Rietmeijer, F.J.M. and Flynn, G.J. (Eds) (1994). Analysis of In-terplanetary Dust. *Amer. Inst. Physics. Conf. Proc.*, 310, 357p., Amer. Inst. Physics, N.Y.

PHYSICAL PROPERTIES OF COMETARY DUST. THE CASE OF CHURYUMOV- GERASIMENKO

A.C. Levasseur-Regourd, E. Hadamcik, J. Lasue
Université Paris VI / Aéronomie CNRS-IPSL, BP 3, 91371, Verrières, France

J.B. Renard
LPCE-CNRS, 3A av. de la recherche scientifique; 45071 Orléans cedex 2, France

J.C. Worms
ESSC-ESF, c/o ENSPS, bd Sébastien Brandt 67400 Illkirch, France

Abstract   Observations of the linear polarization of light scattered by cometary dust (spatial variations, phase and wavelength dependences) provide unique clues to the physical properties of the dust particles. The polarization of 67P/Churyumov-Gerasimenko has, up to now, only been observed at its 1982/1983 return. Data suggest that the coma is rather dust poor, in agreement with the absence of any silicate feature in the infrared spectra. However, outbursts may modify for a while the properties of the dust coma. Laboratory experiments on aggregates of grains representative of cometary particles should take place on board the ISS before the Rosetta rendezvous. Their results, together with complementary remote observations, should provide a better understanding of the morphology, size distribution and porosity properties of the dust particles to be encountered by the spacecraft within the coma.

Keywords: Comets, 67P/Churyumov-Gerasimenko, dust, light scattering, polarization, laboratory measurements.

1.   Introduction

The physical properties of cometary dust particles are still poorly known, although 1P/Halley in-situ observations and interplanetary dust particles collection in the Earth environment suggest that these particles are porous aggregates of smaller grains. Very few data are available for 67P/Churyumov-Gerasimenko. However, interpretation of the observations of the solar light

L. Colangeli et al. (eds.), The New ROSETTA Targets, 111–118.
© 2004 *Kluwer Academic Publishers. Printed in the Netherlands.*

scattered by dust in the coma should provide an insight on the morphology, size distribution, and porosity of its dust particles, and thus on the question of the dust hazard for Rosetta.

This paper summarizes our present understanding of the light scattering properties of cometary and asteroidal dust particles, with emphasis on the phase and wavelength dependence of the linear polarization, and on polarization observations of 67P/Churyumov-Gerasimenko. Interpretation of these observations in term of physical properties is provided by numerical and laboratory simulation. We then present the preliminary results obtained with PROGRA^2 experiment, together with future opportunities given by the ICAPS/IMPF facility on board the ISS.

2.   Light Scattering by Cometary Dust

Solar light scattered by low concentration dust clouds, such as cometary comae, is partially linearly polarized, with the electric field vector either parallel or perpendicular to the scattering plane. For a given wavelength λ, the intensity scattered in a direction corresponding to a given phase angle α is the sum of two polarized components, parallel and perpendicular to the scattering plane. The polarization P, which may be defined by the ratio of the difference to the sum of these two components (see e.g. Hapke, 1993), only depends upon λ, α, and the physical properties of the scattering dust. Polarization can thus be used to compare data obtained at different times and on different objects, and to point out spatial variations, as well as phase angle and wavelength dependences.

Spatial variations within cometary comae have been pointed out by the OPE/Giotto experiment, through local increases in polarization within 1P/ Halley coma and a lower polarization in the innermost coma (see e.g. Levasseur-Regourd et al., 1999a). The excellent agreement between the observed local intensities and dust fluxes has been used to fit the data with a dynamical model, suggesting that the dust particles are very absorbing, with an albedo of about 0.04, and have an extremely low density, of about $100 \; kg \cdot m^{-3}$ (Fulle et al., 2000). More recently, numerous observations of 1995 O1 Hale-Bopp have extensively illustrated the increase of polarization in jets or arcs and its decrease in the inner coma (see e.g. Jockers et al., 1999, Jones and Gehrz, 2000 and Hadamcik and Levasseur-Regourd, 2003a).

The analysis of the phase dependence of polarization data obtained within a given wavelength range, $P_\lambda(\alpha)$, leads to quite smooth phase curves. As illustrated in Figure 1, they present a small negative branch in the backscattering region, an inversion for a phase in the 20° to 23° range, and a wide positive branch with a maximum in the 90° to 100° range. For α greater than 30° to 35°, such phase curves allow to point out different cometary classes, corresponding to comets with a low maximum in polarization (0.10 to 0.15, depending upon the wavelength), comets with a higher maximum in polarization (0.25 to 0.30) and comet Hale-Bopp, the polarization of which was even higher (Levasseur-Regourd et al., 1996; Levasseur-Regourd and Hadamcik, 2003). The comets with a high maximum in polarization seem to be dust rich while the comets with a lower maximum seem to be dust poor and gas rich (Chernova et al., 1993).

Some trends need to be emphasized. First, as noticed by Levasseur- Regourd et al. (1996) and stressed by Hanner (2003), comets with a high maximum in polarization seem to exhibit a silicate emission feature, while comets with a low maximum do not present such a feature. Secondly, high maximum comets usually show numerous jets-like features, with a polarization higher than in the surrounding coma (Hadamcik and Levasseur-Regourd, 2003b). Finally, it may be added that asteroidal regoliths (and thus possibly comets nuclei surfaces) present similarly smooth phase curves, with comparable levels in polarization for C-type asteroids and cometary comae, at least up to 60°. Such curves have been estimated by various authors to be typical of the interaction of light with irregular particles media, with a size larger than the wavelength.

The analysis of the polarization wavelength dependence, $P_\alpha(\lambda)$, suggests that the polarization is a linear function of the wavelength for α between 20° and 90°, at least in the optical domain (Levasseur-Regourd and Hadamcik, 2003). As first noticed for Halley (Dollfus et al., 1988) and later confirmed for other comets, cometary dust polarization increases with the wavelength, the increase being more important for greater phase angles. The rate of increase of the polarization with the wavelength at a given phase angle is actually higher for Hale-Bopp than for Halley. The polarization could nevertheless decrease with increasing wavelength in the innermost coma, as noticed for Halley and suspected for Hale-Bopp (Levasseur-Regourd and Hadamcik, 2003). The wavelength dependence is less documented for asteroids, which are observed on a narrow range of phase angles near backscattering, with the exception of Near Earth Objects. The polarization decreases with the wavelength for S-type asteroids, while it might possibly increase with the wavelength for C-type asteroids.

3.   Significance of 67P/Churyumov-Gerasimenko Data

The linear polarization of 67P/Churyumov-Gerasimenko has, up to now, only been observed from October 1982 to January 1983, i.e. for phase angles in a 10° to 40° range, mostly by Myers and Nordsiek (1984) with a spectropolarimeter operating in the optical domain and, for one night, by Chernova et al. (1993). The corresponding data points are show in Figure 1. The phase angle dependence is nominal. Although a slight increase in polarization could be

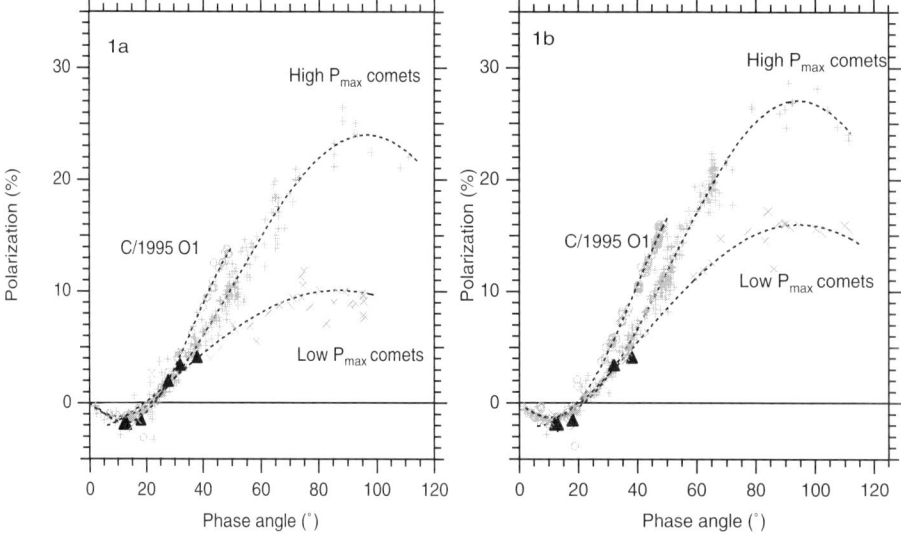

Figure 1. Cometary dust linear polarization as a function of the phase angle, including 67P/Churyumov-Gerasimenko observations (▲). The observations obtained above 30° to 35° show that some comets (×) present a low maximum in polarization, while other ones (+) present a higher maximum, and that Hale-Bopp (○) polarization is the highest. 1a: blue-green wavelengths spectral domain. 1b: red wavelengths spectral domain.

suspected near 30° phase angle, Churyumov-Gerasimenko seems to belong to the low maximum in polarization class. This suggestion is in agreement with the results of a series of fourteen infrared spectral observations, which did not show any silicate emission feature between September 1982 and March 1983 (Hanner et al., 1985).

It is important to add that outbursts, based on amateur monitoring of light curves, have been noticed near perihelion in 2002/2003 and 1996/1997, as well as possibly for the 1982/1983 return (Kidger, 2003; Marsden, 1983). Cometary outbursts usually take place while some small fragments of the nucleus are released, as observed for C/1996 B2 Hyakutake (Desvoivres et al., 2000), or before a complete disruption occurs, as illlustrated for C/1999 S4 LINEAR (Weaver et al., 2001). Such outbursts are immediately followed by an increase in polarization, of 2 to 3 percents. This behaviour has been noticed for Hyakutake (Tozzi et al., 1997; Kiselev and Velichko, 1999) and LINEAR (Hadamcik and Levasseur-Regourd, 2003c), as well as for 1P/Halley (Dollfus et al., 1988) and C/1996 Q1 Tabur (Kiselev et al., 2001). The above-mentioned increase in polarization for Churyumov-Gerasimenko, although almost within the error bars, is related to an increase in brightness (Myers and Nordsiek, 1984) and could be due to an outburst.

It might thus be assumed that 67P/Churyumov-Gerasimenko has a rather dust poor coma, the properties of which may nevertheless be altered during outburst events. The phase angle and wavelength dependence of the polarization (as already observed, and hopefully more documented during future returns) should provide specific constraints on the physical properties of the dust. Such properties can only be derived through appropriate simulations.

4. Laboratory Measurements in Microgravity

Simulations are necessary to interpret the light scattering observations in terms of physical properties. Numerical simulations are most difficult for irregular aggregates: any satisfactory fit may not be unique, the value of the imaginary part of the complex refractive index is always poorly known and, besides, the estimation of the refractive index is extremely difficult for a mixture of particles. A complementary approach is provided by laboratory simulations, which should avoid sedimentation and make use of particles that are representative of cometary dust. Taking into account the fact that the long cruise of Rosetta provides time for elaborate measurements before the encounter, we have developed a modular experimental approach since the mid nineties, with experiments operating under microgravity conditions during parabolic flights, during a rocket flight and tentatively on board the space station (Levasseur-Regourd et al., 1998).

The PROGRA^2 experiment operates during CNES parabolic flight campaigns, as well as in the laboratory for very porous samples. It measures the polarization of the light scattered by a sample contained in a vial, at a given moveable phase angle and for two wavelengths (Worms et al., 2000; Renard et al., 2002). It has demonstrated the feasibility of such measurements on levitating dust clouds (Worms et al., 1999). Besides, a significant data-base is being obtained, and it has been found that mixtures of low-density aggregates of submicron-sized grains provide an excellent fit with cometary observations (Hadamcik et al., 2002).

The CODAG-LSU experiment has operated during ESA MASER-8 rocket flight, to monitor continuously the intensity and polarization of the light scattered by micron-sized spheres contained in a low-pressure chamber where ballistic agglomeration was taking place (Levasseur-Regourd et al., 1999b). It has demonstrated the feasibility of measurements on aggregating particles (Haudebourg, 2000).

The next steps should be the ICAPS precursor experiment and the ICAPS facility on board the ISS. The scientific objectives of ICAPS (Interaction in Cosmic and Atmospheric Particle Systems) are, amongst others, to simulate aggregation processes from micron-sized particles under conditions representative of the proto-solar nebula, to simulate the formation of centimetre-sized regoliths, to interpret light scattering observations in terms of physical properties and to monitor phase changes on icy particles (Blum et al., 1999). A precursor experiment is to be launched in the 2006 time frame; it will test critical subsystems and monitor the evolution of the light scattering properties (measured simultaneously at three wavelengths) of bi-disperse aggregates. The ICAPS facility, now under phase B at ESA, is anticipated to be accommodated on board the ISS on a common rack with the IMPF facility for dusty plasmas.

5.   Conclusions

Light scattering observations indicate that cometary particles are irregular and possibly porous, in agreement with the suspected aggregated morphology of the dust. More specifically, 67P/Churyumov-Gerasimenko observations suggest that the coma is rather dust poor and that outbursts may modify its properties for a while. The dust coma is rather likely, similarly to that of 81P/Wild2 (Tuzzolino et al., 2004) to be heterogeneous, with regions of swarms of fragmenting particles. The precise physical properties of the dust (e.g. morphology, size distribution, porosity) still need to be determined, specially in the innermost coma where some evaporation and breaking-off may take place.

Laboratory experiments on aggregates of grains representative of the cometary dust composition (silicates, carbon) formed under microgravity conditions, as well as on regoliths and icy particles, should be performed on board the International Space Station before the Rosetta rendezvous. The results will provide a better understanding of the physical properties of the dust particles within the coma and on the nucleus surface. It should then be possible to assess more accurately the dust hazard for the Rosetta orbiter and the choice of the landing site, in order to optimize the scientific outcome of the mission.

References

Blum, J. et al. (1999). Research with small particles on board the ISS. *In Proceedings of the 2nd European symposium on the utilization of the space station*, ESA-SP 433, pages 285-289, ESTEC, Noordwijk.

Chernova, E.P., Kiselev, N.N. and Jockers, K. (1993). Polarimetric characteristic of dust particles as observed in thirteen comets, comparison with asteroids. *Icarus*, 103, 144-158.

Desvoivres, E., Klinger, J. and Levasseur-Regourd, A.C. (2000). Modeling the dynamics of fragments of cometary nuclei: Application to comet C/1996 B2 Hyakutake. *Icarus*, 144, 172-181.

Dollfus, A., Bastien, P., Le Borgne, J.L., Levasseur-Regourd, A.C. and Mukai, T. (1988). Optical polarimetry of P/Halley: Synthesis of the measurements in the continuum. *Astron. Astrophys.*, 206, 348-356.

Fulle, M., Levasseur-Regourd, A.C., McBride, N. and Hadamcik, E. (2000). In-situ dust measurements from within the coma of 1P/Halley: first order approximation with a dust dynamical model. *Astron. J.*, 119, 1968-1977.

Hadamcik, E., Renard, J.B., Levasseur-Regourd, A.C. and Worms, J.C. (2002). Polarimetric study of levitating dust aggregates with the PROGRA^2 experiment. *Plan. Space Sci.*, 50, 895-901.

Hadamcik, E. and Levasseur-Regourd, A.C. (2003a). Dust evolution of comet C/1995 O1 (Hale-Bopp) by imaging polarimetric observations. *Astron. Astrophys.*, 403, 757-768.

Hadamcik, E. and Levasseur-Regourd, A.C. (2003b). Imaging polarimetry of cometary dust: different comets and phase angles. *J. Quant. Spectros. Radiat. Transfer*, 79-80, 661-678.

Hadamcik, E. and Levasseur-Regourd, A.C. (2003c). Dust coma of Comet C/1999 S4 (LINEAR): Imaging polarimetry during the nucleus disruption. *Icarus*, 166, 188-194.

Hanner, M.S., Tedesco, M.S., Tokunaga, A.T., Veeder, G.J., Lester, D.F., Witteborn, F.C., Bregman, J.D., Gradie, J. and Lebofsky, L. (1985). The dust coma of periodic comet Churyumov-Gerasimenko. *Icarus*, 64, 11-19.

Hanner, M.S. (2003). The scattering properties of interplanetary dust. *J. Quant. Spectros. Radiat. Transfer*, 79-80, 695-705.

Haudebourg, V. (2000). Propriétés de diffusion des particules en suspension, étude expérimentale de la transition du régime de Mie à des agglomérats. *PhD thesis, Université Paris VI.*

Hapke, B. (1993). Theory of reflectance and emittance spectroscopy. *Cambridge Univ. Press*, Cambridge.

Jockers, K., Rosenbush, V.K., Bonev, T. and Credner, T. (1999). Images of polarization and color in the inner coma of comet Hale-Bopp. *Earth Moon Planets*, 78, 373-379.

Jones, T.J. and Gehrz, R.D. (2000). Infrared imaging polarimetry of comet C/1995 O1 (Hale-Bopp). *Icarus*, 143, 338-346.

Kidger, M.R. (2003). Dust production and coma morphology of 67P/Churyumov-Gerasimenko during the 2002-2003 apparition. *Astron. Astrophys.*, 408, 767-774.

Kiselev, N. and Velichko, F.P. (1999). Aperture polarimetry and photometry of comet Hale-Bopp. *Earth Moon Planets,*, 78, 347-352.

Kiselev, N.N., Jockers, K., Rosenbush, V.K. and Korsun, P.P. (2001). Analysis of polarimetric, photometric, and spectroscopic observations of comet C/1996 Q1 (Tabur). *Solar System Reseach*, 35, 6, 480-495.

Levasseur-Regourd, A.C., Hadamcik, E. and Renard, J.B. (1996). Evidence for two classes of comets from their polarimetric properties at large phase angles. *Astron. Astrophys.*, 313, 327-333.

Levasseur-Regourd, A.C., Cabane, M., Haudebourg, V. and Worms, J.C. (1998). Light scattering experiments under microgravity conditions. *Earth, Moon, Planets*, 80, 343-368.

Levasseur-Regourd, A.C., McBride, N., Hadamcik, E. and Fulle, M. (1990a). Similarities between in situ measurements of local dust scattering and dust flux impact data within the coma of 1P/Halley. *Astron. Astrophys.*, 348, 636-641.

Levasseur-Regourd, A.C., Cabane, M., Chassefière, E., Haudebourg, V. and Worms, J.C. (1999b). The CODAG Light Scattering Experiment for light scattering measurements by dust particles and their aggregates. *Adv. Space Res.*, 23, 1271-1277.

Levasseur-Regourd, A.C. and Hadamcik, E. (2003). Light scattering by irregular dust particles in the solar system : observations and interpretation by laboratory measurements. *J. Quant. Spectros. Radiat. Transfer.*, 79-80, 903-910.

Marsden, B. (1983). Comets in 1983. *Q. J. R. Astr. Soc.*, 27, 102-118.

Myers, R.V. and Nordsiek, K.H. (1984). Spectropolarimetry of comets Austin and Churyumov-Gerasimenko. *Icarus*, 58, 431-439.

Renard, J.B., Worms, J.C., Lemaire, T., Hadamcik, E. and Huret, N. (2002). Light scattering by dust particles in microgravity: polarization and brightness imaging with a new version of the PROGRA^2 instrument. *Appl. Opt.*, 91, 609-618.

Tozzi, G.P., Cimatti, A., di Serego Alighieri, S. and Celino, A. (1997). Imaging polarimetry of comet C/1996 B2 (Hyakutake) at the perigee. *Planet. Space Sci.*, 45, 535-540.

Tuzzolino et al. (2004). Dust measurements in the coma of comet 81P/Wild2 by the dust flux monitor instrument. *Science*, 304, 1776–1780.

Weaver, H.A. et al. (2001). HST and VLT investigations of the fragments of comet C/1999 S4 (LINEAR). *Science*, 292, 1329-133.

Worms, J.C., Renard, J.B., Levasseur-Regourd, A.C. and Hadamcik, E. (1999). Light scattering by dust particles in microgravity: the PROGRA^2 experiment achievements and results. *Advances in space research*, 23, 1257-1266.

Worms, J.C., Renard, J.B., Hadamcik, E., Brun-Huret, N. and Levasseur- Regourd, A.C. (2000). The PROGRA^2 light scattering instrument. *Planet. Space Sci.*, 48, 493-505.

PHYSICAL MODEL OF THE COMA OF COMET 67P/CHURYUMOV–GERASIMENKO

J.F. Crifo
CNRS, Service d'Aéronomie, BP3, F91371 Verrières Cedex, France
crifo@aerov.jussieu.fr

G.A. Lukyanov, V.V. Zakharov
CAS, St–Petersburg State Polytechnical University, 195251 St–Petersburg, Russia
zvv@fn.csa.ru

A.V. Rodionov
TsNIIMASch, Korolev, Moscow Region, 14070 Russia
avrodionov@mtu–net.ru

Abstract   We overview an ongoing program of three–dimensional, time–dependent, physically realistic simulation of the gas and dust coma of the comet, using both fluid equations and Direct Monte–Carlo Simulations.

Introduction

Several Rosetta mission instruments will perform totally new observations which require an extremely realistic physical modelling of the coma – not just approximate orders of magnitude: for instance, the high resolution mm wave sounder MIRO, and the in–situ dust velocity analyzer GIADA. In addition, the deposition of the lander will require an accurate and predictive model of the gas distribution. The presently discussed effort is being developed to meet such requirements. Its essential characteristics is to be based exclusively on plausible physical assumptions (as opposed to unphysical simplifications) and to maximize the use of frontier gasdynamic modelling techniques through the direct implication of gasdynamic experts. Here, for want of space, we do not describe comprehensively the model – the latest description of its time–stationary version is to be found in Rodionov et al., 2002, supplemented by Crifo et al., 2002; 2003 and Crifo et al., 2004 – nor do we give a thorough account of the results

L. Colangeli et al. (eds.), The New ROSETTA Targets, 119–130.

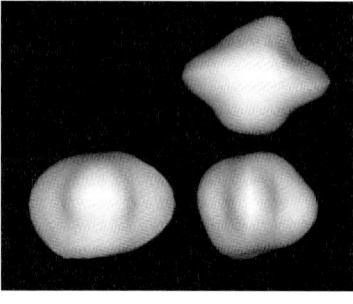 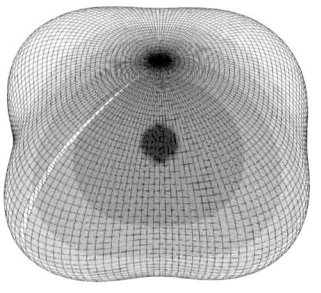

phieye =   45.00 thetaeye = 45.00

Figure 1.   Shape of the nucleus of 67PCG as derived by Lamy et al., 2003 (*left*), and approximation to this shape used in the present study (*right*). The isocontours at right are those of an assumed solar illumination with subsolar point at center of the circular spot.

already obtained on comet Churyumov–Gerasimenko (hereafter designated as 67P/CG). We give only an outline of the model, and present a few illustrative examples of results. One of the most regretful omission will be the absence of any presentation of the spectacular (and instructive) time–dependent simulations of the comet activity.

1.   Nucleus model

Shape

As described in Rodionov et al., 2002, even though the shape and rotational parameters of 67P/CG nucleus were known accurately, it would still be necessary to perform simulations of its coma at increasing levels of spatial filtering of the shape, to fully understand the physical origin of the computed structures of the comas. Hence here we use a set of alternative shapes of increased complexity: (1) a 2km radius sphere; (2) two ellipsoids having, respectively, semi–axes $3 \times 1.5 \times 1.5$ km, and $3 \times 3 \times 1.5$ km; (3) a "starfish" analytical shape with topological similarity to the shape derived by Lamy et al., 2003 from Hubble Space Telescope 2002 observations, and with envelope dimensions $(4.0 \times 4.0 \times 2.5$ km). Fig. 1 compares the two latter shapes: in spite of some differences, they surely exhibit the same kind of coma structuring.

Composition

We assume the nucleus outer layers to be a mixture of porous ice and porous, non–icy dust, with a common specific mass ρ_n, but not necessarily in a uniform proportion. A version of the model in which icy dust is present has been developed, but not yet exploited. The exposed ice (not the exposed non–icy dust!) sublimates when sunlit. We characterize each element of the surface by its icy area fraction f, ratio of its exposed ice area to its total area, related to the total dust–to–ice mass ratio $\Re$ in the nucleus at the same point, by (Crifo, 1997):

$$f = 1/(1 + \Re)$$

Volatile molecules (e.g. CO) are allowed to diffuse across both the ice and the dust, according to some arbitrary pattern. It is assumed that the surface dust mass flux is, at each point, proportional to the total net gas mass flux (sublimated + diffused – condensed).

Gas production model

We have in the past described two versions of gas production model (see Rodionov et al., 2002): surface sublimation of dusty ice (model I), or diffusion of CO from the interior across an inert surface at asteroidal temperature (model II). Here, we use also a new model, in which CO and H_2O are simultaneously produced from the surface (model III).

Model I is fully deterministic, involving only the heliocentric distance r_h, the nucleus orientation angles, and the distribution of surface icy area fraction f. Energy budget equations are written for each icy element of the surface, to compute the upward flux of H_2O. Water molecules backscattered from the coma to icy (or cold enough dusty) parts of the surface are recondensed. The net H_2O flux (upward minus downward) at each point is not function of the ice temperature only, but also of the initial outflow Mach number (to be discovered by solving the gasdynamic equations of the coma gas outflow). We introduce a night side (and shadow) internal heat transfer free parameter κ, to keep the surface temperature there within reasonable limits. To define a homogeneous nucleus, f is taken constant over the whole surface. To define a nucleus with active and inactive regions, in a simplistic way, we set $f = f_B$ in the background, and $f = f_i \gg f_B$ in $\mathcal{N}$ "active regions" $\mathcal{A}_i$. The active regions can have any contour. Since the contours are defined using the grid which defines the nucleus surface, good contour definitions require $\mathcal{N}$ to remain small (e.g. at most 5). In practice, we use circular areas of unequal sizes, randomly placed over the surface (which may assume any of the selected shapes).

Model II is heuristic: it postulates arbitrarily the CO flux at each point of the surface. This is done, of course, with consideration of r_h. As to the surface flux distribution, it can be either uniform, or restricted to active areas (not related to surface ice, as, in this model, there is no surface sublimation). CO molecules backscattered from the coma to the surface are backscattered from it to the coma. We usually write that the upward flux of CO is:

$$Z_{CO}(z_\odot, \vec{u}_\odot) = Q_{CO} f^{CO}\left(\frac{a_0}{A_{ext}} + \frac{(1-a_0)\max[\cos z_\odot, 0.]}{A_\odot(\vec{u}_\odot)}\right) \text{molecule/m}^2\text{s}$$

where $z_\odot$ is the solar zenith angle, Q_{CO} is the desired total CO production rate, $0 \leq a_0 \leq 1$ is a day–night asymmetry parameter, A_{ext} is the external nucleus area, $A_\odot$ is the nucleus solar light capture cross–section when the Sun is in the nucleocentric direction with unit vector $\vec{u}_\odot$, and, finally, the function f^{CO} allows the definition of active areas. It is easy to check that, with the preceding definition, and if $f^{CO} \equiv 1$, or $a_0 \simeq 1$, Q_{CO} does not vary during the nucleus rotation.

In model III, we assume the dusty ice to sublimate as in model I, and CO to diffuse, from the interior, across both the dust and the ice, as in model II. With its preceding definition, Z_{CO} is, if a_0 is small and $f^{CO} \propto f$, about proportional to Z_{H_2O} on the dayside. The ratio $Q(CO)^{night}/Q(CO)^{day}$ is proportional to a_0. Other assumptions can of course be made.

In the limit $f^{CO} \longrightarrow 0$ model III yields model I, whereas in the limit $f \longrightarrow 0$, it yields model II.

2.   Gas Coma model

Our objective is to provide, in a 3D+t model, a gaskinetically exact description of the distribution of CO, H_2O, and their photodestruction products, out to a distance of order 10^5 km. It is necessary to take the photodestruction of CO into account, mostly because, at large r_h, it may provide a heating source greater than that of H_2O. A correct allowance for the coma heating is, in its turn, needed to obtain a correct (often non–Maxwellian) velocity distribution of the molecules, as requested to interpret the high resolution rotational line spectra.

Only a 3D+t DSMC method can meet the above objective. At the time being, however, the construction of such a code is not completed; one of the worrysome difficulties is that the collision sections between fast dissociation products and the molecules are not available. Another is the computer resources needed. Thus, presently we have only developed two complementary approximations to the final objective.

Fluid method

In this case, we solve fluid (Euler or Navier–Stokes) equations everywhere in the coma. The analytical photodissociation model described in Rodionov et al., 2002 (inspired from the old 1–D DSMC result of Bockelée Morvan & Crovisier, 1986) is generalized to the case of the two mother molecules. In view of the weakness of the dissociation rate of CO, the concentrations of the dissociation products of CO are quite small, even at 10^5 km from the nucleus, so that we neglect the radial change in CO mass concentration, as well as the mass concentrations of O and C due to it. For the unknown collision sections of a fast radical of mass m with either H_2O or CO (of mass M) which are needed in the model, we write, heuristically:

$$\sigma_{(m,M)} = \sigma_M/4 \left(1 + (m/M)^{1/3}\right)^2$$

where σ_M is the so–called "variable hard sphere" (VHS) cross–section of either H_2O or CO, deduced from their well–known viscosity.

We use fluid equations for the molecular mixture. That is, the gas molecular mass and specific heat ratio are obtained from mixture laws. For the specific heat ratios of water and CO, we use vibrationally relaxed values. This approach yields a first–order model which usually overestimates the cooling, hence provides too small coma temperatures and slightly too high velocities. But it is computationally very efficient and lends itself to the introduction of a molecular radiative exchange term.

DSMC method

The extension of the DSMC method described in Crifo et al., 2002; 2003 is being done along three lines: (1) allowance for the CO – H_2O mixture, (2) allowance for translational relaxation and (3) allowance for photodissociation of the CO – H_2O mixture. The problem of correctly representing the molecular internal states is momentarily left aside as it is coupled to the problem of developing a 3–D radiation transfer model.

Initial conditions at the nucleus surface

While the DSMC method can be started from the surface, this is computationally inefficient. Near to the nucleus, for models I and II, we solve the Boltzman equation by an extremely accurate analytical approximation, and fluid equations to some distance, as described in Rodionov et al., 2002; in the case of model III, it was necessary to generalize the approach described in Rodionov et al., 2002 to a $H_2O - CO$ gas mixture, as will be described in detail elsewhere: three different algorithms relate the initial mass concentrations C_W and C_{CO} of H_2O and CO, the initial pressure p_0 and the initial temperature T_0 (at the top of the disequilibrium layer), to the nucleus icy element temperature T_I, and to the (initially unknown) initial mixture Mach number M_0. Surface energy budget equations (involving M_0) are written at each point, as in models I and II. In these equations, the heat transfer to the interior is represented by a weak heat source effective at non–sunlit points: the heuristic parameter κ defined in Crifo & Rodionov, 1999a is used for that purpose. The surface energy budget equations are coupled to the gasdynamic equations of mixture outflow, and solved simultaneously with them.

3. Dust Coma Model

We here treat only the region where the nucleus gravity and/or aerodynamic force are non–negligible in the dust dynamics. Outside of it, correct Keplerian models are available (Fulle et al., 2004). The grains are assumed spherical and with the same specific mass ρ_n as the nucleus (the choice of ρ_n greatly affects the dynamics of the heavy dust). Any size distribution can be specified. For definiteness, we use the Halley–like distribution defined in Crifo & Rodionov, 1997a;b. To compute the coma grain velocity and number density, we once more use two alternative techniques: multifluid, and DSMC.

Multifluid method

This method (see Rodionov et al., 2002) is computationally fast and accurate, except in cases where (a) dust flux from different points of the surface, and of comparable magnitude, cross one another, or (b) dust grains are mirrored towards the nucleus, either by the radiation pressure, or by the gravity. In case (a) the method provides only an approximation to the exact solution, and in case (b) the method does not converge. Also the method cannot reveal the distribution of trace components of the distribution (see below). When it is necessary to have a thorough information, DSMC must be used.

DSMC method

The first DSMC method for modelling the near–nucleus dust coma was developed recently (Crifo et al., 2004). The method can, not only obviate the limitations just described, but, as well, take into account mutual grain collisions (case of very active comets) and non–spherical grains (not done presently).

4. Results

A large set of alternative models of comet 67P/CG, all compatible with the available observational constraints, is under production. They are obtained by selecting (1) one of the four defined nucleus shapes, (2) a uniformly active active surface, or a surface with from one to five different circular active areas, (3) a nucleus specific mass ρ_n (in the range $0.03 - 3.$ g cm^{-3}), (4) an heliocentric distance (here always $r_h = 3$ AU) (5) the appropriate gas coma models (I, II or III), in particular the desired total gas production rate $Q(H_2O)$ (here in the range $0 - 10^{26}$ molec/s) and/or $Q(CO)$ (here in the range $0 - 10^{27}$ molec/s), and the CO asymmetry parameter a_0, (6) the dust model described above. The results presented here are extracted from this set.

Pure CO coma around a spherical nucleus

Fig. 2 (top, left panel) shows the gas temperature dispersion along the sun-ward axis in a DSMC computed pure CO coma around a 2km sphere radius nucleus, at an assumed $Q_{CO} = 10^{27}$ molec/s total production rate. The surface production was assumed nearly uniform ($a_0 = 0.9$). One can see that equilibrium conditions prevail only out to about 4 km above the surface. Beyond this distances, the thermal velocity distribution becomes elliptical, with the parallel temperature $T_\parallel$ freezing–out, and the perpendicular temperature $T_\perp$ continuing to decrease. This distance is found to decrease to 2km at the terminator, and then to increase up to 10 km on the antisolar axis.

Fig. 2 also shows the corresponding number density n, flow velocity V and translational temperature T, computed from the same DSMC and from fluid equations. One sees the (expected) excellent agreement, which also persists at larger distances, since above 40 km practically zero–temperature free–molecular effusion prevails. In the following, we therefore show only fluid dynamical results. However, one should keep in mind that, for a proper interpretation of high resolution CO emission spectra, the temperature dispersion must be taken into account, i.e. DSMC results are requested. One will notice from the figure that the CO temperature and velocity are much smaller in the night side coma (as already established in Crifo & Rodionov, 1999a and Crifo & Rodionov, 1999b), whereas the density is significantly greater there, even though the surface gas production was assumed about uniform.

Fig. 3 shows the individual trajectories of large dust grains (i.e., not much below the maximum liftable mass), for two assumed values of ρ_n. One sees that grain turnbacks are frequent, including cases where the grains encircle the nucleus before escaping. In the latter case, these grains provide a minority dayside component, which can only be revealed by DSMC. Also, notice that grain impacts occur on the nucleus surface, here on the night side. Such effects exist in all comets, but are here revealed for the first time, owing to the new DSMC method of Crifo et al., 2004. The figure also shows the density and velocity distribution of such large grains. One will especially notice the large density spike aligned anti–sunwards, and the extremely small values of the ejection velocities.

Mixed $H_2O - CO$ coma around an aspherical nucleus

Fig. 4 shows the H_2O and CO distributions in two orthogonal symmetry planes crossing the coma around the nucleus of Fig. 1, illuminated by the Sun as shown on Fig. 1. The total gas production of CO and H_2O are, respectively, 1.0×10^{27} and 2.5×10^{26} molec/s. The CO production is once more assumed to be about uniform ($a_0 = 0.9$). One can see that, as established in all similar previous works, the surface topography dominantly structures the CO coma,

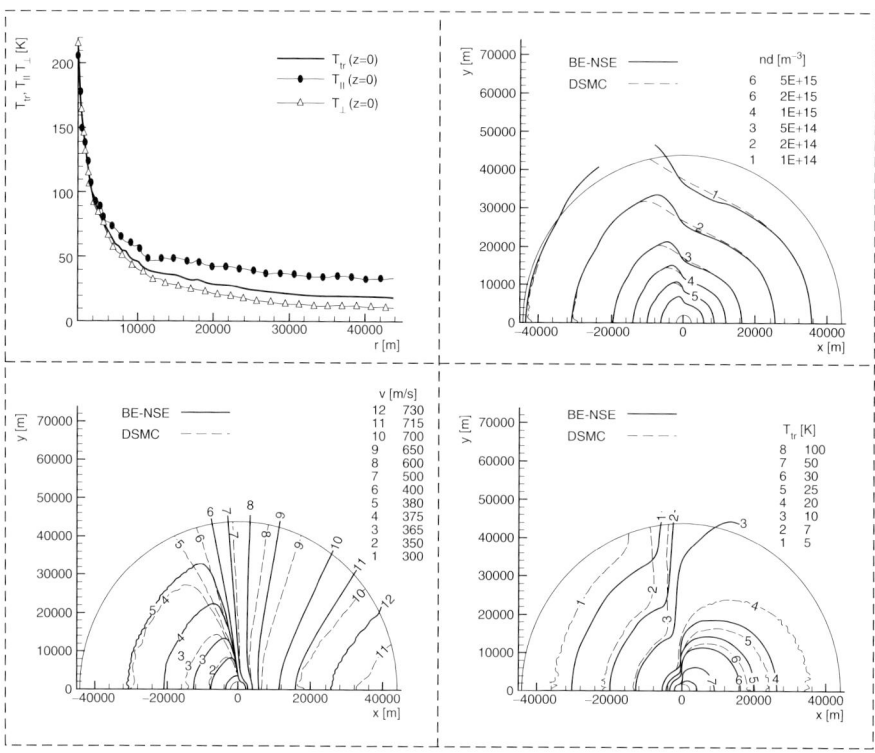

Figure 2. Pure CO coma around a 2 km radius spherical nucleus at $r_h = 3$ AU. The surface flux is assumed nearly uniform ($a_0 = 0.9$). The top left panel shows the translational, parallel, and perpendicular velocities as a function of distance along the comet–to–Sun axis. The other panels show the good agreement of the fluid and DSMC methods for the CO number density (*top right*), velocity (*bottom left*), and translational temperature (*bottom, right*).

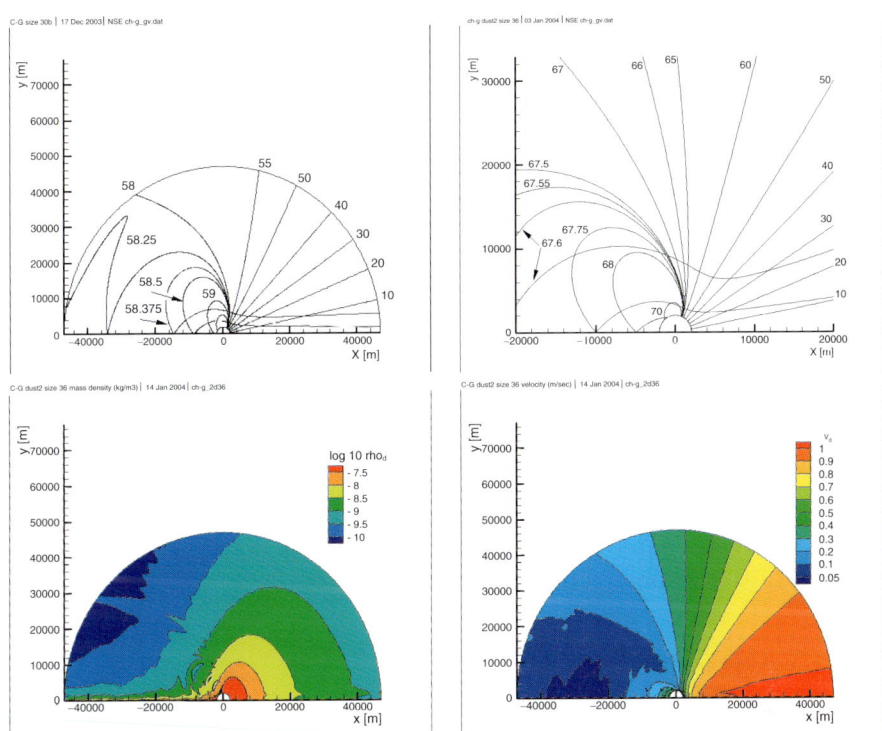

Figure 3. Heavy grains distribution in the CO coma of Fig. 2. The top panels show individual large grains trajectories: *left*: assumed $\rho_n = 1$ g cm^{-3}, grain radius $a_d = 1$ mm; *right*: $\rho_n = 0.1$ g cm^{-3}, $a_d = 9.1$ mm. Notice that the parts of the trajectories lying below the horizontal axis have been mirrored above it. The numbers labelling the curves are the solar zenith angle of grain detachment from the surface. The lower two panels show the resulting mass density (kg m^{-3}) (*left*) and velocity (m/s) (*right*) in the case $\rho_n = 0.1$ g cm^{-3}, $a_d = 9.1$ mm.

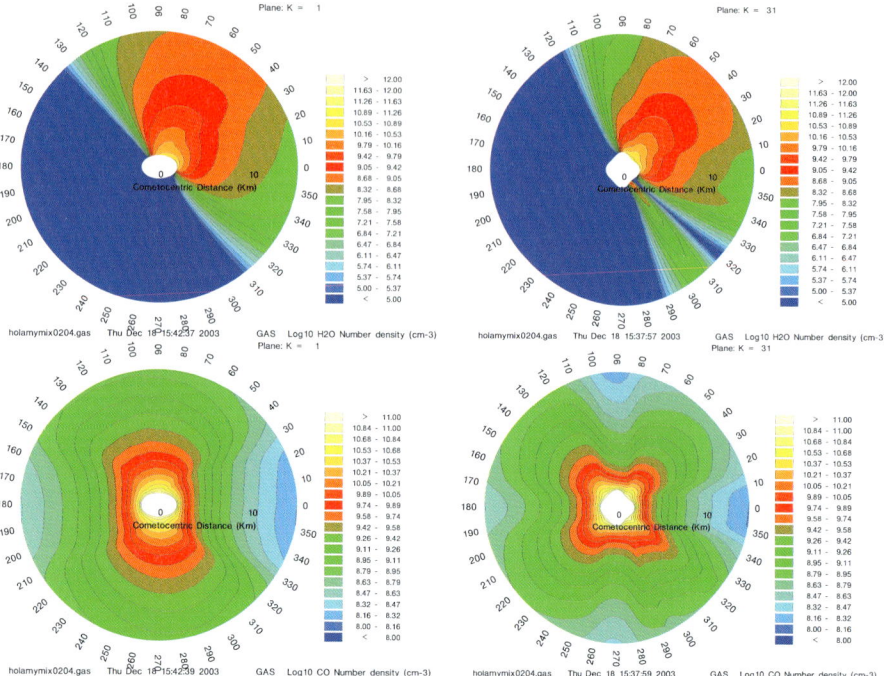

Figure 4. The gas coma around the nucleus of Fig. 1, in two orthogonal symmetry planes (*right and left*). Notice the nucleus intersection at panels centers. The nucleus is assumed sublimating H_2O and diffusing out CO. Shown are $Log_{10}(H_2O$ number density (cm^{-3})) (*top*), $Log_{10}(CO$ number density(cm^{-3})) (*bottom*).

and significantly affects the H_2O coma (notice, in particular, the gas diverted in the angular sector 305 in the equatorial plane).

Fig. 5 shows the 5cm radius dust grains distributions in the same two orthogonal planes, computed by the multifluid method, assuming a nucleus specific mass 0.1 g cm^{-3}, and approximating its gravitational field by a spherically symmetric field. The coma appears entirely structured by the gas topography; in particular, one notices two orthogonal pan–cake distributions of dust, associated with the concave "valleys" of the nucleus, as in Crifo & Rodionov, 1997a;b and Crifo & Rodionov, 1999b. Let us point out that surface inhomogeneities ("active areas") can complicate the coma, but can never suppress the topography induced structures: this is why the automatic attribution of observed coma structures to active areas is a physical nonsense (Crifo & Rodionov, 1999b).

Another evident (and general) property of the computed dust distribution is the high non–uniformity of the terminal dust velocity. Notice also that there

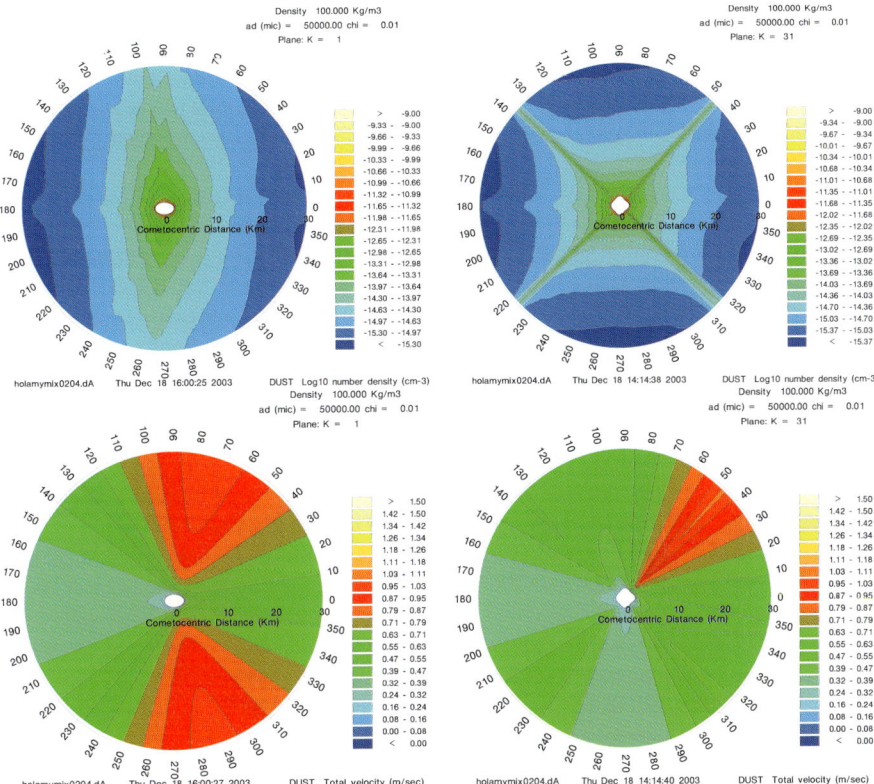

Figure 5.   The 5 cm radius grains coma in the two planes of Fig. 4: Log_{10}(grain number density, cm^{-3}) (*top*), velocity (m/s) (*bottom*). The nucleus and grain specific mass is assumed to be 0.1 g cm^{-3}.

is no correlation between high density and high velocity – contrary to what is often, arbitrarily, postulated.

Acknowledgments

This work was supported by French CNES/CTT 2003 and CNES/DPS 2003 grants.

References

Crifo, J.F. and Rodionov, A.V. (1997a) The dependence of the circumnuclear coma structure on the properties of the nucleus. I. Comparison between a homogeneous and an inhomogeneous spherical nucleus, with applicatio to P/Wirtanen. *Icarus*, 127:319–353; (1997b) The dependence of the circumnuclear coma structure on the properties of the nucleus. II. First investigation of the coma surrounding a homogeneous, aspherical nucleus. *Icarus*, 129:72–93.

Crifo, J.F. (1997). The correct evaluation of the sublimation rate of dusty ices under solar illumination, and its implication on the properties of P/Halley nucleus. *Icarus*, 130:549–551.

Crifo, J.F., Rodionov, A.V. and Bockelée–Morvan, D. (1999a). The Dependence of the Circumnuclear Coma Structure on the Properties of the Nucleus: III. First Modelling of a CO–dominated Coma, with Application to comets P/Wirtanen and P/Schwachmann–Wachmann I. *Icarus*, 138:85–106

Crifo, J.F. and Rodionov, A.V. (1999b). Modelling the Circumnuclear Coma of Comets: Objectives, Methods and Recent Results. *Planetary and Space Sciences*, 47:797–826.

Crifo, J.F., Loukianov, G.A., Rodionov, A.V., Khanlarov, G.R. and Zakharov, V.V. (2002). Comparison between Direct Monte–Carlo Simulation and Gasdynamic Solutions. I. Homogeneous, spherical nucleus. *Icarus*, 156:249–268. (2003) Comparison between Direct Monte–Carlo Simulation and Gasdynamic Solutions. II. Homogeneous, aspherical nucleus. *Icarus*, 163:479–503.

Crifo, J.F., Loukianov, G.A., Rodionov, A.V. and Zakharov, V.V. (2004). Direct Monte–Carlo and Multifluid modelling of the circumnuclear dust coma I. Spherical grains dynamics revisited *Icarus*, submitted.

Bockelée Morvan, D. and Crovisier, J. (1986) The role of water in the thermal budget of the coma. *Symposium on the Diversity and Similarity of Comets*, ESA SP–278:167.

Fulle, M., Barbieri, C., Cremonese, G., Rauer, H., Weiler, M., Milani, G. and Ligustri, R. (2004). The dust environment of Comet 67P/CG: a challenge for the Rosetta probe? *Astron. Astrophys.*, submitted.

Lamy, Ph., Toth, I. and Weaver, H. (2003). *Space News International*, http://www.spacenews.be/art2003/rosetta_050903.html

Rodionov, A.V., Crifo, J.F., Szego, K., Lagerros, J. and Fulle, M. (2002). An advanced physical model of cometary activity. *Planet Space Sci.*, 50:983–1024.

THE DUST ENVIRONMENT OF COMET 67P/CHURYUMOV–GERASIMENKO

Marco Fulle
Istituto Nazionale di Astrofisica (INAF), Osservatorio Astronomico di Trieste, Via Tiepolo 11, I-34131 Trieste, Italy

Cesare Barbieri
Dipartimento di Astronomia, Universita' di Padova, Vicolo dell'Osservatorio 2, I-35122 Padova, Italy

Gabriele Cremonese
Istituto Nazionale di Astrofisica (INAF), Osservatorio Astronomico di Padova, Vicolo dell'Osservatorio 5, I-35122 Padova, Italy

Heike Rauer
Institute of Planetary Research, DLR, Rutherfordstrasse 2, 12849 Berlin, Germany

Michael Weiler
Institute of Planetary Research, DLR, Rutherfordstrasse 2, 12849 Berlin, Germany

Giannantonio Milani
Unione Astrofili Italiani, UAI

Rolando Ligustri
Osservatorio di Talmassons, CAST, Talmassons (Ud), Italy

Abstract   Comet 67P/Churyumov–Gerasimenko passed its last perihelion on August 2002 and was well observed from fall 2002 to spring 2003. Its most prominent feature was a bright thin dust tail, which is best fitted by the Neck–Line model. Fits of the whole tail provide the dust environment of 67P during a year around perihelion, which is dominated by nucleus seasons. The dust mass loss rate appears constant since 2 AU before perihelion at about 200 kg s^{-1}, a factor 100

131

higher than 46P/Wirtanen, the previous target of the Rosetta mission. Neck–Line photometry during winter 2002-03 allows us to conclude that such a dust environment remained constant since 3 AU before perihelion, i.e. the distance at which Rosetta science operations will start and the lander will be delivered to the surface.

Keywords: comets, ROSETTA mission

Introduction

Due to a launch delay, in spring 2003 ESA changed the target of Rosetta Mission (ESA 2003): from 46P/Wirtanen to 67P/Churyumov–Gerasimenko (67P hereafter for brevity). Here we focus on the dust environment of 67P: this is fundamental not only to plan the operations of the many instruments studying the dust ejected from the comet, but also to determine the orbiting strategies, the software robustness of navigation cameras against false nucleus detection, and overall the lifetime of the landing probe against dust pollution.

1.   Neck–Line in the 67P tail

Amateur Astronomers' Databases contain many CCD images following the evolution of a thin comet tail during nine months after August 2002 perihelion (UAI 2003). Since January 2003, the aberration angle (difference with respect to a perfect antisolar direction) became too large to be consistent with a ion tail. Since the structure of the tail remained constant during all the nine post–perihelion months, the most probable explanation is that the same dust tail was observed during all the observations. The Position Angle (PA_T) of the observed tail is sampled for the best subset of available CCD images in Table 1, and is compared with the antisolar direction (PA_{RV} is the Position Angle of the prolonged radius vector projected on the sky plane) and with the expected position of a dust trail (PA_{CO} is the Position Angle of the sky–projected comet orbit along which we observe trails). The observed tail is inconsistent with both computed PA values, while it is consistent with the PA of a Neck–Line (Kimura and Liu 1977), provided by the synchrone ejected at the true anomaly described in Table 1.

2.   Dust Tail: fit by the Inverse Montecarlo Model

Comet 67P was observed as target of opportunity by means of the Galileo Telescope (TNG). We added all the obtained R CCD images to obtain an input image to be processed by means of the inverse tail model (Fulle 1989). The tail image (shown in Figs. 1 and 2) is not polluted by any significant ion tail and is dominated by a prominent antitail, which was undoubtedly composed of dust. The fit of the antitail and of the dust coma by means of the same model

Table 1. The observed Position Angle PA_T of the Tail is compared with the antisolar direction (PA_{RV} of the prolonged radius vector) and the expected position of the dust trail (PA_{CO} of the comet orbit projected on the sky). The observed PA_T values are best fit by preperihelion synchrones ejected at heliocentric Sun–Comet distances $r(\theta - \pi)$ and at the $t(\theta - \pi)$ times (days with respect to perihelion), where θ is the comet true anomaly at observation.

Time	PA_T [°]	PA_{RV} [°]	PA_{CO} [°]	θ [°]	$t(\theta - \pi)$	$r(\theta - \pi)$ [AU]
5 Oct 2002	288 ± 1	291	284	39	−500	4.5
9 Nov 2002	296 ± 1	295	294	61	−280	3.0
12 Jan 2003	300 ± 1	289	304	89	−160	2.2
7 Feb 2003	298 ± 1	279	303	96	−140	2.0
25 Feb 2003	295 ± 1	254	301	101	−120	1.9
12 Mar 2003	294 ± 1	172	299	105	−110	1.8

is important because the input TNG images were not calibrated by means of standard stars, so that the isophote levels shown in Figs. 1 and 2 refer to arbitrary brightness levels. However, the dust tail model provides also the expected $Af\rho_c$ values (A'Hearn et al. 1984), to be compared with the observed $Af\rho_o$ values obtained by means of coma absolute photometry. This comparison allowed us to calibrate in absolute units (meters) the computed $Af\rho_c$ values provided by the inverse tail model. Since the relationship between $Af\rho_c$ and the mass loss rate computed by the inverse tail model depends on the dust albedo only, it was possible to calibrate in absolute units (kg s^{-1}) the dust loss rate provided by the tail model by assuming $A_p = 0.04$.

Fig. 3 shows the dust parameters provided by the inverse tail model after the least square fit of the input image changing the free parameters u and w. These fits and the Neck–Line photometry suggest that the most probable solution is provided by the long–dashed line ($u = -1/2$ and $w = \pi$). All the ouputs are valid within a time–dependent dust size interval shown in the left–lower panel. This fact affects the loss rate and $Af\rho_c$ values, which are surely lower limits of the observed $Af\rho_o$ ones: these depend on all the ejected dust sizes. All the other outputs show the same asymmetry with respect perihelion: larger values before than after.

The temporal evolution of all the quantities shown in Fig. 3 can be easily explained by seasons of the comet nucleus, following a similar successful interpretation of the outputs of the inverse tail model applied to Comet 2P/Encke (Epifani et al. 2001). Gas production rates (A'Hearn et al. 1995; Cochran et al. 1992) during the 1982 perihelion passage were systematically higher after perihelion than before. This fact requires that both the obliquity of the nucleus spin axis and its angle with the perihelion–aphelion line are closer to $\pi/2$ than to zero: in such a case, a less icy nucleus hemisphere (let's name it the north one) will be the only one to be exposed to Sun light before perihelion, while

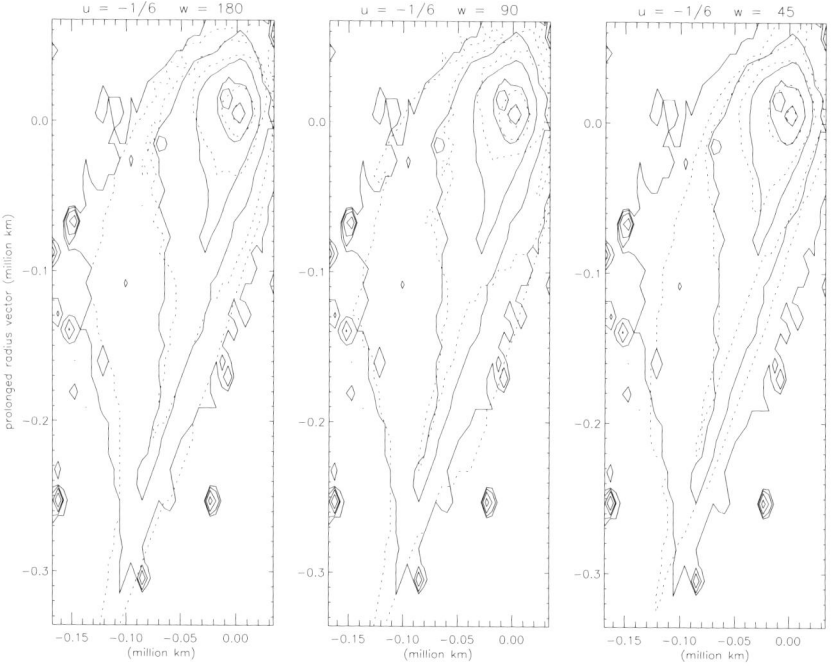

Figure 1. Image of the dust tail of comet 67P obtained on 27.0 March 2003 by means of the TNG Galileo Telescope at La Palma. The isophote steps differ by a factor 2 in intensity. Continuous lines: isophotes of the observed image. Dashed lines: isophotes of the model tail computed by means of the Montecarlo Tail model adopting the dust parameters summarized in Fig. 3. The panels refer to different combinations of the w and u parameters. w refers to the anisotropy of dust ejection: $w = \pi$ for perfectly isotropic dust ejection from the inner coma; $w = \pi/2$ for dust ejection confined to the half coma facing the Sun; $w = \pi/4$ for dust ejection confined within a cone with its axis pointing to the Sun and with half width of 45 degrees. u describes the dependence of the dust ejection velocity v versus the dust radius s: $v = v_o(s/s_o)^u$ (v_o and s_o are shown in Fig. 3). The Sun is at the panel bottom.

after perihelion only the more icy south one will be heated by Sun. The outputs shown in Fig. 3 require that the dust covering the active areas belonging to the north hemisphere is characterized by a particle size distribution and mean shape different with respect to the south one.

In order to evaluate how the finite size range influences $Af\rho_c$, we selected the most probable u and w parameter combination ($u = -1/2$ and $w = \pi$). From the outputs in Fig. 3, we selected sample velocities and size distribution power indices, listed in Table 2. Then we assumed the dust size distribution is a power law with the same index also outside the size interval covered by the model. Then we computed the number loss rates outside this range (Table 2)

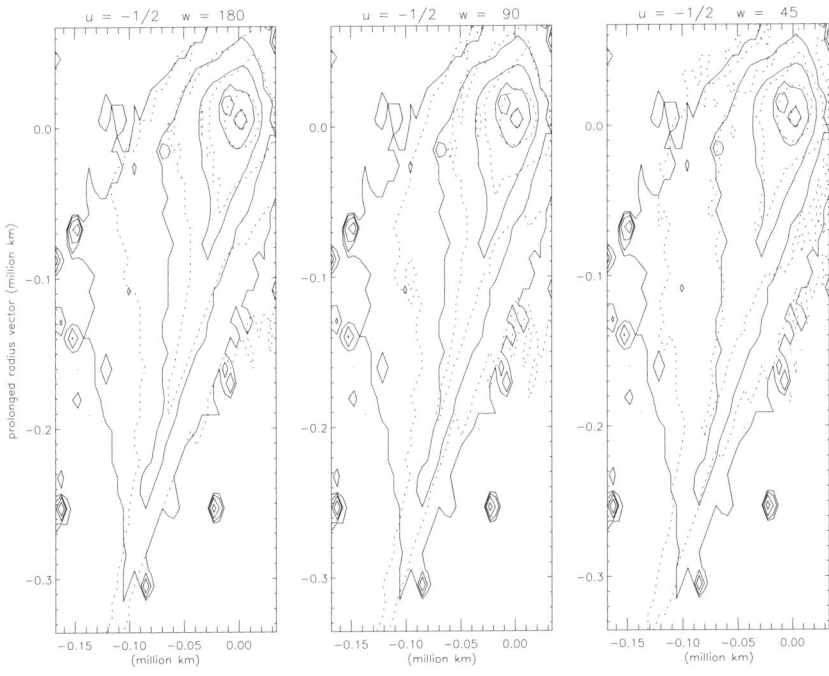

Figure 2. Fits of the same image shown in Fig. 1 for other three combinations of the u and w parameters. See Fig. 1 for further details.

and the $Af\rho_c$ obtained from these new loss rates. In this way we obtained how sensitive the $Af\rho_c$ quantity is on these changes, and how much we can extrapolate the obtained dust parameters outside the size interval considered by the tail model. The results shown in Table 2 suggest that the adopted extrapolations of the dust size range imply deep changes of $Af\rho_c$ around perihelion only: in particular, the rough minimum around perihelion in Fig. 3 becomes a well defined maximum at about 50 days after perihelion. In Fig. 4 we show available $Af\rho_o$ data covering a whole year around two perihelion passages of 67P, provided by CCD coma photometry of UAI amateur observations. The agreement between $Af\rho_c$ values in Table 2 and the $Af\rho_o$ values in Fig. 4 shows that the dust parameters listed in Table 2 offer a schematic summary consistent with all available 67P observations during one year centered on perihelion. Moreover, $Af\rho_c = 0.5$ m at $t = 150$ days in Table 2 is consistent with $Af\rho_o = 0.4$ m observed by Lamy et al. (2003) at $t = 206$ days after perihelion, while $Af\rho_c = 0.1$ m at 2 AU before perihelion is consistent with $Af\rho_o = 0.16$ m observed at 3 AU after perihelion (Schulz 2004).

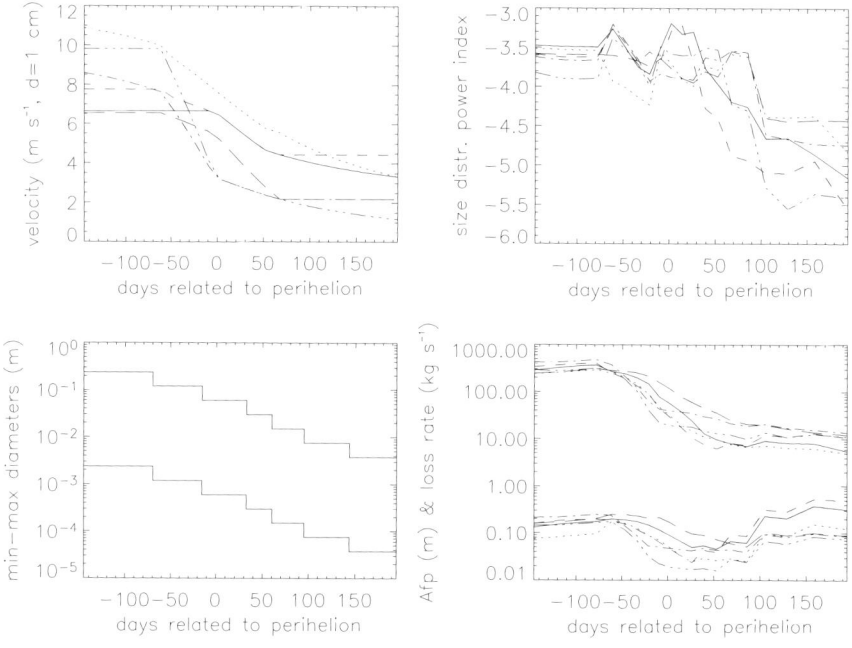

Figure 3. Dust parameters describing the dust environment of 67P during the 2002 perihelion passage. The six curves are related to six combinations of the parameters w and u described in Fig. 1. Continuous lines: $w = \pi$ and $u = -1/6$. Long–dashed lines: $w = \pi$ and $u = -1/2$. Dashed lines: $w = \pi/2$ and $u = -1/6$. Dot–dashed lines: $w = \pi/2$ and $u = -1/2$. Dotted lines: $w = \pi/4$ and $u = -1/6$. Three–dot–dashed lines: $w = \pi/4$ and $u = -1/2$.

3. Dust Tail: Fit by means of the Neck–Line Model

The highest quality images of the Neck–Line of 67P were taken at the Tautenburg Schmidt Telescope during the nights of 27 and 28 March, 2003 (Weiler et al. 2004). Fulle & Sedmak (1988) have developed a photometric theory of the Neck–Lines allowing us to obtain direct information on the dust velocity and size distribution of the Neck–Line dust ejected 105 days before perihelion (see Table 1). Fig. 5 shows the results regarding the dust ejection velocity. The dashed line is the best β power law fitting the data, providing $u = -1/2$. This explains the u value adopted in Table 2. Fig. 6 shows the results regarding the β distribution. Its values can be fitted by any power index γ between –0.5 and –1, implying $-3.5 \leq \alpha \leq -3.0$.

All the CCD images of 67P collected by the UAI amateurs during fall and winter 2002 shows a Neck–Line with almost perfectly constant shape and brightness. Fig. 7 shows the peak mag arcsec^{-2} of the thin tail between August

Table 2. Dust Parameters of 67P at several time intervals related to perihelion. α: power index of the differential dust size distribution (obtained from Fig. 3). v_o: dust ejection velocity from the inner coma at the reference dust radius $s_o = 5$ mm (obtained from Fig. 3). r: Sun–Comet distance. $Af\rho_c$: computed value over the extrapolated dust radius range between 1 cm and 0.5 μm. Q_d [s^{-1}]: cumulative dust number loss rate for all the grains with radius larger than s. No Q_d value is given when the distribution extrapolation is not sustained by any observation.

Time [days]	< -100	-50	0	$+50$	$+150$
α	-3.4	-3.4	-3.7	-4.0	-4.5
v_o [m s^{-1}]	6.5	6.0	5.5	4.0	3.0
r [AU]	> 1.7	1.4	1.3	1.4	2.0
$Af\rho_c$ [m]	0.1	0.2	1.0	1.7	0.5
Q_d at $s > 1\,cm$	$1.6\ 10^{03}$	$3.2\ 10^{03}$	$1.4\ 10^{03}$	$2.1\ 10^{02}$	$3.8\ 10^{00}$
Q_d at $s > 5\,mm$	$8.5\ 10^{03}$	$1.7\ 10^{04}$	$8.9\ 10^{03}$	$1.7\ 10^{03}$	$4.3\ 10^{01}$
Q_d at $s > 1\,mm$	$4.0\ 10^{05}$	$8.1\ 10^{05}$	$6.9\ 10^{05}$	$2.1\ 10^{05}$	$1.2\ 10^{04}$
Q_d at $s > 750\mu m$	$8.0\ 10^{05}$	$1.6\ 10^{06}$	$1.5\ 10^{06}$	$5.2\ 10^{05}$	$3.2\ 10^{04}$
Q_d at $s > 500\mu m$	$2.2\ 10^{06}$	$4.3\ 10^{06}$	$4.5\ 10^{06}$	$1.7\ 10^{06}$	$1.3\ 10^{05}$
Q_d at $s > 250\mu m$	$1.1\ 10^{07}$	$2.3\ 10^{07}$	$2.9\ 10^{07}$	$1.3\ 10^{07}$	$1.5\ 10^{06}$
Q_d at $s > 100\mu m$	$1.0\ 10^{08}$	$2.0\ 10^{08}$	$3.5\ 10^{08}$	$2.1\ 10^{08}$	$3.7\ 10^{07}$
Q_d at $s > 50\mu m$	$5.5\ 10^{08}$	$1.1\ 10^{09}$	$2.2\ 10^{09}$	$1.7\ 10^{09}$	$4.2\ 10^{08}$
Q_d at $s > 30\mu m$	$1.8\ 10^{09}$	$3.7\ 10^{09}$	$8.9\ 10^{09}$	$7.7\ 10^{09}$	$2.5\ 10^{09}$
Q_d at $s > 15\mu m$	$9.5\ 10^{09}$	$1.9\ 10^{10}$	$5.8\ 10^{10}$	$6.2\ 10^{10}$	$2.8\ 10^{10}$
Q_d at $s > 10\mu m$	$2.5\ 10^{10}$	$5.1\ 10^{10}$	$1.7\ 10^{11}$	$2.1\ 10^{11}$	$1.2\ 10^{11}$
Q_d at $s > 4\mu m$	$2.3\ 10^{11}$	$4.6\ 10^{11}$	$2.1\ 10^{12}$	$3.2\ 10^{12}$	$2.9\ 10^{12}$
Q_d at $s > 2\mu m$	$1.2\ 10^{12}$	$2.4\ 10^{12}$	$1.3\ 10^{13}$	$2.6\ 10^{13}$	$3.2\ 10^{13}$
Q_d at $s > 1\mu m$	$6.5\ 10^{12}$	$1.3\ 10^{13}$	$8.7\ 10^{13}$	$2.1\ 10^{14}$	–
Q_d at $s > 0.435\mu m$	$4.8\ 10^{13}$	$9.5\ 10^{13}$	$8.2\ 10^{14}$	$2.5\ 10^{15}$	–
Q_d at $s > 0.203\mu m$	$2.9\ 10^{14}$	$5.9\ 10^{14}$	$6.4\ 10^{15}$	$2.5\ 10^{16}$	–
Q_d at $s > 0.0942\mu m$	$1.9\ 10^{15}$	$3.7\ 10^{15}$	$5.1\ 10^{16}$	$2.5\ 10^{17}$	–
Q_d at $s > 0.0437\mu m$	$1.2\ 10^{16}$	$2.4\ 10^{16}$	$4.1\ 10^{17}$	–	–
Q_d at $s > 0.0203\mu m$	$7.7\ 10^{16}$	$1.5\ 10^{17}$	$3.2\ 10^{18}$	–	–

2002 and March 2003, after we have subtracted $5\log r$ to correct the changing Sun–Comet distance r (the surface brightness does not depend on the changing Earth–Comet distance). For $t > 50$ days after perihelion, the tail surface brightness remains fairly constant, confirming both that the total dust cross section remained constant between August 2002 and March 2003, and the interpretation of the tail in terms of a Neck–Line.

Davidsson (2003) has recently found that at 3 AU the Knudsen layer of water gas around the 67P nucleus is able to lift up dust spheres of 1 cm radius when the dust bulk density is 10^3 kg m^{-3}. This means that surely the Neck–Line ejected at 3 AU before perihelion contains the biggest grains considered in Table 2. Therefore, we must conclude that (i) the dust population typical of the Neck–Line was ejected since 3 AU before perihelion at least; (ii) the dust cross section of such a population remains about constant since 3 AU before

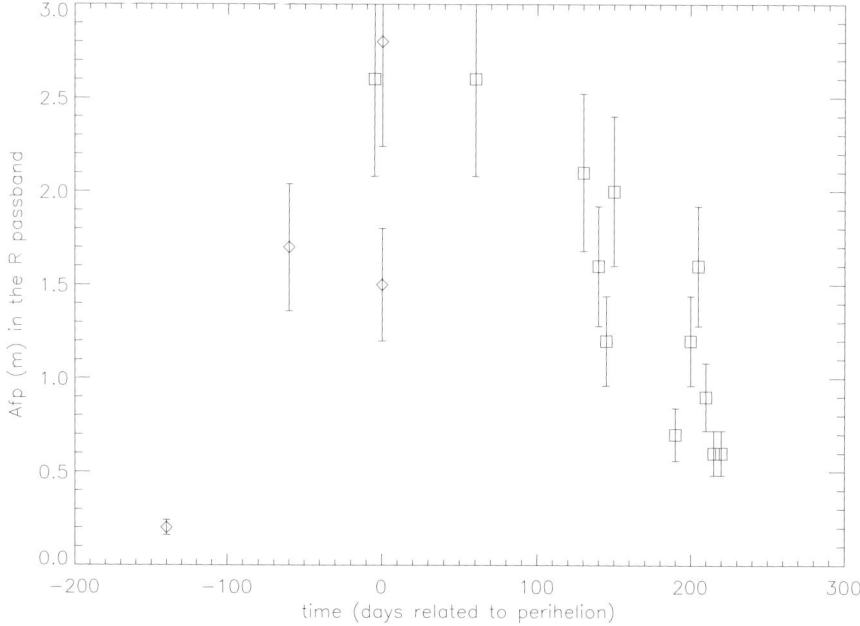

Figure 4. $Af\rho_o$ values obtained by means of CCD R–band photometry provided by the CARA network (http://cara.uai.it). Diamonds: 1996 perihelion passage. Squares: 2002 perihelion passage. The values show a clear asymmetry with respect to perihelion, with higher values after perihelion. This time evolution is consistent with the extrapolated $Af\rho_c$ values computed in Table 2, supporting the outputs of the inverse Montecarlo tail model, and is opposite to the dust mass loss rate evolution, due to the strong changes in time of the dust size distribution.

perihelion; (iii) the dust size distribution remains constant since 3 AU before perihelion. Therefore, also the dust mass loss rates remained constant since 3 AU before perihelion. Both the Rosetta orbiter and lander will face a dust environment dominated both in mass and in brightness by cm–sized grains ejected with a total mass loss rate larger than 100 kg s^{-1} since their approach to the target comet.

References

A'Hearn, M.F., Schleicher, D.G., Feldman, P.D., Millis, R.L., and Thompson, D.T. (1984). Comet Bowell 1980b. *AJ*, 89: 579–591

A'Hearn, M.F., Millis, R.L., Schleicher, D.G., Osip, D.J., and Birth, P.V. (1995). The ensemble properties of comets: Results from narrow band photometry of 85 comets. *Icarus*, 118: 223–270

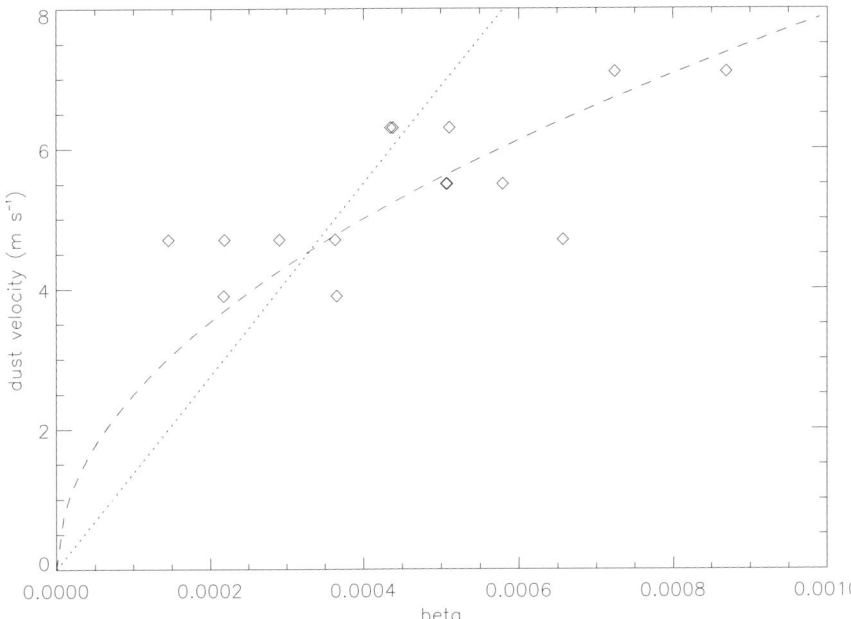

Figure 5. Dust velocity function provided by the Neck–Line model applied to Tautenburg Schmidt CCD images. Diamonds: dust velocity values provided by the width of the Neck–Line along its axis. Dotted line: threshold of validity of the analytical Neck–Line model. Dashed line: $\sqrt{\beta}$ curve, corresponding to $u = -1/2$ in Figs. 1, 2 and 3.

Cochran, A.L., Barker, E.S., Ramseyer, T.F., and Storrs, A.D. (1992). The McDonalds Observatory Faint Comet Survey – Gas production in 17 comets. *Icarus*, 98: 151–162

Davidsson, B.J.R. (2003). Acceleration of Dust Grains in a Cometary Knudsen Layer. *Presentation at Capri Rosetta Workshop*

Epifani, E., Colangeli, L., Fulle, M., Brucato, J., Bussoletti, E., de Sanctis, C., Mennella, V., Palomba, E., Palumbo, P., and Rotundi, A. (2001). ISOCAM Imaging of Comets 103P/Hartley 2 and 2P/Encke. *Icarus*, 149: 339–350

ESA 2003, Proceedings of the 12th RSWT, 13 Feb 2003

Fulle, M. (1989). Evaluation of cometary dust parameters from numerical simulations – Comparison with an analytical approach and the role of anisotropic emissions. *A&A*, 217: 283–297

Fulle, M. and Sedmak G. (1988). Photometrical analysis of the Neck–Line structure of comet Bennet 1970II. *Icarus*, 74: 383–398

Kimura, H. and Liu, C.P. (1977). On the structure of cometary dust tails. *Chin. Astron.*, 1: 235–264

Lamy, P. L., Toth, I., Weaver, H., Jorda, L. and Kaasalainen, M. (2003). The nucleus of Comet 67P/Churyumov–Gerasimenko, the New Target of the Rosetta Mission. *AAS, DPS 35*, 30.04

Schulz, R., Stüwe, J.A., and Boehnhardt, H. (2004). Monitoring comet 67P/Churyumov-Gerasimenko from ESO in 2003. *This volume.*

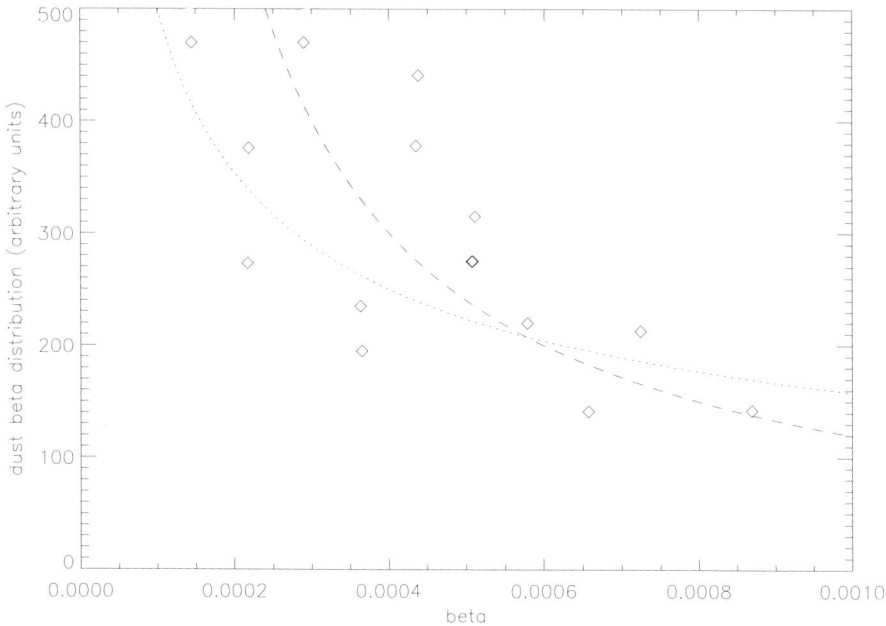

Figure 6. β distributions obtained by the Neck–Line photometry applied to the same Tatuten-
burg Schmidt images already analysed in Fig. 5. Diamonds: β–distribution provided by the peak
brightness along the Neck–Line axis. Dotted line: β distribution with power index $\gamma = -0.5$,
corresponding to $\alpha = -3.5$. Dashed line: β distribution with power index $\gamma = -1$, corresponding to $\alpha = -3$.

UAI 2003 Comet Archive, http://comete.uai.it/67p/index.htm

Weiler, M., Rauer, H., Helbert, J. (2004). Optical observations of Comet 67P/Churyumov–
    Gerasimenko. *A&A*, 414: 749–755

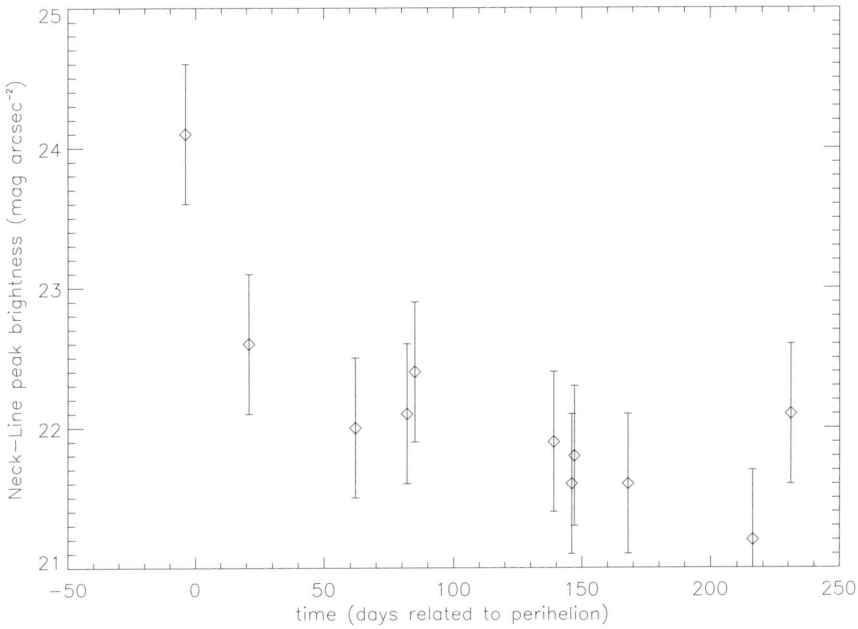

Figure 7. Peak surface brightness of the thin tail observed from August 2002 to March 2003 by UAI amateurs. Before perihelion, the much fainter tail is probably a ion tail. Since October 2002, the tail brightness remains fairly constant, supporting the conclusions the tail is a Neck–Line characterized by a constant cross section of dust ejected at a constant mass loss rate.

MODELLING THE ENVIRONMENT OF 67P/CHURYUMOV–GERASIMENKO

J. Agarwal
Max Planck Institute for Nuclear Physics, Saupfercheckweg 1, D-69117 Heidelberg
and
European Space Operations Centre, Robert-Bosch Str. 5, D-64293 Darmstadt

M. Müller
European Space Operations Centre, Robert-Bosch Str. 5, D-64293 Darmstadt

E. Grün
Max Planck Institute for Nuclear Physics, Saupfercheckweg 1, D-69117 Heidelberg
and
Hawaii Institute of Geophysics and Planetology, HIGP, University of Hawaii, 1680 East West Road POST 512c, Honolulu, HI 96822

Abstract   The ESA Cometary Environment Model is applied to comet 67P/Churyumov-Gerasimenko (C–G). From observed gas and dust production rates the model allows us to determine the conditions in the inner coma of a comet. Assuming a spherical nucleus and an axis symmetric surface activity, the densities and velocities of the gas and of dust particles are calculated as functions of location in the coma. The model is also used to estimate the infrared and visible radiation received by an observer inside the coma due to dust. Observational data for the activity of 67P/C–G are available mainly from the apparitions in 1982/83 and 2002/03. For an insolation-driven activity model corresponding to a homogenous surface composition, the results for 67P/C–G are compared to those obtained earlier for P/Wirtanen. In order to derive an upper limit for the local densities inside a jet, a Gaussian activity profile centred at the subsolar point is considered.

Keywords: 67P/C–G, comet dust and gas, comet environment modelling, jet activity

1.   Model description

The ESA Cometary Environment Model (Müller and Grün, 1997; Müller and Grün, 1998; Müller, 1999; Landgraf et al., 1999) is aimed at providing a quantitative estimate of the state of gas and dust in a cometary coma. Obser-

L. Colangeli et al. (eds.), The New ROSETTA Targets, 143–152.

vational data are used to determine the parameters of the model. The coma is assumed to be in a stationary state at all times and a variation of the comet activity with heliocentric distance or time is achieved through a corresponding variation of the parameters.

To compute the state of the gas in the coma, H_2O and CO activities are taken into account. The water production at the nucleus surface has to be specified as a function of the local zenith angle of the Sun. For a homogeneous surface composition, we use an insolation-driven activity profile where the sublimation rate is determined by the local temperature. On the dayside this is obtained assuming that solar irradiation, thermal emission and water sublimation are in local equilibrium. The nightside temperature is set to a constant value of 150 K. From the surface temperature, the theoretical sublimation rate of pure water ice is calculated and rescaled such that the total water production of the model comet matches the observed value. The assumption of a homogeneously composed surface is in accordance with the available observations for P/Wirtanen. In contrast, images of 67P/C–G show more complicated structures suggesting the presence of one or more jets in the coma emitted from active regions on the surface (Böhnhardt et al., 2003). There, the local density could be much higher than the average density resulting from a homogeneous surface. In order to obtain an upper limit for the density, a jet is simulated using for the surface activity a Gaussian profile centred at the subsolar point (Fulle et al., 1995). The width of the profile is defined from two constraints: the total gas production must match the observed value, and the peak activity is limited by the amount of incoming solar radiation. The insolation-driven model is expected to yield better estimates of quantities averaged over time and space, and it allows us a direct comparison with the results obtained for P/Wirtanen. The Gaussian activity distribution is more suitable to derive upper limits for local quantities. The sublimation temperature of CO is lower than that of water. Sublimation takes place well below the surface where temperature gradients are less distinct. Hence the CO production rate in the model is treated as constant over the nucleus surface. Using the surface activity distribution as initial condition, numerical integration of the Euler equations yields the density, temperature and velocity of the gas as functions of the location inside the coma.

Based on the computed gas flux, the motion of dust in the coma is determined using a test particle approach. The relative rate of dust grains of a given size lifted from a specific surface element depends on the local gas production. The grain motion is determined by the nucleus gravitation and the transfer of momentum from colliding gas molecules. From the production at the surface and the trajectories of single particles, the densities and velocities of dust of all considered sizes are obtained as functions of location inside the coma. The absolute values of the dust production rates are scaled in such a way that the measured $Af\rho$ value is reproduced. For a more detailed description of the model we refer to Landgraf et al. (1999).

Based on a given dust density distribution, the model allows us to investigate the radiation environment inside the coma. Müller et al. (2002) have shown that for small comets, such as P/Wirtanen and 67P/C–G, the assumption of an optically-thin coma is justified, and single scattering can be used to describe the interaction of photons with dust particles. Both scattered solar radiation having its intensity maximum in the visible wavelength range and thermal infrared radiation emitted by the dust and nucleus are considered. The amount of radiation of either type received at a specified position in the coma is calculated.

Input to the model are the production rates of water and CO as well as the $Af\rho$ parameter which is an indicator for the total dust production of the comet. Here it is assumed that the water activity is equal to the OH activity. Spectroscopic measurements of the OH and H_2O production rates of comet 67P/C–G have been performed during its apparition in 1982/83 (LOCD, described by A'Hearn et al., 1995; Osip et al., 1992; Hanner et al., 1985; Feldman et al., 2004). The measured values are shown in Fig. 1 (left) together with a power-law fit for the dependence of the production rate, Q_{H_2O}, on heliocentric distance, r_h. Lacking a significant number of data taken before perihelion passage, it is assumed that Q_{H_2O} depends on r_h only, although the $Af\rho$ curve (Fig. 1, right) indicates that the activity of 67P/C–G is not symmetric about perihelion. The CO-production rate has not yet been determined from observations, but an upper limit of 9×10^{26} s^{-1} at 3 AU is given by Bockelée-Morvan (2004). We use a constant rate of $Q_{CO} = 10^{26}$ s^{-1}.

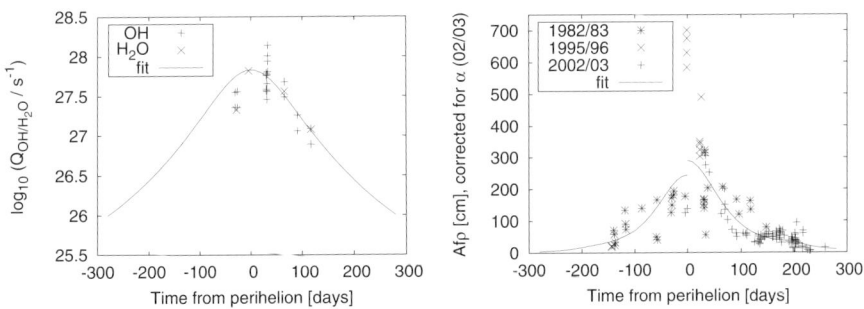

Figure 1. **Left:** Measured OH and H_2O production rates with corresponding power-law fit $Q_{H_2O} = 2.45 \times 10^{28} \, r_h^{-4.97}$. **Right:** Observed $Af\rho$ values, rescaled for geometry of 2002/03 apparition. Power-law fits for phase angle corrected $Af\rho$, separately for in-bound and out-bound motion. $Af\rho(\alpha=0, \text{in}) = 1778 \, r_h^{-5.06}$ cm, $Af\rho(\alpha=0, \text{out}) = 1698 \, r_h^{-4.19}$ cm, r_h being given in AU.

Afρ values are available from 1982/83, 1995/96 and 2002/03 (Storrs et al., 1992; A'Hearn et al., 1995; Osip et al., 1992; CARA; Kidger, 2003; Rauer et al., 2003; Feldman et al., 2004; Schulz et al., 2004). *Afρ* is independent of geocentric distance and of the field of view of the instrument. It does, however, depend on the phase angle, α. This is due, first, to the phase angle dependence of the scattering properties of dust (phase function according to Divine, 1981, and dust geometric albedo of 4%.). Second, it is due to the projection of a not spherically symmetric coma onto the plane of observation. The combined influences of both factors on the *Afρ* parameter are described by the geometric phase function introduced by Müller et al. (2002). In order to compare *Afρ* data taken at different times it is necessary to eliminate the phase angle dependence from the measured values. Fig. 1 (right) shows *Afρ* data from three apparitions corrected to match the orbital geometry of 2002/03. Fits have been made separately for in-bound and out-bound motion assuming a linear relation between $\log(Af\rho(\alpha = 0))$ and $\log(r_h)$. The solid line shows how the fitted *Afρ* curves would appear if observed under the geometry of 2002/03. The bump in the graph at about 200 days after perihelion results from a distinct phase angle minimum at that time.

Dust particle masses covering 20 orders of magnitude are taken into account, distributed over 20 discrete intervals. For a constant particle density, their radii range over nearly seven orders of magnitude. The relative amount of particles in each of the size intervals is characterised by the size distribution function which must be deduced either from *in situ* measurements or indirectly from astronomical data. We have used the function given by Divine and Newburn (1987). It was obtained from the VEGA-2 measurements near comet Halley. For this distribution, the model has been checked to reproduce the observations for the smaller particles.

The nucleus is assumed to be spherical because it is at present hardly possible to constrain the shapes of cometary nuclei from Earth based observations. An average radius of $R_n = 1980 \pm 20$ m for 67P/C–G has been derived by Lamy et al. (2003). For P/Wirtanen, $R_n = 700$ m is used.

2. Results for 67P/C–G and comparison to P/Wirtanen

The described model with an insolation-driven activity profile is applied to 67P/C–G at 1.34 AU and at 1.87 AU. The data indicate that in 1982 and 2002 the activity of 67P/C–G was highest at about 30 days after perihelion passage ($r_h = 1.34$ AU). 1.87 AU (4 months after perihelion) is the largest heliocentric distance for which a measured value of the water production rate, Q_{H_2O}, is available. We use $Q_{H_2O} = 10^{28}$ s^{-1}, $Q_{CO} = 10^{26}$ s^{-1} and $Af\rho = 300$ cm ($\alpha = 36°$) at 1.34 AU and $Q_{H_2O} = 1.2 \times 10^{27}$ s^{-1}, $Q_{CO} = 10^{26}$ s^{-1}, $Af\rho = 65$ cm ($\alpha = 32°$) at 1.87 AU. The results are compared to those obtained earlier for

P/Wirtanen at perihelion (r_h= 1.06 AU) with Q_{H_2O}= 1.3×10^{28} s^{-1}, Q_{CO}= 10^{26} s^{-1} and $Af\rho$=120 cm ($\alpha = 40°$). In addition, the state of 67P/C–G at 1.34 AU with a Gaussian activity distribution is investigated.

Gas density

Schematic plots of the modelled gas densities in the coma of 67P/C–G are shown in Fig. 2. The left-hand image shows the distribution that would result from a homogenous surface composition with the sublimation rate depending only on the amount of incoming solar irradiation. On the right, the density for a jet-like Gaussian surface activity profile is shown.

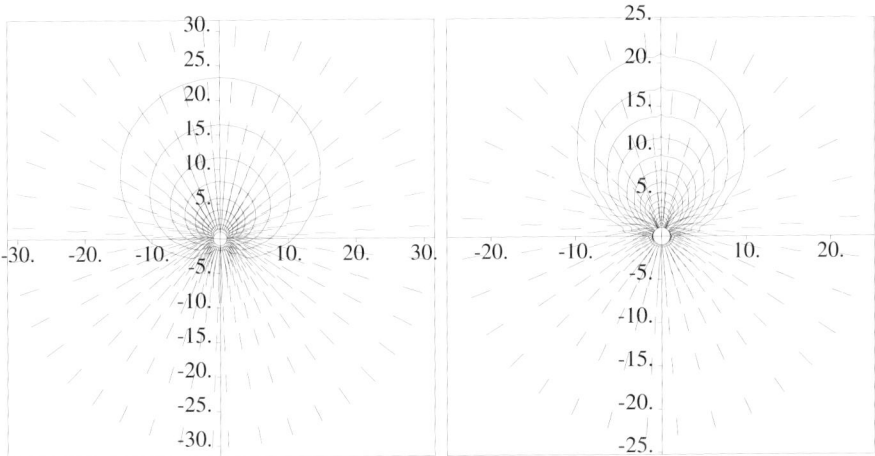

Figure 2. Schematic plots of iso-density lines and velocity field of gas in the coma. The scale is comet radii, Sun direction is up. **Left:** Insolation-driven surface activity profile. **Right:** Jet simulated by a Gaussian activity profile with the sublimation rate of pure water ice in the centre.

For a quantitative comparison, gas densities above the subsolar point are listed in the last line of Table 1. At maximum activity, the gas density inside the coma of 67P/C–G in the insolation-driven model is by one order of magnitude lower than in the coma of P/Wirtanen on the same *relative* length scale ($2 R_n$ from the nucleus centre). However, on *absolute* scales, i.e. at a fixed distance of $2 R_{n,C-G} = 4$ km from the centre, the density in the coma of 67P/C–G is slightly greater due to the higher total gas production rate. At 1.87 AU, the gas density around 67P/C–G has dropped by nearly one order of magnitude. Inside a jet above the subsolar point, the density would be higher than in the insolation-driven model by a factor of five. This value is interpreted as an upper limit for the local gas density.

The ratio of the sublimation that would take place from a sphere of pure water ice to the observed total H_2O/OH production rate is an indicator for the average fraction of active area on a comet. For 67P/C–G at 1.34 AU, this ratio is 9%, whereas for P/Wirtanen it is 49%.

Dust densities and terminal velocities

As an example for the dynamics of dust in the coma, the density of particles having a radius of 200 μm is shown in Fig. 3. The right-hand figure shows that, in contrast to the gas in the jet, dust particles of this size remain in the narrow region from which they are emitted. Due to their inertia they are not much carried away by the gas which expands towards the terminator. Table 1

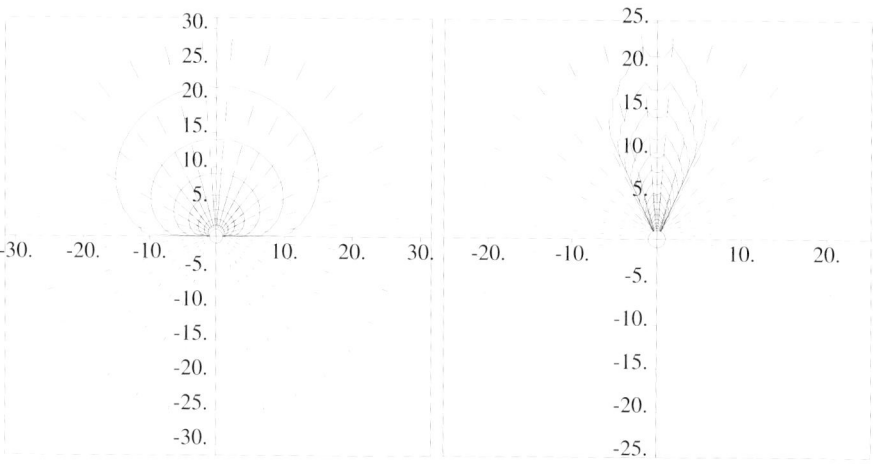

Figure 3. Schematic plots of iso-density lines and velocity field of 200-μm dust particles. The scale is comet radii, Sun direction is up. **Left:** Insolation-driven activity model. **Right:** Jet simulated by a Gaussian activity profile with sublimation rate of pure water ice in the centre.

shows the densities of particles of different sizes at $2 R_n$ from the nucleus centre above the subsolar point for 67P/C–G and P/Wirtanen, respectively, and – for comparison on an absolute scale – at 4 km ($2 R_{n,C-G}$) from the nucleus centre for P/Wirtanen.

Fig. 4 shows the terminal velocities of dust particles above the subsolar point as functions of particle radius. We have plotted the values resulting from the insolation-driven activity model for comets 67P/C–G, P/Wirtanen and P/Halley. In addition, 67P/C–G with the Gaussian surface activity model has been considered. For small particles, the terminal velocity curves converge towards constant values because the dust velocity cannot exceed the gas velocity. In the insolation-driven model, the terminal velocities for 67P/C–G are smaller than for P/Wirtanen and for P/Halley because of the lower average gas density. An upper limit is again derived from the Gaussian activity distribution: the terminal velocities of dust grains in a narrow jet above the subsolar point of

Table 1. Densities [1/m³] of dust and gas above the subsolar point for different scenarios.

Particle radius [m]	C–G 1.34 AU 4.0 km	Wirt. 1.06 AU 4.0 km	Wirt. 1.06 AU 1.4 km	C–G 1.87 AU 4.0 km	C–G (Jet) 1.34 AU 4.0 km
2×10^{-8}	1.46×10^5	7.81×10^4	8.52×10^5	1.72×10^4	1.40×10^6
2×10^{-7}	3.68×10^4	1.57×10^4	1.64×10^5	5.88×10^3	$3.23 \times 10^5 6$
2×10^{-6}	8.69×10^3	3.20×10^3	3.42×10^4	1.58×10^3	6.75×10^4
2×10^{-5}	1.78×10^2	6.16×10^1	6.69×10^2	3.41×10^1	1.27×10^3
2×10^{-4}	4.95×10^{-1}	1.55×10^{-1}	1.69	8.98×10^{-2}	3.15
2×10^{-3}	9.41×10^{-4}	3.11×10^{-4}	3.42×10^{-3}	2.00×10^{-4}	6.32×10^{-3}
2×10^{-2}	1.87×10^{-6}	5.51×10^{-7}	6.06×10^{-6}	0.00	1.12×10^{-5}
Gas	2.94×10^{17}	2.89×10^{17}	3.18×10^{18}	4.01×10^{16}	1.49×10^{18}

67P/C–G would be higher than expected for P/Wirtanen with a homogeneous surface composition.

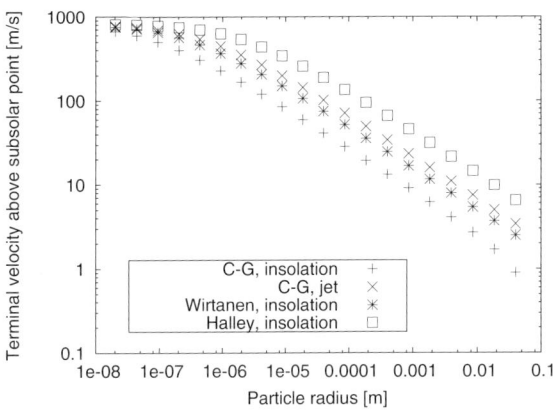

Figure 4 Terminal velocities of dust grains above the subsolar point for comets 67P/C–G, P/Wirtanen and P/Halley.

Total production during one apparition

In order to give an estimate of the total amounts of dust and gas ejected from 67P/C–G during one apparition, the fitted curves for water production and *Afρ* displayed in Fig. 1 have been used. The production rates of particles of different sizes have been calculated for a discrete set of times and integrated over a period spanning from 280 days before perihelion to 280 days after perihelion. By this time, the gas production rate has decreased by about two orders of magnitude according to the fit. The decrease may be even stronger as, in 2003, a major drop of activity between 200 and 260 days after perihelion was reported by Schulz et al. (2004). We conclude that the material ejected by the comet more than 280 days from perihelion does not significantly contribute to the total mass released. The dust mass loss of 67P/C–G during one apparition is then 7×10^8 kg, of which 1.7×10^9 kg are contained in particles larger than 1 mm. The total amount of emitted gas is estimated to 2.7×10^8 kg. With a nucleus density of 1000 kg/m^3, this corresponds – averaged over the surface – to a sublimated layer of 7 centimetres thickness. Note that this estimation is based on a dust mass distribution which has been extrapolated to large particles. If the amount of large particles is higher, more material will be lost.

Radiation inside the coma

Fig. 5 (left) shows the intensity received by an observer at 1 R_n above the subsolar point from a unit solid angle as a function of the direction of the line of sight. Both reflected solar flux (visible) and thermal (infrared) radiation are considered. Only radiation originating from dust particles in the coma is taken into account. When the line of sight intersects with the nucleus (angle >150° in the plot), the nucleus contribution to the total radiation would by far exceed the contribution of the dust. Reflected solar light is most intense for small angles between the line of sight and the Sun direction due to strong forward scattering of the dust. Thermal radiation depends on the integral density of dust particles along the line of sight which is highest when the line of sight just touches the nucleus. Fig. 5 (right) shows the total intensity incident on a surface from all

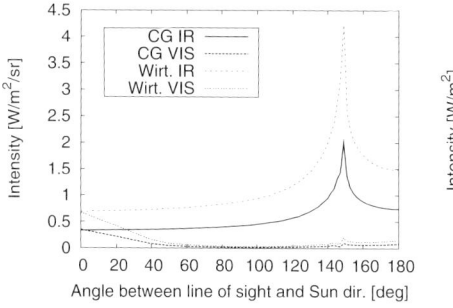 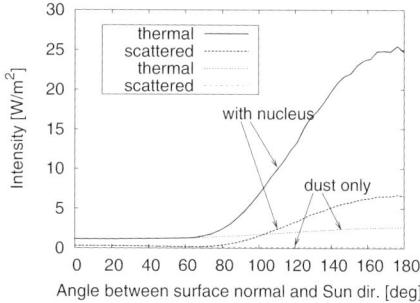

Figure 5. **Left:** Radiation received from a line of sight by an observer at 1 R_n above the subsolar point for 67P/C–G at 1.34 AU and for P/Wirtanen at 1.06 AU. **Right:** Radiation on a surface at 1 R_n above the subsolar point of 67P/C–G at 1.34 AU.

directions as a function of the orientation of the surface. Direct solar irradiation is not considered, whereas the contribution of the nucleus is shown. In all, the radiation intensities in the coma of 67P/C–G are smaller than in the case of P/Wirtanen which is mainly due to the larger heliocentric distance used for 67P/C–G. The perihelion distance of 67P/C–G in 2014 will have decreased, then being comparable to that of P/Wirtanen in 1997 (1.06 AU). The radiation intensity is likely to increase because of the higher solar flux and because of a denser dust environment to be expected. The optical depth near the surface of 67P/C–G at activity maximum is found to be 2-3% for the insolation-driven activity model and about 15% inside a jet.

3.   Summary

We have applied the ESA Cometary Environment Model to comet 67P/Churyumov-Gerasimenko deriving the model parameters from observational data. We have computed gas and dust density distributions, their velocity fields, and the radiation intensity in the coma. In addition to with a homogeneous, insolation-driven surface activity distribution, we have modelled 67P/C–G with a jet-like activity profile and compared the results to those obtained for P/Wirtanen. In the insolation-driven case, the dust density at $2\,R_n$ above the subsolar point of 67P/C–G is smaller by a factor of 3–5 than in the coma of P/Wirtanen; the gas density is about ten times smaller. In the presence of a jet, the highest dust density at $2\,R_n$ that can be reconciled with observational data is twice as high as for P/Wirtanen with insolation-driven activity. However, the gas density is smaller due to lateral expansion. The terminal velocities of dust grains in the homogeneous case are less than those for P/Wirtanen, whereas in a jet they would be higher. From the total mass loss of the comet we estimate that, on average, 7 centimetres of surface material are carried away during each apparition. The radiation inside the coma of 67P/C–G in 2002 is found to be weaker than in the coma of P/Wirtanen. In 2014, it will be enhanced due to 67P/C–G's reduced perihelion distance. To summarise, the average coma environment of 67P/C–G is found to be more benign for the Rosetta spacecraft than that of P/Wirtanen, but is expected to be more severe inside a jet.

Acknowledgments

J. A. is grateful to the Rosetta Project at ESA/ESTEC for bearing the expenses of attending the Capri Rosetta Workshop.

References

A'Hearn, M. F., Millis, R. L., Schleicher, D. G., Osip, D. J., and Birch, P. V. (1995). The ensemble properties of comets: Results from narrowband photometry of 85 comets, 1976-1992. *Icarus*, 118:223–270.

Bockelée-Morvan, D., Moreno, R., Biver, N., Crovisier J., Crifo J.-F., Fulle M. and Grewing M. (2004). CO and dust productions in 67P/Churyumov–Gerasimenko at 3 AU post–perihelion. *This volume*.

Böhnhardt, H., Stüwe, J., and Schulz, R. (2003). Dust Coma and Tail Structures in Comet 67P/CG. *Presentation at Capri Rosetta Workshop*.

CARA. Cometary Archive for Amateur Astronomer, by the Italian Comet Section.
http://cara.uai.it.

Divine, N. (1981). A Simple Radiation Model of Cometary Dust of P/Halley. *ESA SP-174*, pages 25–30.

Divine, N. and Newburn, R. L. (1987). Modeling Halley before and after the encounters. *A&A*, 187:867–872.

Feldman, P. D., A'Hearn, M. F., and Festou, M. C. (2004). Observations of Comet 67P (Churyumov-Gerasimenko) with the International Ultraviolet Explorer in 1982. *This volume*.

Fulle, M., Colangeli, L., Mennella, V., Rotundi, A., and Bussoletti, E. (1995). The sensitivity of the size distribution to the grain dynamics: simulation of the dust flux measured by GIOTTO at P/Halley. *A&A*, 304:622.

Hanner, M. S., Tedesco, E., Tokunaga, A. T., Veeder, G. J., Lester, D. F., Witteborn, F. C., Bregman, J. D., Gradie, J., and Lebofsky, L. (1985). The dust coma of periodic Comet Churyumov-Gerasimenko (1982 VIII). *Icarus*, 64:11–19.

Kidger, M. R. (2003). Dust production and coma morphology of 67P/Churyumov-Gerasimenko during the 2002-2003 apparition. *A&A*, 408:767–774.

Lamy, P. L., Toth, I., Weaver, H., Jorda, L., and Kaasalainen, M. (2003). The Nucleus of Comet 67P/Churyumov-Gerasimenko, the New Target of the Rosetta Mission. *AAS/Division for Planetary Sciences Meeting*, 35.

Landgraf, M., Müller, M., and Grün, E. (1999). Prediction of the in-situ dust measurements of the stardust mission to comet 81P/Wild 2. *Planet. Space Sci.*, 47:1029–1050.

LOCD. Lowell Observatory Cometary Data Base.
http://pdssbn.astro.umd.edu/sbnhtml/comets/comparative.html.

Müller, M., Green, S. F., and McBride, N. (2002). An easy-to-use Model for the Optical Thickness and Ambient Illumination within Cometary Dust Comae. *Earth Moon and Planets*, 90:99–108.

Müller, M. (1999). A Model of the Inner Coma of Comets with Applications to the Comets P/Wirtanen and P/Wild 2. PhD Thesis, Universität Heidelberg.
http://www.mpi-hd.mpg.de/dustgroup/~mmueller.

Müller, M. and Grün, E. (1997). An Engineering Model of the Dust Environment of the Inner Coma of Comet P/Wirtanen, Part 1. *ESA-RO-ESC-TA-5501*.

Müller, M. and Grün, E. (1998). An Engineering Model of the Dust Environment of the Inner Coma of Comet P/Wirtanen, Part 2. *ESA-RO-ESC-TA-5501*.
http://www.mpi-hd.mpg.de/dustgroup/~mmueller.

Osip, D. J., Schleicher, D. G., and Millis, R. L. (1992). Comets - Groundbased observations of spacecraft mission candidates. *Icarus*, 98:115–124.

Rauer, H., Weiler, M., and Helbert, J. (2003). Optical Observations of Comet 67P/Churyumov-Gerasimenko. *Presentation at Capri Rosetta Workshop*.

Schulz, R. Stüwe J.A. and Böhnhardt H. (2004). Monitoring comet 67P/Churyumov–Gerasimenko from ESO in 2003. *This volume*.

Storrs, A. D., Cochran, A. L., and Barker, E. S. (1992). Spectrophotometry of the continuum in 18 comets. *Icarus*, 98:163–178.

PLASMA ENVIRONMENT OF COMET CHURYUMOV-GERASIMENKO – 3D HYBRID CODE SIMULATIONS

T. Bagdonat, U. Motschmann
Inst. for Theoret. Physics, TU Braunschweig

K.–H. Glassmeier
Inst. for Geophysics and Extraterrestrial Physics, TU Braunschweig

E. Kührt
DLR Inst. for Planetary Research, Berlin

Abstract   Recent measurements indicate the H_2O production rate of comet Churyumov–Gerasimenko to be of the same order as for comet Wirtanen. From a plasma physical point of view, this classifies both comets as "weak comets", i.e. the plasma structures known from larger comets differ much from the MHD picture. We present two 3D hybrid model simulation results for Churyumov–Gerasimenko for heliocentric distances of 3.25 AU and 2.8 AU. We focus on the kinetic effects of the ions, which are retained by the hybrid model. The results should be directly applicable to the interpretation and prediction of data which will be obtained by the ROSETTA plasma instruments during the early encounter. It is found, that the "classical" plasma boundaries are not present at this early stage, because their characteristic scales are smaller than the gyroradius of the solar wind protons as well as the heavy ions. Moreover, complex particle energy distributions are found.

Keywords: 3D hybrid simulation, Churyumov–Gerasimenko, plasma, particle distributions

1.   Introduction

Although Churyumov–Gerasimenko (C–G) is larger in diameter than comet Wirtanen (Tancredi et al., 2000), earlier (Crovisier et al., 2002; Osip et al., 1992) and recent observations by Kidger, 2003 indicate the water production rate of C–G to be almost the same as that of Wirtanen. Fig. 1 shows a comparison with data collected from Schwehm and Schulz 1999b; Schwehm, and Schulz, 1999a for Wirtanen and from Kidger, 2003; Crovisier et al., 2002

L. Colangeli et al. (eds.), The New ROSETTA Targets, 153–166.

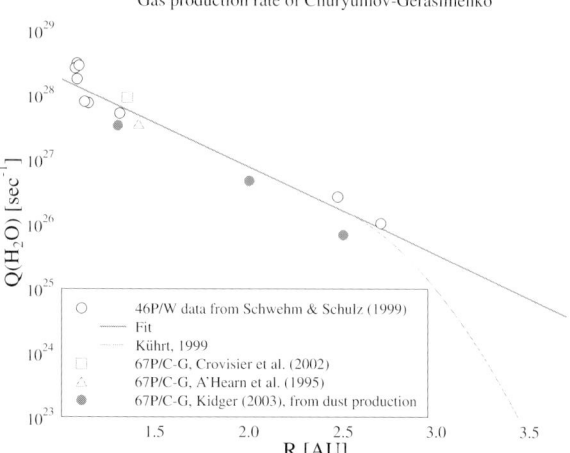

Figure 1. Water production rate of comet Churyumov–Gerasimenko compared to comet Wirtanen. The continuous line is a fit for C–G, while the dashed line is a qualitative extrapolation according to the model by Kührt, 1999. Note, that the values for Wirtanen and C–G are almost the same.

and A'Hearn et al., 1995 for C–G. Therefore, plasma simulation results for Wirtanen by Sauer et al., 1996, as well as the hybrid results of Bagdonat and Motschmann, 2002 should be qualitatively valid for C–G too. However, the results of Bagdonat and Motschmann, 2002 and Bagdonat and Motschmann, 2004 have shown that pronounced kinetic effects occur at weak comets. However, besides the qualitative study of Lipatov et al., 2002 there are no further 3D hybrid model simulations for weak comets.

The mass–loading process at strong comets, i.e. the incorporation of the cometary ions into the solar wind flow, proceeds in three main steps. First, the cometary ions starting with almost zero velocity in the cometary rest frame experience the interplanetary electromagnetic field and start a gyrating motion. This forms the typical ring distribution around the solar wind velocity, cf. Motschmann and Glassmeier, 1993, found at Halley (Balsiger et al., 1986). Now, as shown by Motschmann et al., 1997, this non–gyrotropic ring distribution is unstable and decays relatively quickly into a spherical shell distribution around the solar wind bulk velocity, cf. Szegö et al., 2000 for a review on these processes. This distribution, however, is relatively stable, and decays only slowly into a fully thermalized distribution, where the cometary ions are finally Maxwellian and flow with the same velocity as the solar wind. The MHD picture neglects all details of this mass–loading process, because no non–Maxwellian distributions can be modeled. For strong comets, this already provides a remarkably good agreement of theory with the data obtained from Halley, cf. Israelevich et al., 1999. However, as already bi–ion fluid simulations (Bogdanov et al.,; Sauer et al., 1996) have shown, the situation is completely different at weak comets. The reason for this is the large gyroradius of the cometary ions as well as the solar wind protons compared to the scales of the interaction region. Consequently, also the interaction times, i.e. the time before the involved ions stay in the interaction region is also much smaller than at strong comets. The thermalization of the cometary ions is therefore different from what was described above. Furthermore, in this different regime of mass loading the solar wind is not influenced as in the stronger case, hence, no bow shock is formed, as was shown by the results of Bagdonat and Motschmann, 2002; Lipatov et al., 1997; Sauer et al., 1996; Lipatov et al., 2002. This implies a supersonic solar wind flow in the whole interaction region which leads to a complex ion dynamics which we described in detail in Bagdonat and Motschmann, 2004.

This paper focuses on the global 3D picture of the ion tail and the field configurations and the particle energy distributions resulting from the weak mass loading interaction. The results shown here should serve as a reference of what is to be expected of the ROSETTA plasma measurements at the early encounter of Churyumov–Gerasimenko.

2. Numerical model

Code. The numerical investigations are done using a newly developed hybrid code described by Bagdonat and Motschmann, 2002. Our code is able to use an arbitrary, curvilinear grid in up to three spatial dimensions. This feature is used here to enhance the resolution near the cometary nucleus. For an introduction to the hybrid simulation technique cf. Winske, 2003 or Lipatov, 2002.

As an enhancement of the model used by Bagdonat and Motschmann, 2002 we included some enhancements described by Bagdonat and Motschmann, 2004 for the results presented here. Moreover, instead of using a finite anomalous resistivity η to describe magnetic diffusion, it turned out that for 3D simulations, $\eta = 0$ strongly enhances numerical stability, when using a simple five–point smoothing procedure for the magnetic field to reduce noise.

Coma model. The coma model is the same as used in Bagdonat and Motschmann, 2004.The density of the neutral atoms as function of radial distance r from the nucleus is given by Lipatov et al., 2002

$$n_n(r) = \frac{G}{4\pi r^2 v_0} \quad , \tag{1}$$

where G is the H_2O production rate and v_0 the outgassing velocity, taken to be $v_0 = 1 \text{ km s}^{-1}$. The values for the H_2O production rate were taken from the fit in Fig. 1. The ionization frequency is given as $\nu = 10^{-6}\mathrm{s}^{-1}$ by Mendis et al., 1985.

Simulation geometry.   In all runs the simulation box has three spatial dimensions, with the undisturbed solar wind flowing in positive x–direction, and the initial magnetic field pointing in positive z–direction, i.e. perpendicular to the solar wind flow. The interplanetary electric field points towards the negative y–direction. Thus, the plane $y = 0$ corresponds to the ecliptic in the case of nominal solar wind conditions. In order to have a higher resolution in the vicinity of the nucleus, we use the same grid as used by Bagdonat and Motschmann, 2004 generalized for three dimensions.

3.   Results

Simulation parameters.   Two runs for the heliocentric distances of 3.25 AU and 2.8 AU have been performed. The parameters for these runs are listed int Tab. 1. The background values for the undisturbed solar wind, i.e. B_0 and n_0 were taken from a fit according to the model by Parker, 1958 to data given by Richardson et al., 1995; Richardson et al., 1996. From these two values various normalization constants are derived as described by Bagdonat and Motschmann, 2002. These are the Alfvèn velocity v_{A0}, the length scale $x_0 = c/\omega_{p,i}$ (ion inertia length) and the timescale $t_0 = \Omega_i^{-1}$ (inverse gyrofrequency), where $\omega_{p,i}$ and Ω_i are the solar wind proton plasma frequency and gyrofrequency, respectively. The parameters labeled as A^* are normalized to these values.

Basic physical mechanisms.   For the understanding of the results shown in this section it is convenient to recall the different forces acting on the ions and the magnetic field lines.

   The electron bulk velocity can be expressed by the bulk velocities of the solar wind protons $\vec{v}_p$ and cometary ions $\vec{v}_h$, respectively, to yield

$$\vec{E} = - \left(\frac{n_p \vec{v}_p}{n_p + n_h} + \frac{n_h \vec{v}_h}{n_p + n_h} \right) \times \vec{B} + \frac{\left(\nabla \times \vec{B} \right) \times \vec{B}}{\mu_0 e (n_h + n_p)} - \frac{\nabla p_{e,p} + \nabla p_{e,h}}{e (n_h + n_e)} \quad (2)$$

Because the cometary ion density is low compared to the solar wind density in most places, the first term in Eq. 2 is dominated by the solar wind bulk velocity $\vec{v}_p$ and balances the $\vec{v}_p \times \vec{B}$–term. Therefore, the solar wind protons experience the force resulting from this term only, where n_h becomes large. Where this is the case, the electric field becomes smaller, thus, the solar wind protons are deflected upwards. The cometary ions, however, who enter the simulation box with $\vec{v}_h = 0$ experience the full force from this term and are therefore strongly accelerated downwards in regions, where n_h is small. This

Table 1. Simulation parameters for the two runs according to heliocentric distances of $R_\odot = 3.25$ AU and $R_\odot = 2.8$ AU.

Parameter	$R_\odot = 3.25AU$	$R_\odot = 2.8AU$
H$_2$O production rate G	$7.5 \cdot 10^{25} \text{s}^{-1}$	10^{24}s^{-1}
outgassing velocity v_0	1km/s	1km/s
IMF B_0	1.13nT	1.23nT
solar wind density n_0	0.66cm^{-3}	0.88cm^{-3}
Alfvèn velocity v_{A0}	30km/s	28.6km/s
length scale x_0	280km	240km
time scale t_0	9.2s	8.5s
Alvénic Mach number	10	10
solar wind ion plasma beta $\beta_{i,p}$	0.4	0.4
solar wind electron plasma beta $\beta_{e,p}$	0.4	0.4
cometary ion plasma beta $\beta_{i,h}$	0^a	0^a
cometary electron plasma beta $\beta_{e,h}$	0.05	0.05
simulation box dimensions $L_x^* \times L_y^* \times L_z^*$	$4 \times 4 \times 4$	$30 \times 30 \times 30$
number of grid cells $N_x \times N_y \times N_z$	$100 \times 100 \times 100$	$90 \times 90 \times 90$
time step Δt^*	0.001	0.005
average number of particles per cell^b	≈ 10	≈ 10

^{a}The small cometary ion temperature is neglected.
^{b}Due to the use of curvilinear coordinates, the number of particles per cell varies with the cell size. The value given here is therefore taken as an average.

force will be denoted as the **"pickup–force"** in the following. It always acts perpendicular to the magnetic field and plasma velocity vector.

The electric field arising from the pressure gradient terms in Eq. 2 is experienced by both species in the same way. This force mainly acts radially outward from the nucleus, due to the dominant radial density profile of the cometary ions. This accelerates the cometary ions outwards and deflects the solar wind protons around the obstacle. This force will be simply referred to as **"electron pressure"**. The second term in Eq. 2 can be further split into the two parts

$$(\nabla \times \vec{B}) \times \vec{B} = (\vec{B} \cdot \nabla)\vec{B} - \frac{1}{2}\nabla(B^2) \quad . \qquad (3)$$

The first term in Eq. 3 acts as a **"magnetic tension"**, which tends to shorten the magnetic field lines, when they become curved. Due to this term in Eq. 2, also the plasma is accelerated whenever the field lines are curved. The second term in Eq. 3 is called **"magnetic pressure"**, which tends to expand the magnetic field lines, where their density becomes higher. Again, this magnetic pressure also acts on the plasma pointing in the direction of the magnetic pressure gradient. Finally, the first term of Eq. 2 substituted in Faraday's law, is responsible

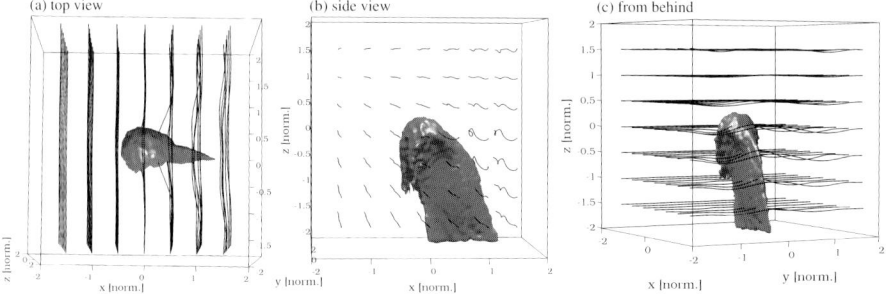

Figure 2. Three different 3D views of the global plasma tail configuration for Churyumov–Gerasimenko at 3.25 AU. (a) Top view. (b) Side view. (c) View from behind. The isosurface corresponds to a cometary ion density of $n/n_0 = 0.2$. The black lines are the magnetic field lines. The solar wind flows in positive x–direction, the IMF is oriented in positive y–direction. The ion tail does not point anti–sunwards but is cycloidal, following the direction of the interplanetary electric field. It is "flat", i.e. it is much broader in the plane perpendicular to the IMF. Only very few fieldlines are draped around the obstacle, as can be seen in (a). They are also bent upwards due to the deflection of the solar wind, as can be seen in (c). Behind the tail, the field lines shorten again, which excites a circularly polarized Alfvèn wave, as visible in (b).

for the **"convection"** of the magnetic field along the plasma flow ("frozen–in" magnetic field). In regions with high n_h and low $\vec{v}_h$ (inner coma) this convective term becomes smaller, thus, the magnetic field lines cannot be transported through this region and are "draped" around the obstacle. This effect will be denoted simply as **"draping"**.

Churyumov–Gerasimenko at 3.25 AU. Fig. 2 shows an overview of the global three–dimensional ion tail and magnetic field configuration for C–G at 3.25 AU, shown from three different viewing angles. In Fig. 2(a) and (b) the solar wind flows from left to right, whereas it comes out of the figure in (c). The magnetic field is oriented along the shown field lines pointing in positive y–direction, i.e. from bottom to top in (a), inwards in (b) and from left to right in (c).

The magnetic field lines are draped weakly around the nucleus, as can be seen in Fig. 2(a). Behind the obstacle, the field lines shorten themselves due to the magnetic tension. It appears from Fig. 2(a) that this relaxation produces an overshoot, which excites an oscillation in the field lines in the wake. Also, due to the slight upwards deflection of the solar wind, and the radial outwards motion above the nucleus, the field lines are also bent upwards a bit, as can be seen in Fig. 2(c). Together, this results in an excitation of circular polarized Alfvèn waves in the wake as can be best seen from Fig. 2(b).

The ion tail exhibits the typical cycloidal form already described by **Zitat: diverse** for the 2D case. Considering the three dimensional situation, however,

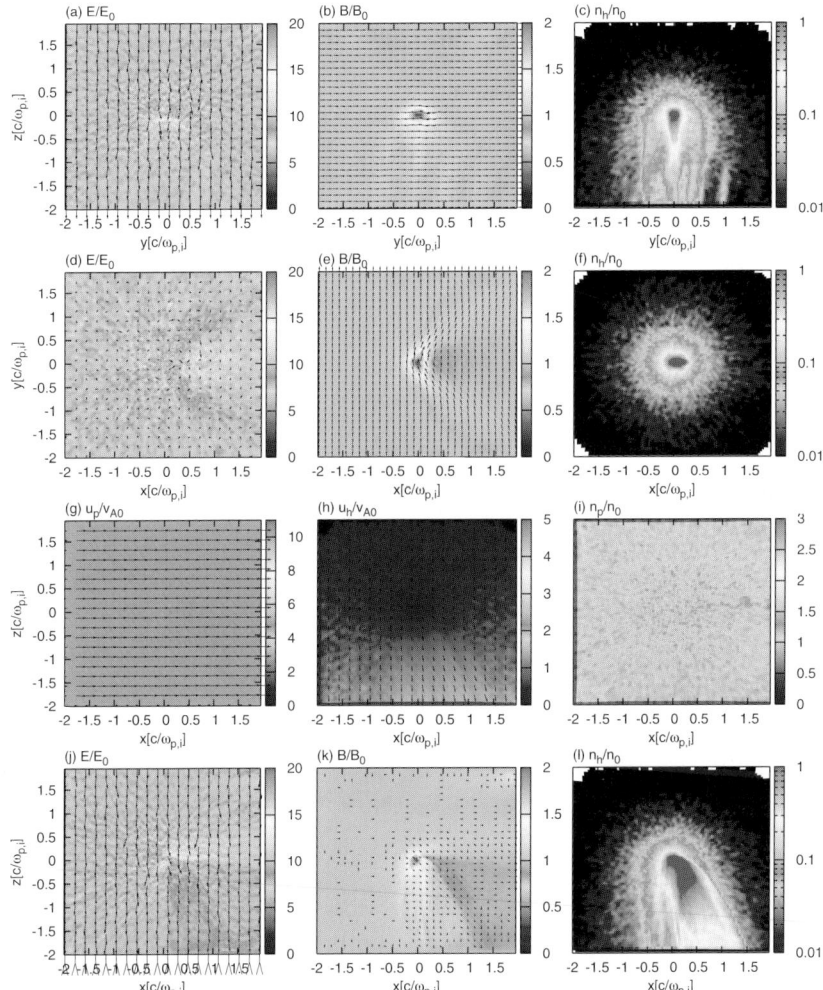

Figure 3. C–G at 3.25 AU, 2D cuts. (a)–(c) E/E_0, B/B_0 and cometary ion density n_h/n_0 in the plane $x = 0$. (d)–(f) same for the plane $z = 0$ (ecliptic). (g)–(l) SW bulk velocity v_p/v_{A0}, cometary ion bulk velocity v_h/v_{A0}, SW density n_p/n_0, E/E_0, B/B_0 and n_h/n_0 in the plane $y = 0$. The electric and magnetic field as well as the bulk velocity panels show the amount of the according quantity color coded and the direction with small arrows. (k) and (l) show the formation of a magnetized, cycloidal ion tail, with an overall downward motion, seen in (h). The magnetic field is piled–up inside the whole coma, as can be seen from (b), (e) and (k). The piled–up magnetic field forms a weak Mach cone in (e) and (k). In (c) "tail rays" are visible. The solar wind flow does not react much in this regime, as can be seen from (g) and (i). (a), (d) and (j) show, that the electric field is decreased in the inner coma.

it can be seen that the cometary ion plasma forms an almost spherical "head", which is slightly pulled downwards by the pickup force. The main cycloidal part, however, is "flat", i.e. it has only a very small extension in the direction perpendicular to the field lines. This is due to the ion pickup acting perpendicular to the draped field lines, which focuses the pick–up ions into the midplane $y = 0$. Moreover, the cycloidal tail is pushed backwards somewhat directly underneath the nucleus due to the magnetic tension force.

The two–dimensional cuts shown in Fig. 3 exhibit some more details of the mass–loading process. First of all, panels (g) and (i) show, that the solar wind flow is almost undisturbed, besides a very small density enhancement, which indicates the slight upwards deflection. The magnetic field, shown in panels (b), (e) and (k) is enhanced in the inner coma due to the draping. Attached to this pile–up region is a faint Mach cone visible in (e) and (k). The most interesting feature, however, is the existence of at least three cycloidal tails as can be seen in panel (c) of Fig. 3. The focusing effect described above resembles the wave structure of the magnetic field lines shown in Fig. 2. If this tail structuring is in any way connected to the observed tail rays at stronger comets is questionable. The electric field shown in panels (a), (d) and (k) exhibits a slight decrease in the inner coma, as was expected from Eq. 2. The cometary plasma "shields" itself from the interplanetary electric field, comparable to a Faraday cage. However, in this regime, the pickup force is still the dominant mechanism. The overall direction of motion of the cometary ions therefore coincides mostly with the electric field lines as can be seen from Fig. 3(h).

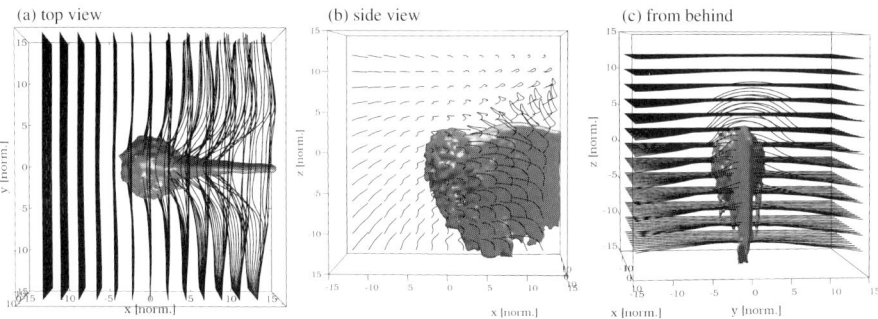

Figure 4. Three different 3D views of the global plasma tail configuration for Churyumov–Gerasimenko at 2.8 AU. (a) Top view. (b) Side view. (c) View from behind. The isosurface corresponds to a cometary ion density of $n/n_0 = 0.5$. The black lines are the magnetic field lines. The solar wind flows in positive x–direction, the IMF is oriented in positive y–direction. The tail is still "flat" as can be seen in (a) and (c). However, besides the cycloidal part, moving downwards, also a anti–sunward main tail is formed visible in (b). At the top a LF wave is excited. (a) and (c) show, that the draping of the magnetic field lines as well as their upwards bending is much more pronounced compared to Fig. 2. Also the Alfvèn wave excitation in the wake above the tail can be seen in (b).

Churyumov–Gerasimenko at 2.8 AU. Fig. 4 shows the global 3D tail and magnetic configuration for C–G at 2.8 AU. All features already visible at 3.25 AU in Fig. 2 can also be found in this figure. However, the ion tail is no longer only cycloidal, but directed anti–sunward more pronounced. The draping and upwards bending of the field lines has become stronger, due to the increased cometary ion densities. In the wake, the wave excitation has also increased.

The reason for the more anti–sunward tail can be found in Fig. 5. First of all the magnetic pile–up is strongly enhanced, as can be seen from panels (b), (h) and (n). The Mach cone in the $y = 0$ plane is steeper compared to Fig. 3(k). In the $z = 0$ plane, the draping is attached to the anti–sunward main tail. The most striking difference, however, compared to the 3.25 AU case is the strong shielding of the electric field in the inner coma and along the main tail region, as is visible in panels (a), (g) and (m), whereas it is enhanced directly above in (m) and besides the main tail in (g). This is simply due to the increased magnetic field strength. The sharp boundary between enhanced electric field outside and almost vanishing electric field inside keeps the cometary ions from crossing the sharp boundary visible in Fig. 5(o). On the other hand, the solar wind protons do not cross this boundary from the outside either, because they are deflected at the obstacle by the electron pressure gradient force, as can be seen in (d) and (j). This forms a "proton cavity" visible in (f) and (l). This boundary can be identified as the "ion composition boundary" or "cometopause". Both ion species "avoid" being mixed, because the cometary ions experience the strong pickup force focusing them to the midplane, whereas the solar wind protons only experience the electron pressure force pushing them away.

The ion tail is still very "flat" due to the focusing effect described above, see panel (i), which is now even stronger due to the stronger field line draping. Whereas the cycloidal part going downwards still reaches high velocities, the anti–sunward part is formed by rather slow drifting cometary ions, as can be seen in (k). This indicates, that the main mechanism is no longer the pickup force acting on the cometary ions, but the magnetic tension, which does not accelerate the cometary ions that strong. In Fig. 5(c) some tail structuring is visible again, resembling the wavy magnetic field line structure. Although no shock is developed and the solar wind flow is still supersonic in the whole interaction region, one can already spot the onset of a diamagnetic cavity formation in Fig. 5(n). This is because of the missing convection of magnetic field flux, due to the deflection of the solar wind around this area.

Particle energy distributions. In Fig. 6 some energy distributions are compiled. These diagrams show the energy of all particles which entered a certain "sample cell" in the simulation during an interval of 400 time steps. This rather long interval was taken to enhance the statistics. The position of these sample

162

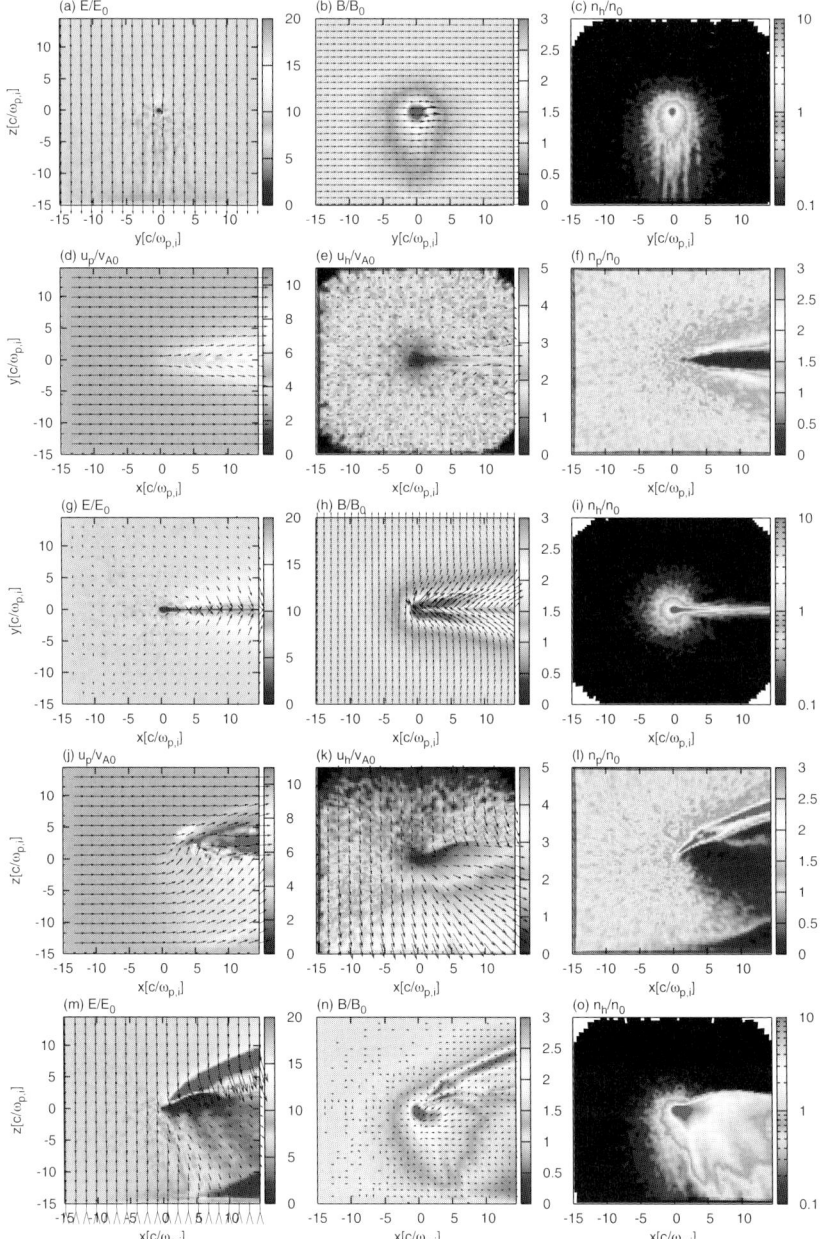

Figure 5. C–G at 2.8 AU, 2D cuts. (a)–(c) E/E_0, B/B_0 and n_h/n_0 in the plane $x = 0$. (d)–(i) SW bulk velocity v_p/v_{A0}, cometary ion bulk velocity v_h/v_{A0}, SW density n_p/n_0, E/E_0, B/B_0 and n_h/n_0 in the plane $z = 0$ (ecliptic). (j)-(o) same as (d)–(i) for the plane $y = 0$. See text for a description of this figure.

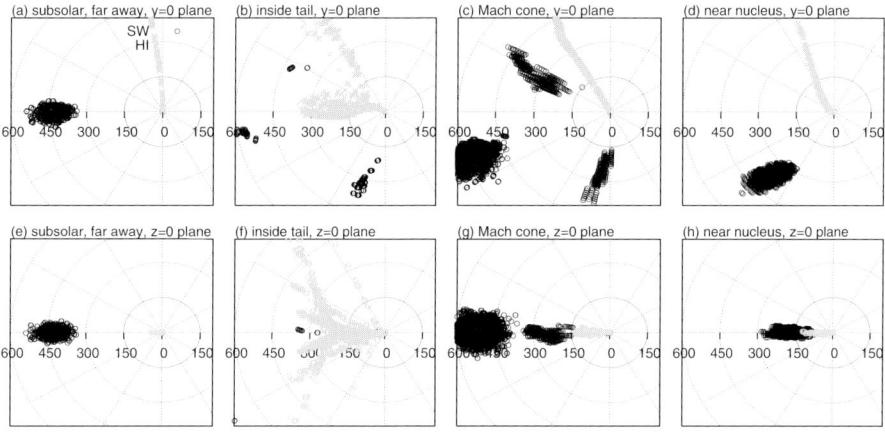

Figure 6. Energy spectra for C–G at 2.8 AU. Polar plots in the planes (a)–(d) $y = 0$ and (e)–(h) $z = 0$. Each dot represents a cometary ion (light) or solar wind proton (dark) respectively. The distance from the center gives the particles energy in eV perpendicular to the respective plane. The angular position indicates the direction from which the particle moves towards the point, where the distribution is taken. All distributions were taken in the midplane $y = 0$. (a),(e) Subsolar point, far away from the nucleus at $x = -10.0$, $z = 0.0$. (b),(f) Inside the tail, far away from the nucleus at $x = 10.0$, $z = 0.0$. (c),(g) At the Mach cone area at $x = 5.0$, $z = 3.0$. (d),(h) Inside the tail, near the nucleus at $x = 0.5$, $z = 0.0$. See Fig. 5 for the location of these positions.

cells are given in the figure caption. All four sample cells are located in the plane $y = 0$. Each point represents a particle with its energy in eV and the direction from where it enters the respective location, where the distribution was taken. However, because this direction is a three–dimensional vector, two different "cuts" in velocity space are shown in Fig. 6 for each location, namely the planes $y = 0$ and $z = 0$. Only the energy component perpendicular to that plane is plotted. For example, Fig. 6(a) shows the energy $m/2(v_x^2 + v_z^2)$ for each particle, whereas Fig. 6(e) shows $m/2(v_x^2 + v_y^2)$.

Fig. 6(a) and (e) were sampled at the subsolar point, far away from the nucleus. The Maxwellian distribution of the solar wind with an average energy of about 400eV coming from the "left", i.e. from the sunward direction can be seen. The pickup ions form the typical ring distribution, coming from "above", i.e. moving downwards along their cycloidal pickup trajectory. These pickup ions, originating from far away of the nucleus can reach energies up to some keV. Fig. 6(b) and (f) were sampled inside the tail, far downstream of the nucleus. As can be seen from Fig. 5(f) and (l), in this area, the solar wind protons almost vanish completely. Consequently, these panels show mainly cometary ions. From (b) it can be seen, that there is still a high energetic component originating from pickup ions coming from above. This is overlayed by a low energetic component with energies up to 300 eV which comes from the "left",

164

i.e. from the nucleus. These cometary ions were accelerated not by the electric pickup field, but by the magnetic pressure and tension forces, which are less effective. Moreover, panel (f) indicates that the low energetic particles also enter the main tail from the sides, which is due to the focusing effect of the draped field lines, described above. The ray structure in this panel originates probably from the wave structure of the magnetic field lines visible in Fig. 4.

Fig. 6(c) and (g) were sampled at the Mach cone region above the ion tail. The solar wind is deflected upwards rather strongly there due to the enhanced magnetic field. This leads to a gyromotion, with some particles performing loops and getting reflected. Therefore the solar wind particles are split into several populations moving in different directions. This indicates the transition of the linear Mach cone to the non–linear shock structure.

Finally, Fig. 6(d) and (h) were sampled near the nucleus, about 120 km behind it in the tailward direction. It can be seen, that the acceleration process by the magnetic pressure and tension has not yet effectively accelerated the cometary ions at that distance. Therefore, the main population has very low energies, below 50 eV. However, still the high energetic population of the pickup ions is overlayed. Note, that for that case, although being located in the main tail, a particle detector would register almost no particles when oriented towards the nucleus.

Almost all of these distribution differ strongly from simple Maxwellian distributions as assumed by fluid theories. Therefore, Fig. 6 clearly indicates a complex kinetic behavior.

4. Summary

Full 3D hybrid code simulations for the plasma interaction of the solar wind with comet Churyumov–Gerasimenko at heliocentric distanced of $R_\odot = 3.25$ AU and $R_\odot = 2.8$ AU have been performed.

The global ion tail configuration for $R_\odot = 3.25$ AU shown in Fig. 2 exhibits the cycloidal ion tail form typical for weak comets. The tail becomes more erected and pointing anti–sunward at $R_\odot = 2.8$ AU as can be seen from Fig. 4. The tail is very thin in the direction perpendicular to the magnetic field due to the focusing effect of the draped field lines.

The magnetic field lines are not only draped around the obstacle, but also bent upwards due to the upward deflection of the solar wind. When the field lines have crossed the inner coma and tail regions, they relax and shorten themselves. This relaxation results in an overshoot which excites a circularly polarized Alfvèn wave in the wake. This, in return, gives rise to some tail structuring visible in Fig. 3(c) and Fig. 5(c).

The anti–sunward tail portion for $R_\odot = 2.8$ AU is not formed by the pickup force, but mainly by the magnetic pressure and tension forces of the draped and piled–up magnetic field lines. The interplanetary electric field, important for this pickup process, is "short–circuited" in the inner coma and the main tail region by the high cometary ion densities. This also gives rise to a sharp boundary at the "top" of the ion tail. The cometary ions cannot cross this boundary because they are held in by the strong pickup field outside. On the other hand, the solar wind protons do not feel this "pickup field", but only the electron pressure force from the cometary plasma, which keeps them from crossing this boundary from the outside. Thus, a solar wind proton cavity is formed and both species tend to not mix each other. This is probably the onset of the formation of an "ion composition boundary" or "cometopause".

Instead of a bow shock, a Mach cone is formed. This Mach cone, however, is already capable of reflecting solar wind particles. Near the nucleus, where the pickup process is not effective, the cometary ions produced near the nucleus reach only low energies up to 50 eV. They are accelerated by the electron pressure, as well as the magnetic pressure and tension forces. The main contribution near the nucleus comes from the high energetic pickup ions, coming from "above". These can reach an energy up to some keV. Further away from the nucleus (about 2500 km) inside the tail, however, the mentioned forces have already accelerated the heavy ions up to an energy of about 200 eV. Still the high energetic pickup component can be found there.

Acknowledgments

This work is supported by the Deutsche Forschungsgemeinschaft through the grant MO 539/10-2 and by the DLR.

References

A'Hearn, M. F., Millis, R. L., Schleicher, D. G., Osip, D. J., and Birch, P. V. (1995). The ensemble properties of comets: Results from narrowband photometry of 85 comets, 1976-1992. *Icarus*, 118:223–270.

Bagdonat, T. and Motschmann, U. (2002). 3D hybrid simulation code using curvilinear coordinates. *Journal of Computational Physics*, 183:470–485.

Bagdonat, T. and Motschmann, U. (2002). From a weak to a strong comet – 3D global hybrid simulation results. *Earth, Moon and Planets*, 90:305–321.

Bagdonat, T. and Motschmann, U. (2004). Plasma boundaries and ion dynamics at comet churyumov-gerasimenko. *Ann. Geophysicae.* in press.

Balsiger, H., Altwegg, K., Bühler, F., Fuselier, S. A., Geiss, J., Goldstein, B. E., Goldstein, R., Huntress, W. T., Ip, W.-H., Lazarus, A. J., Meier, A., Neugebauer, M., Rettenmund, U., Rosenbauer, H., Schwenn, R., Shelley, E. G., Unstrup, E., and Young, D. T. (1986). Ion composition and dynamics at comet Halley. *Nature*, 321:330–334.

Bogdanov, A., Sauer, K., Baumgärtel, K., and Srivastava, K. (1996). Plasma structures at weakly outgasing comets–results from bi-ion fluid analysis. *Planet. Space Sci.*, 44(6):519–528.

Crovisier, J., Colom, P., Gérard, E., Bockelée-Morvan, D., and Bourgois, G. (2002). Observations at Nançay of the OH 18-cm lines in comtes. the data base. observations made from 1982 to 1999. *Astron. & Astrophys.*, 393:1053–1064.

Israelevich, P. L., Gombosi, T. I., Ershkovich, A. I., DeZeeuw, D. L., Neubauer, F. M., and Powell, K. G. (1999). The induced magnetosphere of comet halley. 4. comparison of in situ observations and numerical simulations. *J. of Geophysical Research*, 104:28,309–28,319.

Kidger, M. R. (2003). Dust production and coma morphology of 67P/Chuyrumov-Gerasimenko during the 2002–2003 apparition. *Astron. & Astrophys.*, 408:767–774.

Kührt, E. (1999). H_2O-activity of comet hale-bopp. *Space Science Rev.*, 90:75–82.

Lipatov, A. S. (2002). *The Hybrid Multiscale Simulation Technology*. Springer.

Lipatov, A. S., Motschmann, U., and Bagdonat, T. (2002). 3D hybrid simulation of the solar wind-weak comet interaction. *Planet. Space Sci.*, 50:403–411.

Lipatov, A. S., Sauer, K., and Baumgärtel, K. (1997). 2.5D hybrid code simulation of the solar wind interaction with weak comets and related objects. *Adv. Space Res.*, 20(2):279–282.

Mendis, D. A., Houpis, H. L. F., and Marconi, M. L. (1985). The physics of comets. *Fundamentals of Cosmic Physics*, 10:1–380.

Motschmann, U. and Glassmeier, K.-H. (1993). Nongyrotropic distribution of ions at comet P/Grigg-Skjellerup: A possible source of wave activity. *J. of Geophysical Research*, 98: 20977–20983.

Motschmann, U., Kafemann, H., and Scholer, M. (1997). Nongyrotropy in magnetoplasma: Simulation of wave excitation and phase space diffusion. *Ann. Geophysicae*, 15:603–613.

Osip, J. D., Schleicher, D. G., and Millis, R. L. (1992). Comets – Groundbased observations of spacecraft mission candidates. *Icarus*, 98: 115–124.

Parker, E. N. (1958). Dynamics of the interplanetary gas and magnetic fields. *Astrophys. J.,* 128: 664–676.

Richardson, J. D., Belcherand, J. W., Lazarus, A. J., Paularena, K.I., and Gazis, P. R. (1996). Statistical properties of the solar wind. In Winterhalter, D., Gosling, J.T., Habbal, S.R., Kurth, W.S., and Neugebauer, M., editors, *AIP Conference Proceedings*, volume 382, pages 483–486.

Richardson, J. D., Paularena, K. I., Lazarus, A. J., and Belcher, J.W. (1995). Radial evoultion of the solar wind from imp 8 to voyager 2. *Geophysical Research Letters.*

Sauer, K., Bogdanov, A., Baumgärtel, K., and Dubinin, E. (1996). Plasma environment of comet Wirtanen during its low-activity stage. *Planet. Space Sci.*, 44:715–729.

Schwehm, G. and Schulz, R. (1999a). Coma composition and evolution of rosetta target comet Wirtanen. *Space Science Rev.*, 90:321–328.

Schwehm, G. and Schulz, R. (1999b). Roestta goes to comet Wirtanen. *Space Science Rev.,* 90:313–319.

Szegö, K., Glassmeier, K.-H., Bingham, R., Bogdanov, A., Fischer, C., Haerendel, G., Brinca, A., Cravens, T., Dubinin, E., Sauer, K., Fisk, L., Gombosi, T., Schwadron, N., Isenberg, P., Lee, M., Mazelle, C., Möbius, E., Motschmann, U., Shapiro, V. D., Tsurutani, B., and Zank, G. (2000). Physics of mass loaded plasmas. *Space Science Rev.*, 94:429–671.

Tancredi, G., Fernández , J. A., Rickman, H., and Licandro, J. (2000). A catalog of observed nuclear magnitudes of jupiter family comets. *Astron. & Astrophys. Suppl.*, 146:73–90.

Winske, D. (2003). *Space Plasma Simulation*, chapter Hybrid Simulation Codes: Past, Present and Future – A Tutorial. Springer.

MODELLING OF THE STRUCTURE AND LIGHT SCATTERING OF INDIVIDUAL DUST PARTICLES

Ivano Bertini
CISAS – University of Padova
Physikalisches Institut – University of Bern
bertini@pd.astro.it
ivano.bertini@phim.unibe.ch

Nicolas Thomas
Physikalisches Institut – University of Bern
nicolas.thomas@phim.unibe.ch

Cesare Barbieri
Astronomy Department – University of Padova
barbieri@pd.astro.it

Abstract   Cometary dust environment modelling is required to make predictions for the European Space Agency's Rosetta mission. Following results from pre–planetary dust aggregation experiments both on Earth and in space (CODAG), we have developed a code to generate and study the morphological properties of dust chains characterized by a fractal mass dimension $D_f \sim 1.3$. We are now analysing the scattering properties of these chains and comparing their behaviour with the dust scattering properties seen in the cometary comae. Preliminary results and future prospects are described in this paper.

Keywords: Comets, dust, light scattering

Introduction

Cometary dust environment modelling is an essential tool needed for the future operation and data evaluation of the European Space Agency's Rosetta mission to comet 67P/Churyumov–Gerasimenko. Comets are commonly thought to be primitive objects, containing information about the early formation phase of our Solar System. Ground–based observations of light scattered

L. Colangeli et al. (eds.), The New ROSETTA Targets, 167–176.

in cometary comae show that dust scattering properties are typical of non–spherical particles or fractal aggregates. Although a morphological metamorphism of pre–planetary dust might take place inside comets, it is usually assumed that cometary dust maintains its pristine nature (Kimura et al., 2002). Although we are somewhat sceptical of this ourselves, it implies that laboratory experiments and theoretical modelling of dust grains growth are vital instruments to derive information about the morphology of cometary dust aggregates.

Ground–based experiments with pre–planetary dust aggregation have shown that the resulting aggregates, grown in a turbulent rarefied gas, are morphologically characterized by: a fractal mass dimension $D_f \sim 1.91$, reproducible with Ballistic Cluster Cluster Aggregation (BCCA) models; an asymmetry value, or maximum to minimum radius of gyration ratio for a single chain, around 3 for small aggregates and 2 for large aggregates; and a preferred alignment along the vertical axis, i.e. the direction of gravity, or gas drag direction (Wurm and Blum, 1998; Wurm and Blum, 2000).

The Cosmic Dust Aggregation Experiment (CODAG), flown on STS–95 (Oct/Nov 1998) and on ESA Maser 8 Rocket (May 1999), shows instead a different behaviour of dust aggregation in a microgravity environment. Dust chains were highly elongated, with an asymmetry between 3 and 4, randomly oriented, i.e. unaligned, and characterized by a low fractal dimension $D_f \sim 1.3$ (Blum et al., 2000; Blum et al., 2002). These structures, not reproducible with usual BCCA ($D_f \sim 2$), were explained by Blum et al. as the result of a quasi–ballistic Brownian motion–driven aggregation and a high rotation of aggregates that can lead to non–central collisions. They obtained this result modelling the observed chains with a BCCA with impact parameter greater than 0.65 to simulate aggregate rotation.

Their latest work about CODAG has led to a new method of determining D_f, plotting the mass of a large sample of aggregates as a function of their linear size. They found that up to the biggest detected aggregates, comprising 100–200 monomer grains, the fractal dimension of the aggregates is $D_f \sim 1.4$, close to the value derived with the previous method (J. Blum, private communication, 2004).

On the other hand, theoretical works about light scattering by fractal dust aggregates have shown that it is possible to reproduce some coma observed scattering properties of the cometary dust. Petrova et al. (2001) used randomly oriented aggregates of spherical monomers positioned in a tetrahedral or cubic lattice and the N–sphere scattering method. More recently similar fits, using large BCCA aggregates consisting of optically dark submicron grains and the T–matrix method have been obtained (Kimura et al., 2003). Analogous studies, using the Discrete–Dipole Approximation (DDA) method, have been made by Xing and Hanner (1997) and Yanamandra–Fisher and Hanner (1999).

Our scientific goal is to study the morphological and scattering properties of dust aggregates characterized by a low fractal mass dimension. We have therefore constructed a simulation of the aggregation process which successfully reproduces the fractal mass dimension observed by CODAG. We now seek to determine if this kind of structure can reproduce observed scattering properties of dust in cometary comae. In the immediate future, we will study if they can be used, together with more compact structures and the presence of ice, to model scattering properties of nuclei surfaces. In this paper we present our preliminary results.

1. The Model

In order to model and study dust aggregates with a low fractal mass dimension, we have built a code to generate dust chains randomly using spherical particles (monomers) as constituents. We are using a Particle Cluster Aggregation (PCA) mode under some assumptions. The generation mechanism is quite straightforward. The first particle of the chain is placed in a fixed point in a 3D grid. The next particles are then randomly generated on a sphere centred at the first particle's coordinates and having a radius bigger than the maximum theoretical chain size (in order to avoid any compression of the chain). After generation the particles are forced to move in the direction of the closest one and they stick with it at the contact point, along the line which connects the two spheres centres. In this way we obtain random chains with an elongated structure.

Examples of the generated chains, with a different number, N, of monomers are shown in Fig.1.

After the generation of the entire chain, we study the morphological characteristics of the dust agglomerate. We calculate the chain orientation, through the analysis of the direction of the line which minimises the radius of gyration of the system. We then determine the anisotropy or asymmetry and the fractal mass dimension, D_f, which is the exponent of the relation between the mass and the size of the aggregate.

Monomers with different radii have been used both to analyse the effects of different sizes and to prepare the scatterers for scattering calculations. We have taken into account monomers of radii R = $0.1\mu m$ (used in Kimura et al., 2003), $0.15\mu m$ (used in Petrova et al., 2001), $0.50\mu m$ and $0.95\mu m$. The latter are the sizes of SiO_2 spherical dust analogues monomer grains used in the CODAG experiment.

The analysis of the orientation of the minimum radius of gyration line in 3D space shows that our aggregates are randomly oriented, i.e. they are unaligned with respect to a preferred direction (Fig.2).

170

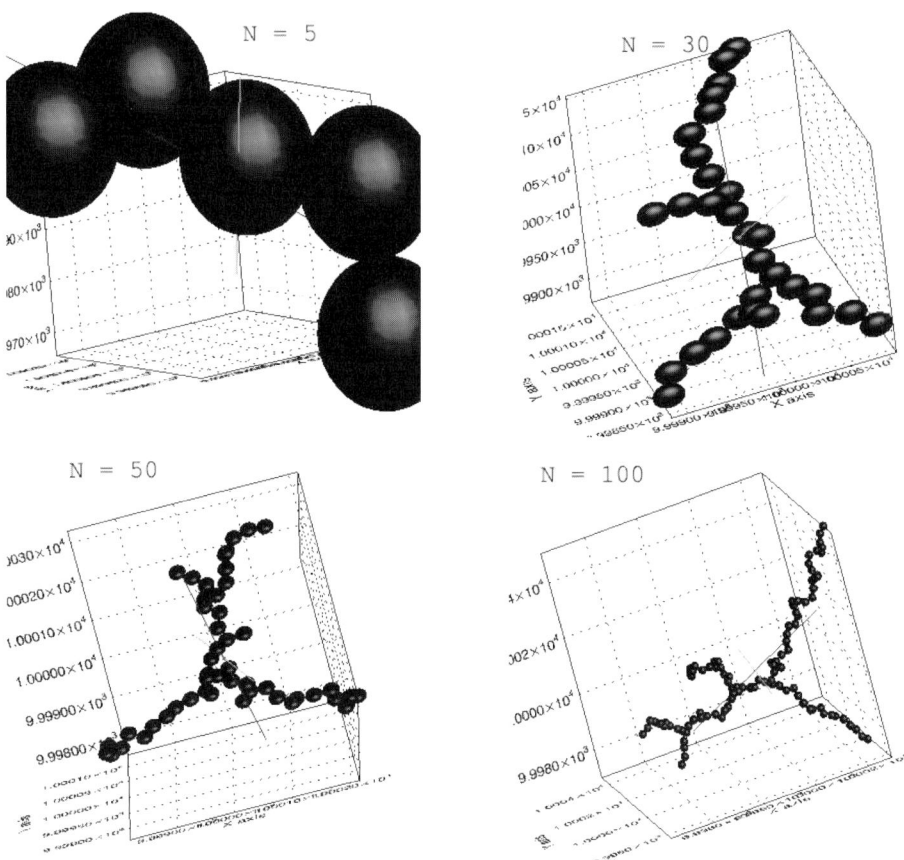

Figure 1.   Examples of chains generated by the code. N is the number of constituent monomers. N values of 5, 30, 50 and 100, with radius of monomers, R=0.15μm are taken into account. The two lines are useful additional information generated by the code regarding the radius of gyration of the chain calculated in respect to a line. They are the two perpendicular lines which minimise and maximise the radius of gyration. They both pass through the center of mass of the aggregate. The two lines give the minimum and maximum chain cross section on their perpendicular planes.

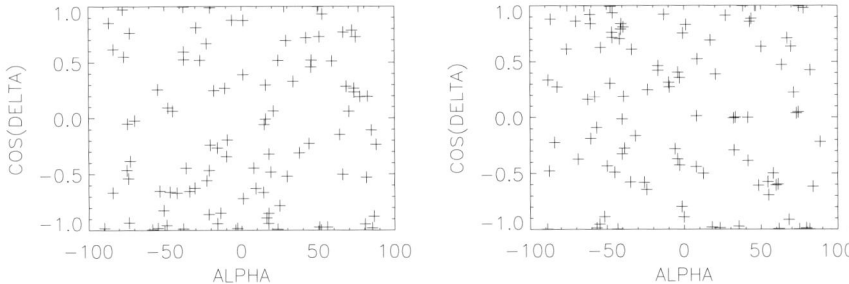

Figure 2.   3D orientations of 100 chains with N=5 (left), 30 (right) and R = 0.95μm. On axes the two angular coordinates which describe the 3D position of the alignment line for each chain in a reference frame centred in its center of mass are plotted. As can be seen the agglomerates are randomly oriented.

Anisotropy for agglomerates with a number of monomers N between 5 and 300 and R = 0.95μm is plotted in Fig.3. Each point is the mean value of 100 chains. Despite the dispersion of its values (shown through the standard deviation error bar in the figure) we can see that anisotropy decreases with increasing N. It has a mean value of ~3 for small aggregates and ~2 for N~100. This behaviour is similar to the result obtained with dust aggregation laboratory experiments on Earth (Wurm and Blum, 2000).

The fractal dimension for aggregates constituted by a different number N of monomers is derived plotting the aggregate mass, in units of monomer mass (i.e. the N number), versus the chain radius of gyration, in units of monomer radius, and fitting an exponential curve. The exponent is the fractal dimension, D_f. Each theoretical point obtained from calculations and used to fit the curve is the mean value of 120000 chains with same N.

In Tab.1 values of D_f for N=5, 15, 30, 50, 100 and R=0.10, 0.15, 0.50, 0.95μm are shown.

Table 1.   Values of the fractal mass dimension D_f(N) for chains with different number N of monomers and different radius of monomers, R(μm). Each point, fitted by the curve, is the mean value of 120000 chains for each N. Results are almost independent from monomer size.

R	$D_f(5)$	$D_f(15)$	$D_f(30)$	$D_f(50)$	$D_f(100)$
0.10	1.270±0.011	1.315±0.001	1.291±0.001	1.263±0.008	1.234±0.009
0.15	1.267±0.010	1.327±0.006	1.281±0.022	1.253±0.008	1.203±0.015
0.50	1.267±0.011	1.324±0.002	1.304±0.006	1.268±0.018	1.177±0.002
0.95	1.266±0.009	1.324±0.007	1.305±0.003	1.286±0.016	1.212±0.001

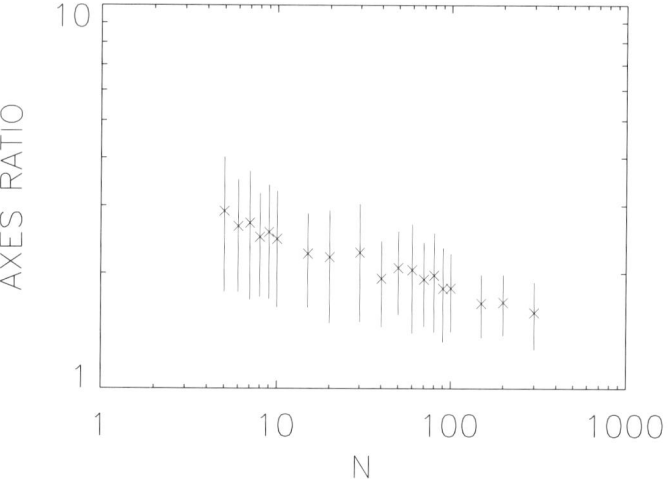

Figure 3. Anisotropy for aggregates with N between 5 and 300 and R=0.95μm. Each point is the mean value of 100 chains with same N. The decreasing behaviour of anisotropy with increasing N has to be noticed.

As can be seen, aggregates constituted by N in the range [5–50] are characterized by $D_f \sim 1.3$, while bigger aggregates have a slightly smaller fractal dimension $D_f \sim 1.2$. The results are moreover almost independent of the size of the monomers, being intrinsic characteristics of the generation mechanism. Hence, we have produced a code which can be used to simulate dust aggregates characterized by a low fractal dimension.

2.   Scattering calculations – first results

Our goal is to determine if dust aggregates characterized by a low fractal dimension can reproduce some observational properties of dust in cometary comae. We will focus our attention on the intensity and polarization curves of scattered light. The intensity of dust in the coma shows a strong rise at large phase angles α (forward direction); a flat profile at intermediate α and a weak increase toward small α (backscattering direction). Cometary dust is also mainly red in color at intermediate α. These properties have been summarized in Kimura et al. (2003). Observed polarization curves of many comets show a bell shape curve, with a peak of 10–30% around α=90-100° and a negative branch at small α, with an inversion angle $\sim 20^o$, as can be seen in Levasseur–Regourd and Hadamcik (2001). Observations with $\alpha < 120^o$ are not available.

Scattering calculations of our chains have been performed using the DDA method in the program DDSCAT.5a10 by Draine and Flatau (2000).

Different approaches for the scattering calculations are possible, because of the DDSCAT structure. We can study the scattering properties of a huge amount of chains with the same number of monomers N, fixed in 3D space with respect to the incident light direction. As explained above, each chain has its own random orientation. Using this method we therefore obtain a mean result averaged over many orientations and many different chain shapes (generated by our code and characterized by the mean D_f value showed in Tab.1). Alternatively, we can use a single chain and rotate it in 3D space, i.e. varying its orientation in respect to the incident light, calculating the averaged scattered intensity and polarization. Although both methods are computational time consuming, the last one is faster and, even if in that case we have only one chain shape taken into account, we have used this to obtain our first results. With unlimited computing resources one can, of course, combine both approaches.

In the future a comparison of the results obtained using methods described above, considering also many different monomer sizes, complex refractive indices and incident light wavelengths, has to be carried on. In our initial calculations, we have used chains with R=0.15μm and complex refractive index m=1.65+0.01i at λ=0.60μm, typical of astronomical silicates. We also used the refractive index m=1.5+0.0032i at λ=0.6708μm, derived from the Imager for Mars Pathfinder (IMP) measurements of optical properties of the dust suspended in the martian atmosphere (Markiewicz et al., 1999). Chains with a number of constituent monomers equal to 5 and 15 have been analysed up to now. For each considered chain we calculated: the effective radius of the chain, a_{eff}; the size parameter of the monomer, X_i and the size parameter of the chain, X_V. Scattering calculations with martian dust have been made with N=5, a_{eff}=0.256μm, X_i=1.41 and X_V=2.40. Silicate chains have common X_i=1.57. The values of the other parameters are: X_V=2.69, a_{eff}=0.256μm (N=5); X_V=3.87, a_{eff}=0.370μm (N=15). Calculations with bigger aggregates are planned in the immediate future.

In Fig.4 and Fig.5 the S_{11} term of the scattering, or phase, matrix and the polarization curves of our sample aggregates, versus the phase angle, are plotted. The S_{11} term is proportional to the scattered intensity (Bohren and Huffmann, 1983).

These first code runs show promising and interesting features both in the intensity and the polarization curve, using astronomical silicate chains. A negative polarization is visible at small phase angles even for small chains and the intensity profile shows a weak increase in the backscattering direction and a bigger increase in the forward direction. These features reproduce only qualitatively the scattering properties of cometary dust. We are at present going on deriving data to complete our calculation plans. This will tell us at the end if

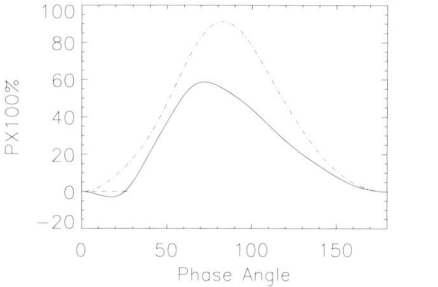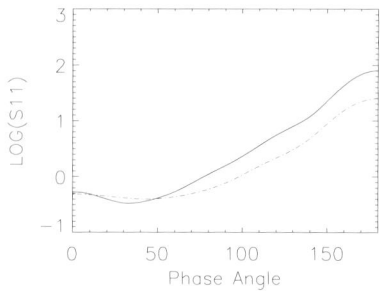

Figure 4. Polarization curve (left) and S_{11} term profile (right) versus the phase angle of a N=5 aggregate. The continuous lines are calculated with m of astronomical silicates. The dash–dotted lines with m of dust suspended in the martian atmosphere. The dotted line is the zero level of the polarization for $\alpha<30°$. Results are averaged over 729 target orientations in the 3D space.

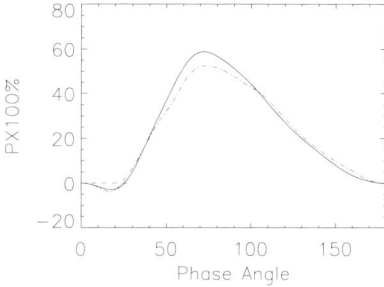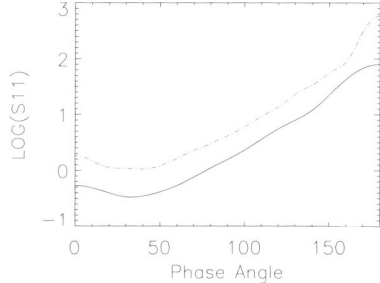

Figure 5. Polarization and S_{11} profiles for silicate chains with different N. Results calculated for single aggregates with N=5 (continuous line) and N=15 (dash–dotted line) are plotted.

the dust particles characterized by low fractal dimension can be representative of the dust in cometary comae.

3. Future prospects

A future important point will be the study of the different scattering properties obtained changing the monomer size and the refractive index. Because observational evidence shows that cometary dust in the coma has a size distribution, we also plan to average results over different size distributions. All our calculations will also be compared by analogue ones regarding BCC aggregates with $D_f \sim 2$. Future goals include also using aggregates with different fractal dimensions (from extremely elongated to spherical ones) in order to build a dust surface that can be representative of cometary nuclei insulating dust mantle and studying its scattering properties, both toward outside and inside the nucleus. The presence of ice, mixed with dust, in the perspective of modelling also the surface active regions will be another key point to take into account.

Acknowledgments

Authors whish to thank Prof. J. Blum for his kind and precious explanations about CODAG experiment results and latest news about his work. We are also grateful to B. T. Draine and P. J. Flatau for providing their fortran scattering code DDSCAT.5a10, available at:
http://arxiv.org/abs/astro–ph/000815v4.

This work is supported by an Italian Space Agency (ASI) grant about theoretical modelling of the ROSETTA mission targets.

References

Blum, J., et al. (2000). Growth and Form of Planetary Seedlings: Results from a Microgravity Aggregation Experiment. *Physical Review Letters*, Vol.85,12:2426–2429.

Blum, J., Wurm, G., Poppe, T., Kempf, S. and Kozasa, T. (2002). First Results from the Cosmic Dust Aggregation Experiment CODAG. *Advances in Space Research*, Vol.29,4:497–503.

Bohren, C. F. and Huffman, D. R. (1983). Absorption and Scattering of Light by Small Particles. *John Wiley and Sons, New York*.

Kimura, H., Mann, I., Biesecker, D. A. and Jessberger, E. K. (2002). Dust Grains in the Comae and Tails of Sungrazing Comets: Modeling of Their Mineralogical and Morphological Properties. *Icarus*, 159:529–541.

Kimura, H., Kolokolova, L. and Mann, I. (2003). Optical properties of cometary dust - Constraints from numerical studies on light scattering by aggregate particles. *Astronomy and Astrophysics*, 407:L5–L8.

Levasseur–Regourd, A. C. and Hadamcik, E. (2001). Clues to the Structure of Meteoroids, from Dust Light Scattering Properties. *Proceedings of the Meteoroids 2001 Conference. Sweden*, ESA SP-495:587–594.

Markiewicz, W. J., Sablotny, R. M., Keller, H. U., Thomas, N., Titov, D. and Smith, P. H. (1999). Optical properties of the Martian aerosols as derived from Imager for Mars Pathfinder midday sky brightness data. *Journal of Geophysical Research*, Vol.104,E4:9009–9018.

Petrova, E. V., Jockers, K. and Kiselev, N. N. (2001). Light Scattering by Aggregate Particles Comparable in Size to Wavelenght: Application to Cometary Dust. *Solar System Research*, Vol.35.1:57–69.

Wurm, G. and Blum, J. (1998). Experiments on Preplanetary Dust Aggregation. *Icarus*, 132:125–136.

Wurm, G. and Blum, J. (2000). An Experimental Study on the Structure of Cosmic Dust Aggregates and Their Alignment by Motion Relative to Gas. *The Astrophysical Journal*, 529:L57–L60.

Xing, Z. and Hanner, M. S. (1997). Light scattering by aggregate particles. *Astronomy and Astrophysics*, 324:805–820.

Yanamandra-Fisher, P. A. and Hanner, M. S. (1999). Optical Properties of Nonspherical Particles of Size Comparable to the Wavelength of Light: Application to Comet Dust. *Icarus*, 138:107–128.

GAS AND DUST ACTIVITY OF THE NUCLEUS OF 67P/CHURYUMOV-GERASIMENKO: SOME PRELIMINARY RESULTS

M. Teresa Capria
Istituto di Astrofisica Spaziale e Fisica Cosmica - CNR, ARTOV, Roma, Italy
capria@rm.iasf.cnr.it

Angioletta Coradini
Istitituto di Fisica dello Spazio Interplanetario - CNR, ARTOV, Roma, Italy
coradini@rm.iasf.cnr.it

M. Cristina De Sanctis
Istituto di Astrofisica Spaziale e Fisica Cosmica - CNR, ARTOV, Roma, Italy
cristina@rm.iasf.cnr.it

Marco Fulle
Osservatorio di Trieste - INAF, Trieste, Italy
fulle@ts.astro.it

Abstract

The planning of planetary missions and the operations preparation of the instruments payload require a knowledge as much as possible accurate of the target. This knowledge cannot be obtained only from ground based observations and can be integrated by theoretical models. Comet 67P/Churyumov-Gerasimenko is the current selected target for the ESA ROSETTA mission. Presently, little is known about this comet, but the successful design of the mission, and in particular of the operations of the on-board scientific instruments, requires some preliminary knowledge of the comet physical parameters, such as surface temperature, percentage of active surface, intensity of gas and dust fluxes, and so on. These quantities cannot be determined only through ground-based observations, so predictive models of the thermal evolution and differentiation of a cometary nucleus can be used. We present here the results of the application of the nucleus evolution model developed at the IASF in Rome to the simulation of 67P/Churyumov-Gerasimenko. This nucleus model, used as a

L. Colangeli et al. (eds.), The New ROSETTA Targets, 177–184.

help in the planning of the VIRTIS instrument operations on Rosetta, is a one-
dimensional model giving gas and dust fluxes, evolution of temperatures and
composition along the orbit.

Introduction

The comet 67P/Churyumov-Gerasimenko is the target of ROSETTA, the
ESA cornerstone mission devoted to the in-situ study of a comet over a period
of several months. The spacecraft will start to observe the comet from the beginning of its Sun-driven activity, then will follow and orbit it until at least the
perihelion. At a distance from the Sun of 3 AU a probe will land on the surface
of the nucleus. The preceding target of ROSETTA, dismissed due to a one year
delay in the launch of the spacecraft, was the smaller comet 46P/Wirtanen.
The successful design of this mission requires some preliminary knowledge of
comet status and activity: the onset of activity (dust and gas emission), the diurnal and nocturnal surface temperatures, the percentage of active surface, the
coma development and the intensities of gas and dust fluxes are some of the
data needed to select safe orbits around the comet and to optimize the scientific
return.

1.   Comet 67P/Churyumov-Gerasimenko

Comet 67P/Churyumov-Gerasimenko, discovered on 1969 September 11,
has been observed during 6 apparitions, of which the latest was in 2002-2003.
The comet is a Jupiter family comet with a perihelion distance of 1.29 AU and a
period of 6.57 years; both values are slightly larger than the corresponding ones
for 46P/Wirtanen (1.06 AU and 5.44 years). From their observations, Tancredi
et al. (2000) estimated a radius of 2.5 km for 67P/Churyumov-Gerasimenko,
making it one of the largest comets in the Jupiter family. Radar observations
(Kamoun et al., 1998) show an upper limit of 3.7 km for the radius of the
nucleus and a probable upper limit of 3 km, consistent with the estimate of
Tancredi et al. (2000). Similarly, the observations by Mueller (1992) give a
radius of 3.2 km with an albedo of 0.03. The estimated minimum axis ratio is
1.7, in good agreement with the previously cited values.

The dynamic history of 67P/Churyumov-Gerasimenko shows two significant drops in the perihelion distance in the last 160 years: Beliaev et al. (1986)
and Carusi et al. (1985) found that the comet had close encounters with Jupiter
in 1840 and 1959 that reduced the perihelion distance significantly. The encounter with Jupiter in 1959 reduced the perihelion distance from q = 2.75 AU
to the present value. The perihelion distance has remained stable since the discovery.

67P/Churyumov-Gerasimenko is considered to be a dusty comet (Kiselev, 1998); combining the estimated radius of the nucleus (Tancredi et al., 2000) with the production rates (Osip et al., 1992) it seems that the active area of the nucleus of 67P/Churyumov-Gerasimenko should be about 6% (Kidger, 2003), that means an effective active area of $\sim 2 km^2$. The nucleus rotation period should have a value of 12.3 hrs.

The comet lightcurves show outbursts at perihelion with strong similarities in the 1982-83, 1996-97 and 2002-2003 apparitions, suggesting that the light curve is quite the same over several returns. The comet shows a pre/post-perihelion asymmetry with a peak water production of $\sim 1.0 \times 10^{28}$ mol/s. Moreover a peak value of $Af\rho = 450$ cm was recorded.

Due to the differences, particularly in size and activity level, between 46P/Wirtanen and 67P/Churyumov-Gerasimenko, a careful characterization of the new comet target is important for the success of the ROSETTA mission and even more for the lander operations.

2.   Comet nucleus evolution model

The results presented in this paper have been computed with a numerical code solving the heat conduction and gas diffusion equations through an idealized spherical comet nucleus. Here we will not describe the model in details: for further details see our previous papers (Capria et al., 2000; De Sanctis et al., 1999; Coradini et al., 1997a,b).

At the beginning of the computations, the comet nucleus is homogeneous and uniformly porous. It is composed by a mixture of dust and ices of H_2O, CO_2 and CO. The dust grains are spherical, and distributed between different size classes. The numerical code computes how the heat diffuses in the porous cometary material, inducing the H_2O ice phase transition and the sublimation of the volatile gas. When temperature rises, ices start to sublimate, beginning from the more volatile ones, and the initially homogeneous nucleus differentiates giving rise to a layered structure, in which the boundary between different layers is a sublimation front. The model takes into account amorphous-crystalline transition with the release of gases trapped in the amorphous ice. Surface erosion due to ice sublimation, particles ejection and dust mantle formation and compaction is computed at each step.

When the ices begin to sublimate the refractory particles become free and are subject to the drag exerted by the escaping gas, so that they can either be blown off or accumulate on the surface to form a crust. To determine how many particles can be blown off and how many can be accumulated on the surface, the different forces acting on the single grain are compared, obtaining for each distribution a critical radius that represents the radius of the largest particle that can leave the comet. At each time step we compute the number of free dust

Table 1. Initial parameters describing the nucleus characteristics

dust-ice	1
radius	2 km
porosity	0.6
rotation period	12.3 h
obliquity	0.
initial temperature	30 K
CO_2/H_2O	0.01
CO/H_2O	0.01

particles and the value of the critical radius: we consider ejected the grains whose radius is smaller than the critical one. The existence of different dust grain size classes allows for differential deposition and differential dust emission.

The model is able to describe the nucleus rotation (day/night effects) and the effects of the nucleus spin axis obliquity.

3.   A reference model for the ROSETTA mission

The composition and structure of comet nuclei are to the moment poorly known, and cannot be easily determined from ground observations: to assign a value to the model parameters describing the initial nucleus status and composition we added to the few known data about this comet our best current knowledge about cometary nuclei in general.

The chosen values are listed in Tab.1.

The initial grains distribution is exponential with power index -4 and five size classes from 1 μm to 1 cm (Fulle et al., 2004). The initial dust grain size distribution used in the computations is shown in Fig. 1.

The thermal conductivity of a dust grain is assumed to be 3 W/K/m and the average density is 1000 kg/m^3. A value of 1000 kg/m^3 for the dust density ρ_{dust} simulates the fact that the grains are the result of an accumulation process, and are therefore highly porous, like the Brownlee particles. The H_2O ice is initially considered to be in the amorphous phase, due to the very low temperatures at which comets are thought to have been formed (Rickman and Huebner, 1990), but it can undergo an irreversible, exothermic phase transition to crystalline form. We have chosen to include the ices of H_2O, CO_2 and CO because the first one is the most prominent molecule in the gas emission of comets, while the other two are representative of the more volatile species,

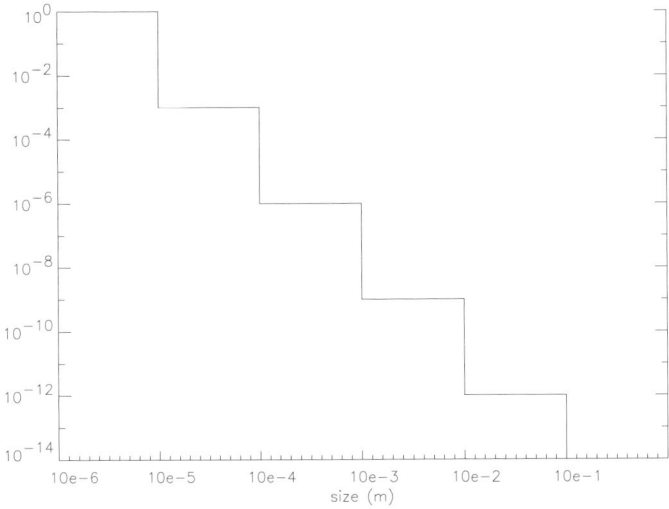

Figure 1.   Initial dust distribution in the comet nucleus.

including also NH_3, CH_4 and CH_3OH, that have been observed in the coma of comets.

At the beginning of computations, the nucleus model is following an orbit in the Kuiper Belt, and then it is moved until it reaches the actual short period orbit. This is done in the attempt to simulate the fact that the comet it is not injected in a short period orbit without having been undergone a differentiation process at least in the layers closer to the surface.

3.1   The activity of 67 P/Churyumov-Gerasimenko from nucleus modelling

Gas production from the nucleus is plotted in fig. 2. Water production shows a perihelion peak whose value is higher than the one estimated by observations, but it should be noted that the gas production plotted in fig. 2 refers to the production at the equator and assumes the whole surface is active, so the results should be scaled in order to make a comparison. The other gases, CO_2 and CO, are coming from layers under the surface and this explains the perihelion asymmetry they show: the heat wave is reaching the layers where those ices are still present (few meters for the CO_2 and more than fifty for the CO) with some delay. The presence of CO, a very volatile gas, introduces some activity along the part of the orbit closer to aphelion.

Dust activity along the orbit, particularly important for many instruments onboard Rosetta, is plotted in fig. 3; the figure shows also how the escaping particles are distributed in the different size classes. Obviously, dust flux is strongly influenced from the initial dust distribution and in general from the assumptions on the physical properties of the grains. Dust production, being mainly the effect of water flux, has a peak at perihelion. A much weaker dust activity, due to the flux of gases more volatile than water, is present far from perihelion. Due to the exponential nature of the initial distribution, close to perihelion most of the ejected particles belong to one class (but not only to this one!).

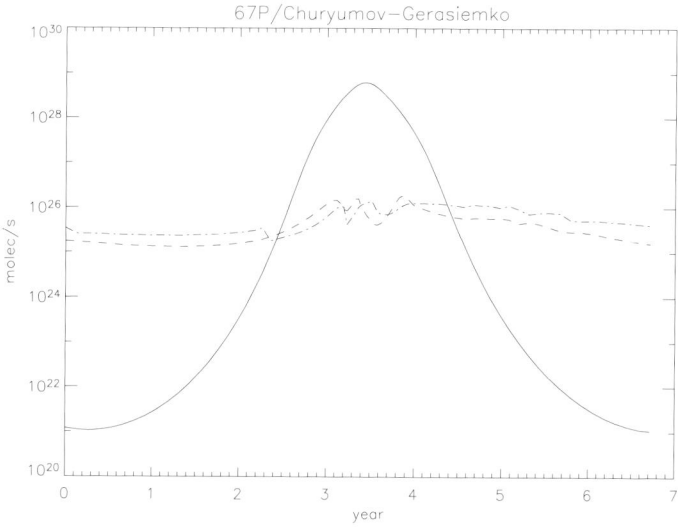

Figure 2. Gas production rate along one orbit. Water is plotted with a continuous line, CO_2 with a dash-dotted line and CO with a dashed line.

4.   Conclusion

It is difficult to give conclusions, both because the results described above are very preliminary and because at the moment too few results from observation are known (and most of them should be confirmed). Anyway, comparing the results of modelling with existing observations, we should note that: 1) the agreement between production rates from the model and from observations is not bad, taking into account the incertitude about observations and the fact that gas and dust production are simulated for a nucleus at the equator, active on the whole surface, so they must be scaled in order to be compared with measured productions; 2) some CO has been considered in the initial nucleus composition, so there is a certain level of activity along the whole orbit; to the moment, we do not know if this is true or not, it must be checked with future observations; 3) as a next step in refining our computations, a spin axis obliquity will

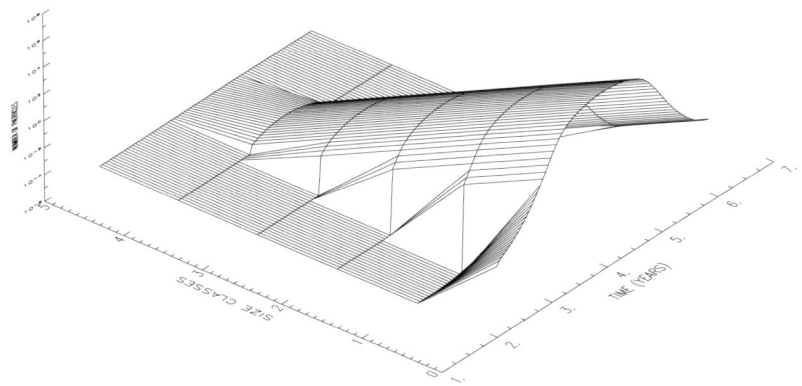

Figure 3. Dust production vs. size distribution along one orbit. Vertical axis reports the number of particles from 10^{-6} to 10^8 in log scale

be introduced, in order to possibly explain the asymmetry effects that seems to have been observed.

References

Beliaev, N.A., Kresak, L. and Pittich, E.M. (1986). *Catalogue of short-period comets*, Bratislava: Slovak Academy of Sciences, Astronomical Institute.

Capria, M.T., Coradini, A., De Sanctis, M.C. and Orosei, R. (2000). CO emission mechanisms in C/1995 O1 (Hale-Bopp). *A&A*, 357: 359.

Carusi, A., Kresak, L., Perozzi, E. and Valsecchi, G.B. (1985). *Long-term evolution of short-period comets*. Bristol (England and Accord, MA, Adam Hilger, Ltd.): 272.

Coradini, A., Capaccioni, F., Capria, M.T., De Sanctis, M.C., Espinasse, S., Orosei, R., Salomone, M. and Federico, C. (1997a). Transition elemets between comets and asteroids. I: Thermal evolution models. *Icarus* 129: 317.

Coradini, A., Capaccioni, F., Capria, M.T., De Sanctis, M.C., Espinasse, S., Orosei, R., Salomone, M. and Federico, C. (1997b). Transition elemets between comets and asteroids. II: From the Kuiper Belt to NEO orbits *Icarus* 129: 337.

De Sanctis, M.C., Capaccioni, F., Capria, M.T., Coradini, A., Federico, C., Orosei, R. and Salomone, M. (1999). Models of P/Wirtanen nucleus: active regions versus non-active regions. *Planet. Space. Sci.* 47: 855.

Fulle, M., Barbieri, C., Cemonese, G., Rauer, H., Weiler, M., Milani, G. and Ligustri, R. (2004). The dust environment of comet 67P/Churyumov-Gerasimenko. *A&A*, submitted.

Kamoun, P., Campbell, D., Pettengill, G. and Shapiro, I. (1998). Radar observations of three comets and detection of echoes from one: P/Grigg-Skjellerup. *Planet. and Space Sci.*, 47: 23.

Kidger, M.R. (2003). Dust production and coma morphology of 67P/Churyumov-Gerasimenko during the 2002-2003 apparition. *Astronomy and Astrophysics*, 408: 767.

Kiselev, N.N. and Velichko, F.P. (1998). Polarimetry and Photometry of Comet C/1996 B2 Hyakutake. *Icarus*, 133: 286.

Mueller, B.E.A. (1992). "CCD-photometry of comets at large heliocentric distances" *Asteroids, Comets, Meteors 1991* : 425.

Osip, D.J., Schleicher, D.G. and Millis, R.L. (1992). Comets - Groundbased observations of spacecraft mission candidates. *Icarus*, 98: 115.

Rickman, H. and Hübner, W.F. (1990). Comet formation and evolution. *Physics and chemistry of comets*, ed. W.F. Hübner, Springer Verlag, Berlin: 245.

Tancredi, G., Fernandez, J.A., Rickman, H., and Licandro, J. (2000). A catalog of observed nuclear magnitudes of Jupiter family comets. *A&A Suppl.*, 146: 73.

BIG PARTICLE EMISSION FROM COMET 67P/CHURYUMOV-GERASIMENKO

E. Grün
Max Planck Institute for Nuclear Physics, Saupfercheckweg 1, D-69117 Heidelberg
Hawaii Institute of Geophysics and Planetology, University of Hawaii, 1680 East West Road,
Honolulu, HI 96822
eberhard.gruen@mpi-hd.mpg.de

J. Agarwal
Max Planck Institute for Nuclear Physics, Saupfercheckweg 1, D-69117 Heidelberg
European Space Operations Centre, Robert-Bosch Str. 5, D-64293 Darmstadt
jessica.agarwal@mpi-hd.mpg.de

Abstract   Big (millimeter-sized) particles released from a comet are displayed in a comet
trail rather than in the tail. A comet tail is an extended ephemeral phenomenon
that is caused generally by sub-millimeter sized particles that were released during the ongoing apparition of the comet in the inner solar system. A dust trail
is defined as a concentration of big (mm-sized) particles close to the orbit of
the parent comet. The distribution along the orbit comes from the fact that big
particles are emitted from the nucleus at low emission speeds (a few m/s) and
that the radiation pressure force only weakly modifies their orbits from the orbit
of the parent comet. A dust trail of 67P/Churyumov-Gerasimenko (C-G) — the
target comet of the Rosetta mission — has been found in IRAS observations in
the infrared and in recent ground based observations at visible wavelength. Simulations of micron- to centimeter-sized particle emissions from C-G have been
performed. Our results show that mm-sized particles that were emitted during
previous perihelion passages have a distribution along the orbit of the comet and
in vertical extent which is compatible with trail observations in spring 2003.

Keywords: 67P/C–G, big comet dust, dust tail, dust trail

1.   Introduction

Comet 67P/Churyumov-Gerasimenko (C-G) is a short period comet of the
Jupiter family with perihelion distance $q = 1.29$ AU, semi major axis $a = 3.5$ AU corresponding to a period of 6.57 years, eccentricity $e = 0.631$, and
inclination $i = 7.1$ deg. It was discovered in September 1969 by the Kiev as-

L. Colangeli et al. (eds.), The New ROSETTA Targets, 185–196.
© *2004 Kluwer Academic Publishers. Printed in the Netherlands.*

tronomers K. I. Churyumov and S. I. Gerasimenko. The comet has an unusual history. Up to 1840 its perihelion distance was at 4.0 AU and the comet was totally unobservable from Earth. That year there was an encounter with Jupiter and the orbit shifted inwards to a perihelion distance of 3.0 AU. From there it slowly decreased further to 2.77 AU, from which, in 1959, a further Jupiter encounter (Marsden 1970) moved it into an orbit with the present perihelion distance. Since the last close Jupiter encounter it has completed seven orbits around the sun.

The comet has been extensively observed during the favorable apparition in 1982 and during the current apparition (perihelion passage August 2002). Recent interest in this comet is caused by the fact that C-G has been selected the new target comet of ESA's Rosetta mission. The selection of a new target comet became necessary because of a launch delay from January 2003 to February 2004. For the 2004 launch window the old target comet 46P/Wirtanen was no longer reachable with the available launcher. There are significant differences between the two comets. The nucleus size of 46P/W is about 0.7 km compared to 1.98 ± 0.02 km for C-G (Lamy et al., 2003). 46P/W displayed an unstructured coma during its perihelion passage while CCD observations of C-G by Boehnhardt et al. (2003) showed jet like coma structures and an extended tail. But most importantly for studies of big cometary dust this comet displayed a dust trail. Already in 1983 the Infrared Astronomical Satellite (IRAS) observed a dust trail associated with C-G. More recently, groundbased observations of the C-G comet trail have been reported by Reach et al. (2003) and Ishiguro (2003).

From the analysis of coma and tail observations information on the dust emission (like the dust size distribution) can be obtained. However, cometary comae and dust tails normally display only the distribution of micron-sized particles, whereas for spacecraft missions like Rosetta millimeter-sized and bigger particles are of special interest.

The purpose of the paper is to discuss some general properties of tails and trails with special emphasis on the C-G trail. It is not our intention to replace full photometric models of comet tails like the ones by Beisser (1989) or by Fulle et al. (2004), but to draw conclusions on the distribution of big particles in the vicinity of the nucleus based on a simple model. Using a simple model has the advantage that some elementary facts become obvious which may be obscured in a full fledged model.

The structure of the paper is as follows. In the next section we discuss methods which provide information on big particle emissions from comets. In section three we present a simple model which we employ in order to derive some results for the large particle distribution in the vicinity of the nucleus of C-G which are discussed in the last section.

2.   Observations of big cometary particles

Traditionally, big cometary dust is studied in the meteor shower phenomenon. Most prominent meteor showers are related to known periodic comets. Visible meteors originate from centimeter-sized meteoroids entering the atmosphere at high speeds (> 20 km/s). Since C-G does not cross the orbit of the Earth we have no direct meteor information for this comet.

The most direct way to analyze big cometary particles is by in-situ spacecraft measurements. Three comets have been visited by spacecraft with dedicated dust instrumentation: 1P/Halley by Giotto and the Vega spacecraft, 26P/Grigg-Skjellerup by Giotto, and recently 81P/Wild 2 by the Stardust spacecraft. From measurements of the Halley encounters a dust mass distribution reaching from 10^{-18} to 1 g has been derived (McDonnel et al. 1991; Divine and Newburn 1987).

An astronomical approach to big cometary particles is the observation of comet trails. A cometary dust trail is an extended structure of long and narrow shape stretching to either side of a parent comet along its orbital path. It is thought that trails consist of mm-sized and larger particles ejected from the comet near perihelion during previous apparitions. Due to their mass, such particles are but weakly accelerated by the gas flowing from the nucleus and leave the coma with low velocities (a few m/s) relative to the parent comet. The effect of radiation pressure on these particles is small as well. Therefore, the orbital elements of their trajectories are similar to those of the parent comet. Assembling along or near the parent's orbital path, large particles remain in the vicinity of the nucleus on timescales of tens to hundreds of years. Smaller particles, on which the influence of both gas drag and radiation pressure is stronger, are set onto trajectories more different from their parent body. After forming the dust tail, bent away from the Sun and observed only around perihelion passage, they vanish from the vicinity of the comet nucleus within a couple of months.

IRAS observed in 1983 trail structures for eight different short-period comets, including C-G (Table 1, Sykes and Walker 1992). Since C-G's close encounter with Jupiter in 1959 this comet was able to eject large particles during four perihelion passages that formed the trail observed by IRAS.

After IRAS, trails were observed by the Infrared Space Observatory (ISO) more than a decade later. ISO observed segments of the 22P/Kopff (Davies et al. 1997) and 2P/Encke trails (Reach et al. 2000). Several years later, a trail associated with 4P/Faye was accidentally detected in the visible by the Spacewatch Survey (Rabinowitz and Scotti, 1991). More recently a program to observe trails from the ground at visual wavelengths has been successful. Ishiguro et al. observed the trails of 22P/Kopff (Ishiguro et al. 2002), of 81P/Wild

Table 1. Cometary dust trail information from Sykes and Walker 1992. This includes the observed angular extent in mean anomaly, θ; the width, W; emission velocity perpendicular to orbit plane, Δv_p; an estimate of the age of the oldest emissions observed.

Name	θ (deg.)	W (10^3 km)	Δv_p (m/s)	Age (years)
67P/Churyumov-Gerasimenko	1	50	2	3–11
2P/Encke	93	680	40	19–21
65P/Gunn	6	111	3	27–74
22P/Kopff	17	47	3	47–158
7P/Pons-Winnecke	3	40	3	6–21
29P/Schwassmann-Wachmann 1	10	769	5	114–148
9P/Tempel 1	7	68	4	14–38
10P/Tempel 2	65	31	2	140–665

2 (Ishiguro et al. 2003), and of C-G (personal communication), Lowry et al. (2003) observed the 2P/Encke trail, and Reach et al. (2003) the C-G trail.

From the IRAS observations of the C-G trail in 1983 at 2.6 AU heliocentric distance Sykes and Walker (1992) determined a length of 1.2 degrees in mean anomaly and a width of about 50000 km corresponding to a normal emission speed of 2 m/s. From the length these authors estimate a trail age of 3 to 11 years. A trail length of several 10 arcmin on the sky has been reported by Reach et al. (2003) from visible observations in May/June 2003.

3.   The model

For an observation time in March 2003 we have calculated positions of big particles of various sizes (radius 0.1, 1, and 10 mm) that were released near perihelion passage during previous apparitions of comet C-G (1963 to 2002). For the present apparition (2001-2004) we have calculated additionally positions of particles in the size range from 0.5 to 10 microns. We have assumed that particles were emitted form the nucleus in different directions: towards the sun, and in 4 directions perpendicular to the sun direction, i.e. north, south, apex (direction of motion of the comet), and anti-apex directions. For the emission speed we have used the values calculated with the C-G Environment Model (Agarwal et al., 2004).

The size dependent emission speeds have been calculated taking into account gas drag consistent with gas production rates determined by astronomical observations at various heliocentric distances. In order to estimate the maximum spatial volume that big particles can occupy, we have assumed that particles emitted in all directions acquire the same speed v_{em} (Table 2). For the determination of this speed Agarwal et al. (2004) assumed that all sun lit nuclear

Table 2. Particle radius, radiation pressure constant β, and perihelion emission speed (Agarwal et al. 2004). We assumed a density 10^3 kg/m^3.

s (mm)	0.0005	0.001	0.002	0.0033	0.005	0.01	0.1	1	10
β		2.0	1.0	0.5	0.303	0.2	0.1	0.01	0.001	0.0001
v_{em} (m/s)		255	191	140	116	95	68	22	7.1	2.0

surface emits gas proportional to the radiation received from the sun. In case the active surface area is smaller, higher dust speeds may be achieved above these active areas, but lower speeds elsewhere. After the particles have been released from the comet we calculate their positions for the respective observation date taking into account solar gravity and radiation pressure. Radiation pressure has been estimated by the radiation pressure constant $\beta = F_{rad}/F_{grav}$ (Table 2).

In order to derive a more realistic estimate of the spatial extent that particles occupy we have taken into account variable speeds for different emission directions and various heliocentric distances. These were determined by the hydrodynamic model and consistent with the observed gas production rates. These emission speeds are mostly significantly lower (Fig. 1) than the maximum ones that are reached only at perihelion and consequently particle positions stay closer to the nucleus. We consider these conditions the "nominal" case. Nevertheless the initially calculated particle positions are extreme cases which may represent very inhomogeneous gas and dust emission which we will call the "maximum" case.

It can be seen in Fig. 1 that mm-sized and bigger particles are not released at large heliocentric distances. This is due to the fact that the gas drag force on the particles is smaller than the gravitational attraction from the nucleus — such particles will not leave the vicinity of the nucleus. Under nominal conditions particles of 1 mm radius will not be emitted outside 200 days from perihelion passage and particles of 10 mm radius will not be emitted outside 60 days from perihelion passage.

In order to demonstrate the different behavior of small (micron-sized) versus big (mm-sized) particles we show the positions of all particles considered in an ecliptic view. In Fig. 2 dust positions for an observation time of 24 March 2003 are shown. The positions of micron-sized particles ($\leq$ 10 microns radius) span the dust tail outside the orbit of the nucleus which is characterized by synchrones (solid lines connecting positions of particles released at the same time) and syndynes (dashed lines connecting positions of particles that experience the same radiation pressure force). 0.1 mm-sized and bigger parti-

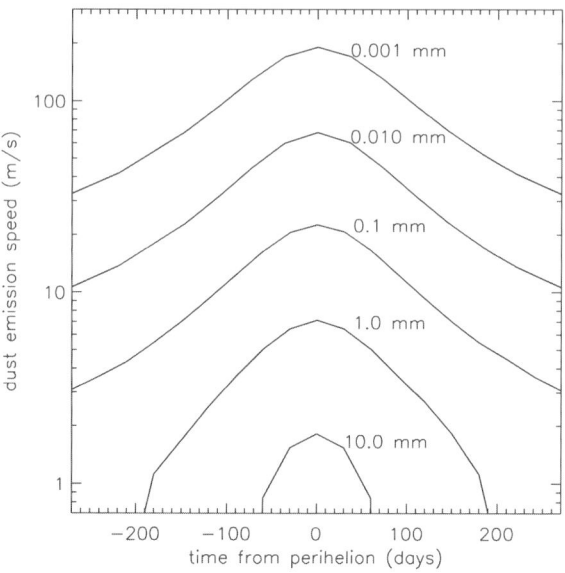

Figure 1. Sunward dust emission speed as function of time from perihelion.

cles delineate closely the orbit of the comet with mm-sized and bigger particles being still in vicinity of the nucleus.

There is another difference between small and big cometary particles. Small particles in the dust tail comprise only particles of this apparition, the same particles from previous apparitions are no more anywhere near positions in the tail and most of them have even left the solar system. Because of the low inclination ($i = 7.1$ deg.) of the orbit of C-G for an observer in the ecliptic plane the tail of this comet will appear only very narrow (bottom of Fig. 2) and both tail and trail will merge. In the following section we will have a closer look at the distribution of big particles in the vicinity of the nucleus as seen from Earth.

4. Distribution of big particles

Observations of comet trails show that they can be found both in front of and behind the nucleus. The first question we ask is: what is the size of particles that lead the nucleus and what sized particles are trailing it? In order to answer this question we consider two particles that are released from the nucleus at perihelion. One particle is emitted at the speed given in Table 2 in the apex direction and the other one in the anti-apex direction. The first particle will initially lead the nucleus but because of its higher speed and because of the action

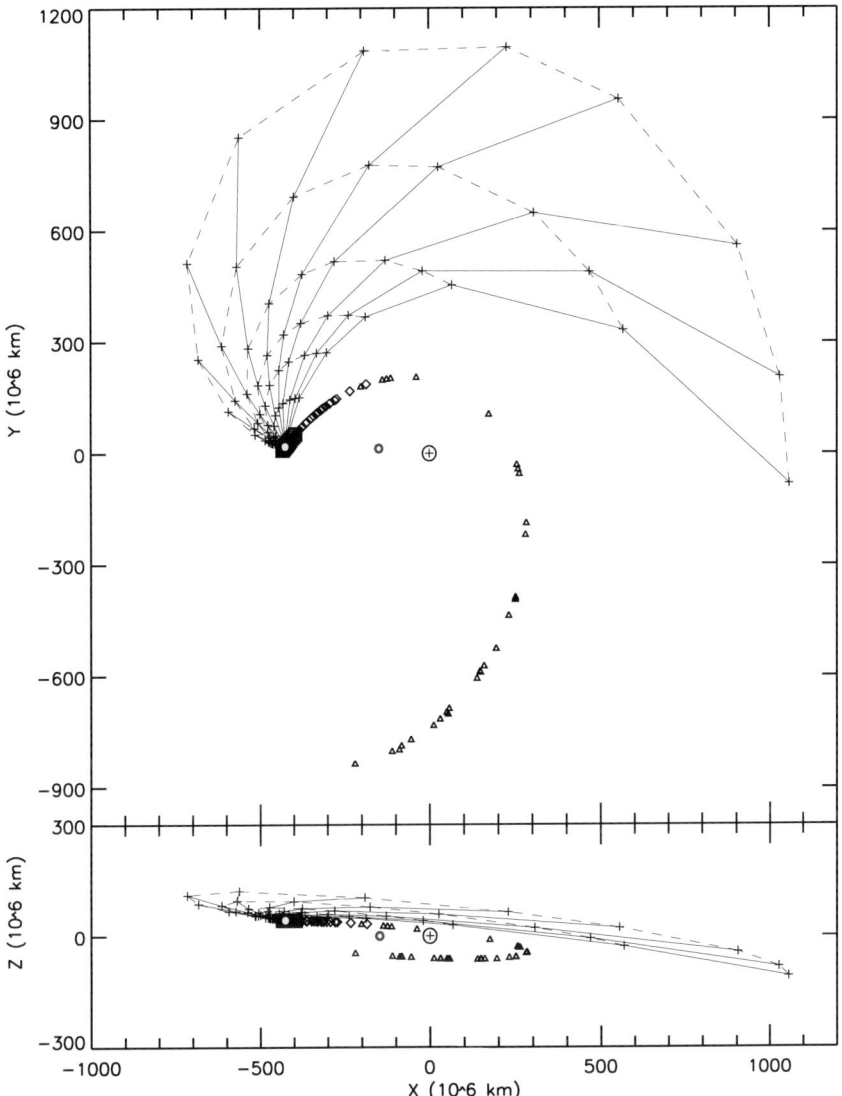

Figure 2.   Positions in ecliptic coordinates of dust particles emitted from C-G observed on March 24, 2003 when the comet was at 2.60 AU from the sun and 1.62 AU from the Earth. Dust tail of micron sized particles (0.5, 1, 2, 3.3, 5, and 10 microns, crosses) and the trail of big particles (0.1, triangles; 1, diamonds; 10 mm, squares) which outlines the orbit of comet. Trail particles were released during seven last perihelion passages, tail particles only during current apparition — others are lost. Synchrones are solid lines, syndynes are dashed, Sun is big circle and Earth is small circle.

192

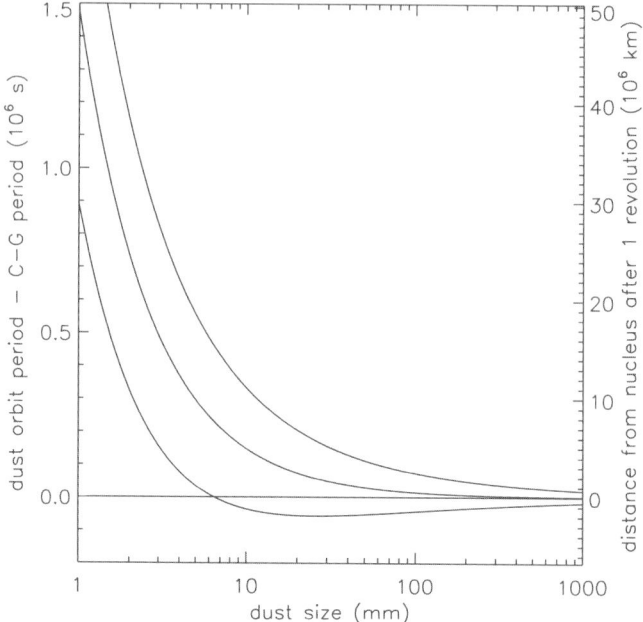

Figure 3.   Orbital periods of particles relative to the orbital period of C-G. Particles of different sizes were emitted at perihelion in apex (upper curve), anti-apex (lower) direction with max emission speed and at zero emission speed (middle). The right scale gives the particle distance from the nucleus along the orbit after one revolution.

of radiation pressure its aphelion distance will increase. Therefore, the orbital period of the particle will be longer than that of the comet. As a consequence after one revolution it will be behind the nucleus. The particle emitted in anti-apex direction will get on an orbit with reduced orbital energy and can after one revolution reach perihelion before the comet. Whether this is true depends on the magnitude of radiation pressure affecting the particle. When we assume the size dependent β-value and the emission speed as indicated in Table 2, we obtain the dependence of the orbital period as shown in Fig. 3. Only particles bigger than 7 mm can be found in front of the comet at the assumed conditions.

Our second question pertains to the trail thickness. Trail observations by Sykes and Walker (1992) report a width of 5×10^4 km for the C-G trail. In order to calculate a trail thickness we assume again that dust emission with speed v_{em} takes place at perihelion, however, this time perpendicular to the orbit plane. At this condition the orbit plane of the dust particle will be slightly inclined to the orbit plane of the comet. The height of the trail at maximum distance from the line of apsides is given by

$$h_{trail} = 2a_d \cdot v_{em}/v_{peri} \cdot \sqrt{1 - e_d^2}, \tag{1}$$

where v_{peri} is the comet's speed at perihelion. For small beta values (which is the case for mm-sized and bigger particles) the semi-major axis a_d and eccentricity e_d of the dust particle can be approximated by the orbital elements of the parent comet. For 0.1, 1, and 10 mm particles the trail widths are 5×10^5 km, 1.3×10^5 km, and 4×10^4 km, respectively. These values are compatible with the trail width observed by IRAS which suggests that this sized particles contribute most to the C-G trail.

The third question relates to the distribution around the nucleus of mm- and cm-sized particles as function of age. In order to study the spatial distribution of old particles that were emitted during previous perihelion passages in contrast to that of particles that were emitted during this apparition we have calculated the positions of particles for an observation time of 24 March 2003. This date refers to a period during which several recent ground based observations of the coma, tail, and trail of C-G have been performed.

We calculated the positions relative to the Earth. A result is shown in Fig. 4 for the maximum model (left) and for the nominal model (right). In three panels we have highlighted positions of differently sized particles (0.1, 1, and 10 mm). Old particles (from previous apparitions) are located in the gray shaded band. Only cm-sized old particles are still close enough to the nucleus in order to be seen in the image section shown. Smaller old particles are lagging further behind (cf. Fig. 3). While for the maximum model old cm-sized particles are found in front and behind the nucleus in the nominal model the closest old cm-sized particles are more than 20 arcmin behind the comet. Only even bigger particles are in front of the nucleus. This is due to a reduced emission speed (by about a factor 3) in directions that are 90 degrees from the sun direction in the nominal model.

Big particles that were emitted during the current apparition are still close to the nucleus, e.g. cm-sized particles are still within 10 arcmin, smaller particles are distributed over a much wider extent, especially in the tail-ward direction (to the right). In the nominal model big particles stay much closer to the orbit of the comet because the emission speed at right angle to the sun direction (apex, anti-apex, north, and south) is significantly reduced. Minimum vertical distance from the comet's orbit plane have particles that were released 180 degrees in true anomaly from the comet's position at observation time, i.e. about 100 days before perihelion passage in August 2002. This is the case for particles of 0.1, 1, 10 mm in size that are located at 140 (outside the image), 14, 1.4 arcmin from the nucleus, respectively (cf. Fig. 4). This phenomenon is called neckline (Fulle et al., 2004). The vertical extent of particles from this apparition is much bigger for the maximum model than for the nominal model.

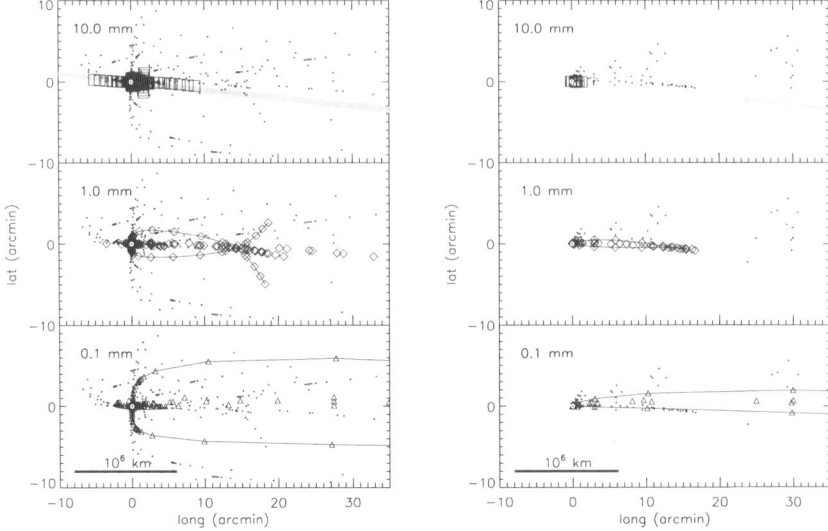

Figure 4. Positions of different sized particles in the vicinity of the nucleus (at the origin) on 24 March 2003 as seen from the Earth. On the left side positions are shown that were calculated with the maximum model, while on the right side positions were calculated with the nominal model. In three panels, particles of a specific size are highlighted by specific symbols, also positions of the other particles and those of micron-sized particles are shown. The gray diagonal band is the region where particles of the respective size can be found that were emitted during previous apparitions. Lines show the maximum vertical extent of the respective sized particles.

These results show that mm-sized and bigger particles are distributed over a range of several tens of arcmin from the nucleus, which agrees well with the observed trail lengths. These mm-sized particles were emitted around the last perihelion passage some 200 day before the observation.

In order to support planned trail observations we have calculated the expected trail brightness in April 2004 when the comet is at 4.66 AU from the sun and 3.68 AU from the Earth. Based on the Environment Model of C-G (Agarwal et al., 2004) and using the size distribution function measured by the Vega-2 spacecraft at comet 1P/Halley (Divine and Newburn, 1987), the total production of particles bigger than 1 mm during a single apparition is estimated to be 1.7×10^{10} kg, corresponding to a total scattering area of 7.4×10^9 m^2. Assuming that the scattering area is homogeneously distributed over the trail extent given above (Fig. 3) and using the orbital geometry of April 2004 and a geometric albedo of 4%, we estimate a trail brightness of 28.5 mag/arcsec^2. This figure may represent a lower limit because only dust of one apparition has been taken into account. Using the same geometry and albedo and a nucleus radius of 1980 m (Lamy et al. 2003), we obtain a nuclear brightness of 23.1 mag.

In spring 2004 the activity of C-G will be very weak and hence the emission rate of small particles will be very small, which further completes the separation of large particles from small ones. This allows us to study of the amount of large grains emitted by the comet. Such an investigation will contribute to the determination of size distribution function of dust from C-G. This function is an important and controversial parameter of all models used to predict the state of the comet environment.

Acknowledgments

J. A. is grateful to the Rosetta project at ESA/ESTEC for bearing the expenses of attending the Capri Rosetta Workshop.

References

Agarwal J., Müller M., and Grün E. (2004). Modelling the environment of 67P/Churyumov-Gerasimenko. *This volume.*

Beisser, K. (1989). Dynamics and Structures of Cometary Dust Tails. *Reviews in Modern Astronomy*, 2:221–228.

Boehnhardt, H., Stüwe, J., and Schulz, R. (2003). Dust coma and tail structures in comet 67/CG. *Presentation at workshop "The NEW Rosetta targets - observations, simulations and instrument performances"*, Capri, 13-15 Oct. 2003.

Davies, J.K., Sykes, M.V., Reach, W.T., Boulanger, F., Sibille, F., and Cesarsky, C.J. (1997). ISOCAM observations of the comet P/Kopff dust trail. *Icarus*, 127:251–254.

Divine, N. and Newburn, R.L. (1987). Modeling Halley before and after the encounters. *A&A*, 187:867–872.

Fulle M., Barbieri C., Cremonese G., Rauer H., Weiler M., Milani A., Ligustri, R. (2004). 67P dust environment: a challenge for the Rosetta probe? *A&A.*, submitted.

Ishiguro M. (2003). *Private communication.*

Ishiguro, M., Watanabe J., Usui F., Tanigawa T., Kinoshita D., Suzuki J., Nakamura R., Ueno M., and Mukai T. (2002). First detection of an optical dust trail along the orbit of 22P/Kopf. *Astrophys. J.*, 572:L117-L120.

Ishiguro, M., Kwon S. M., Sarugaku Y., Usui F., Nishiura S., Nakada Y., and Yano H. (2003). Discovery of the Dust Trail of the Stardust Comet Sample Return Mission Target: 81P/Wild 2. *Astrophys. J*, 589:L101-L104.

Lamy P., Toth I., Weaver H., Jorda L., Kaasaleinen M. (2003). The nucleus of comet 67P/Churyumov-Gerasimenko, the new target of the Rosetta mission. *BAAS*, 35:970.

Lamy, P.L., Toth, I., Weaver, H., Jorda, L., and Kaasalainen, M. (2003). The Nucleus of Comet 67P/Churyumov-Gerasimenko, the New Target of the Rosetta Mission. *AAS/Division for Planetary Sciences Meeting*, 35.

Lowry, S.C., Weissman, P.R., Sykes, M.V., and Reach, W.T. (2003). Observations of periodic comet 2P/Encke: Properties of the nucleus and first visual-wavelength detection of its dust trail. *LPSC*, 34:2056.

Marsden B.G. (1970). Reports on the Progress of Astronomy: Comets. *Quarterly Journal of the Royal Astronomical Society*, 11:221.

McDonnell, J.A.M.; Lamy, P.L.; Pankiewicz, G.S. (1991). Physical properties of cometary dust. *Comets in the post-Halley era.* (eds. Newburn R.L., Neugebauer, M., Rahe J.), 2:1043-1073.

Rabinowitz, D., and Scotti J. (1991). Periodic comet Faye (1991n). *IAUC*, 5366.

Reach, W.T., Sykes M.V., Lien D., and Davies J.K. (2000). The formation of Encke meteoroids and dust trail. *Icarus*, 148:80–94.

Reach W. T., Hicks M. D., Gillam S., Bhattacharya B., Kelly M. S., Sykes M. V. (2003). The debris trail and near-nucleus dust environment of the ROSETTA mission target 67P/Churyumov-Gerasimenko. *BAAS*, 35:970.

Sykes M.V., and Walker R.G. (1992). Cometary dust trails I. Survey. *Icarus*, 95:180–210.

THE SOLAR WIND REMOVAL OF VOLATILES FROM COMETS – PERSPECTIVES ON THE ROSETTA MISSION

R. Lundin and H. Nilsson
Swedish Institute of Space Physics, IRF, Box 812, S–981 28 Kiruna, Sweden

Abstract   The Rosetta–mission is a mission on cosmogony – the evolution of celestial bodies in the solar system. One aspect of this is the solar wind interaction with small bodies approaching the inner solar system, such as comets. A comet will be subject to strong forcing by the Sun near perihelion. Solar radiation, electromagnetic as well as corpuscular, creates an extensive atmosphere and subsequently a removal of matter at a rate kilotons per day for an average–sized comet. The visible evidence for the erosion of matter from a comet is the long tail, specifically the plasma tail (ion tail). The outflow in the plasma tail is considered driven by a combined action of different plasma acceleration process, denoted nonthermal escape.

Nonthermal escape is an example of a plasma process leading to the loss of matter from celestial bodies in the solar system. However, plasma forcing may also govern the agglomeration of matter onto a celestial object, as first suggested by Alfvén and Arrhenius (1974). The balance of forces determining source and loss and the prospects of the plasma consortium to provide measurements that lead to a better understanding of the cometary loss and agglomeration of matter will be discussed in this report.

Keywords: Solar wind, cometary atmosphere, plasma escape

1.   Introduction

Biermann proposed in 1951, before the space age, that the cometary tail is not caused by solar radiation pressure as first suggested by Johannes Kepler, but rather by corpuscular radiation from the Sun. This represents the first indirect proof for the existence of a solar wind in interplanetary space. After the *in–situ* discovery of the solar wind, solar–planetary relations became an important issue in space plasma physics.

The cometary tail is the signature of comets. In fact the word itself, comet, originates from the greek *"Aster cometes"*, *"hairy stars"*. Hairyness is indeed characterising cometary tails, in particular the plasma tail. Comets have two

197

tails, a dust tail and a plasma tail, the latter sometimes referred to as the "ion tail". Ion tail is in part a misnomer because the tail contains an equal amount of electrons and ions. Hence, plasma tail is more appropriate. The plasma tail of a comet has many commonalities with the plasma tail of other non-magnetized bodies in the solar system such as the magnetotails of Mars and Venus. The main difference is that cometary tails are denser and visible, while the magnetotails of Mars and Venus are tenuous and invisible. The plasma tail represents the "windsock" of a *magnetosphere*, the cellular structure that characterises volatile rich celestial objects, with or without an intrinsic magnetic field. The strength of the intrinsic magnetic field, the atmospheric extension, and the strength of the solar wind forcing are factors that define the overall size of the magnetosphere. In fact, the notion magnetosphere is applicable to all volatile–rich and/or magnetized celestial objects subject to external plasma forcing (stellar, interstellar and intergalactic).

The distinguishing difference between the plasma tail and the dust tail of a comet is that the dust tail comprises debris and neutral gas from the comet nucleus while the plasma tail is either a mixture of neutral gas and plasma, or is simply dominated by plasma of comet origin. While the dust tail is primarily subject to radiative and gravitational processes, the plasma tail is dominated by plasma processes. Besides direct *in-situ* measurement proofs (e.g. Johnstone et al., 1986) the filamentary structure of cometary plasma tails (Fig. 1) are typical signatures of magnetized plasmas, including the plasma near non-magnetized planets such as Venus and Mars (e.g. Brace et al., 1987; Lundin and Dubinin, 1992). In fact the strong ionization of tenuous (cometary) atmospheres near the Sun, leading to an ionization rate of $> 10^{-6}$ (comparable to the E–region in the Earth's ionosphere), is sufficient for a gas to become strongly affected by electric and magnetic forces.

The comet tail, specifically the plasma tail, is the signature of outflow of matter being lost from the comet. The plasma outflow reach high speed, eventually the speed of ambient solar wind plasma (typically $\approx$300-900 km/s). The processes accelerating the initially cold (T< 1 eV) cometary plasma is similar to those acting near Mars, i.e. plasma/ion pickup (e.g. Luhman and Schwingenschuh, 1990) and mass–loaded pickup processes (e.g. Lundin et al., 1991). The outflow from Hyakutake, a comet dominated by the plasma tail, is depicted in Fig. 1. Hyakutake displays typical properties of a plasma tail: narrow jets/beams, wavy (Kelvin–Helmholtz) structures and disconnection events induced by the temporal–spatial characteristics of the solar wind. Therefore, beside the general hairiness of comets, the tail characterizes strong variability. The variability depends on local conditions in the coma of a comet and on the general properties of the solar wind (magnetic field, pressure, plasma density and velocity etc).

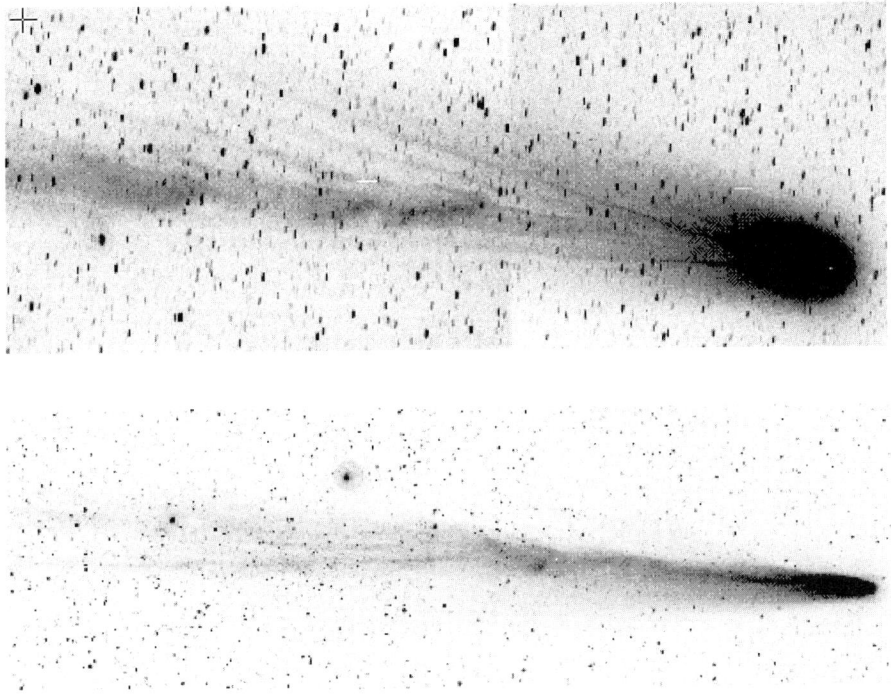

Figure 1. Outflow structures from the long tail of comet Hyakutake: narrow beams/jets, Kelvin–Helmholtz waves and disconnection events. (Courtesy Herman Mikuz, Crni Vrh Observatory, Slovenia)

A textbook example of a comet with a distinguishable dust tail and a plasma tail, Hale–Bopp, is shown in Figure 2. The close–up view demonstrates that filamentary (hairy) properties may connect to the coma as well. Notice that filamentary structures may also be observed in the antisolar direction downstream of the dust tail. This may be interpreted as regions with increased ionization (due to solar UV or solar electrons) above the "critical" value (10^{-6}) and a corresponding direct interaction with the solar wind plasma leading to momentum transfer and the formation of plasma beams.

Figure 3 shows the diagrammatic picture of the magnetosphere of a comet near the Sun. The outflow of gas from the comet creates an expanded atmosphere and ionosphere leading to a large target for solar wind interaction. The diagram shows typical magnetospheric feautures expected from a volatile–rich non-magnetized celestial object, e.g. a bow shock, a magnetic flux pile–up region (cometopause) and an ionopause. Intense plasma acceleration and out-

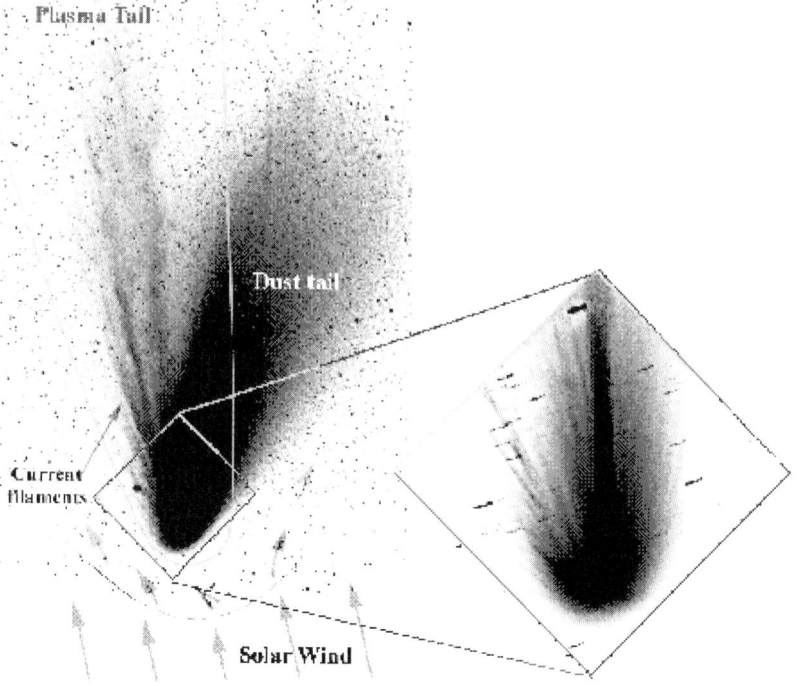

Figure 2. Hairiness is the characteristic of a magnetotail. Here the comet Hale–Bopp (adapted from H. Mikuz, Crni Vrh Observatory, Slovenia). The close–up view to the right (adapted from Tim Puckett, Puckett Observatory, Georgia) illustrates the hairiness down to the small–scale features

flow are expected to take place within a region accessible by impinging solar wind plasma, i.e. at least out to the cometopause.

The objective of the Rosetta Plasma Consortium (RPC) investigation is to perform plasma measurements near the prime target of the Rosetta mission, comet 67P/Churyumov-Gerasimenko (67P/C–G). The RPC instruments will measure hot and cold plasma, magnetic fields and waves (see Trotignon et al., 1999, for a review). The goal is to determine the physical properties of the solar wind interaction with a comet, to characterize the plasma environment and to estimate the net escape of matter (volatiles). In this report we discuss the general conditions for the outflow processes based on the direct momentum exchange between the solar wind and the ionized gas cloud surrounding a comet. We analyze and model the escape rate in the plasma tail from 67P/C-G based on recent estimates of the water escape from the cometary surface (Mäkinen, 2004). We conclude that the outflow in the plasma tail may match the net outflow of neutral gas from the surface inferred by remote means. The relatively

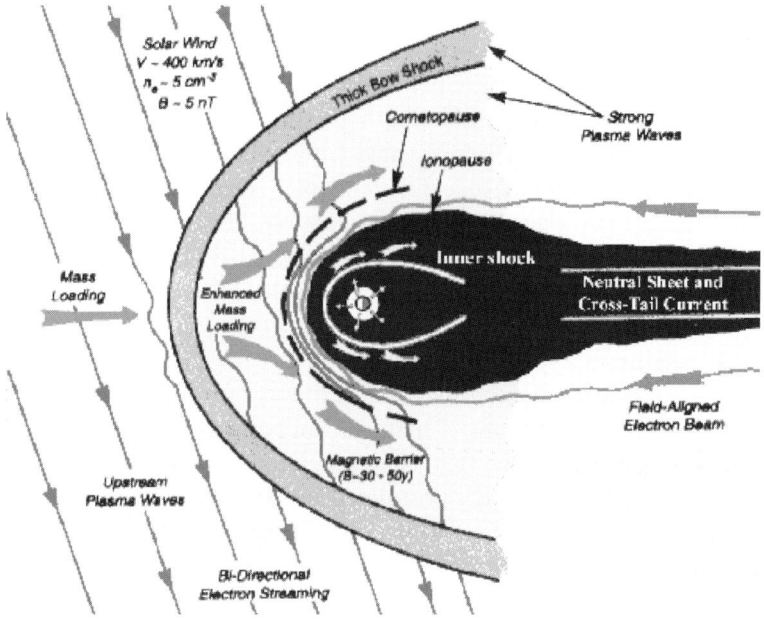

Figure 3.   Magnetosphere of a comet at ≈ 1 AU (Courtesy of D.A. Mendis)

low neutral gas density near the comet implies the build–up of an ionosphere extending down to the comet surface near perihelion. This suggests that plasma processes may prevail even inside the coma of the comet.

2.   Model of the solar wind impact on 67P/Churyumov-Gerasimenko

The solar impact on the surface and atmosphere of a comet depends on the solar irradiance and the solar corpuscular radiation (e.g. the solar wind). The solar effect is therefore essentially inversely proportional to the square of the distance to the Sun. The atmosphere of a comet results from solar heating as the comet approaches the Sun, becoming most dense and voluminous shortly after perihelion, after which it contracts and may condense again onto the surface. The volume of the atmosphere and correspondingly also the atmospheric loss is largest near perihelion. Two major atmospheric loss processes are expected to operate near a comet: (1) Thermal escape and (2) Non-thermal escape. Thermal escape results from solar heating of atmospheric particles. Particles acquiring velocities higher than escape velocity may eventually escape the gravitational binding. Since the escape velocity from a comet is typically in the m/s range, a thermalized gas may quite easily expand out from the weak gravitational binding field of a comet. However, the gas will still form a trail along the orbit unless the gas reaches a speed comparable to the orbital speed, i.e. typically tens of km/s. For a comet like 67P/C–G the velocity near perihelion is >30 km/s. To reach such high speeds require much higher gas temperatures ($> 10^5$ K) than conceivable from solar irradiance only. On the other hand, a combination of solar UV and solar wind impact ionization will create an ionized atmosphere which is susceptible to electromagnetic forcing by the solar wind. Electromagnetic forcing may easily cause acceleration to velocities well above hundreds of km/s, i.e. well above the orbit velocity of a comet. The escape caused by electromagnetic forcing is generally not associated with heating in a traditional sense, it is therefore denoted non–thermal escape. Thermal heating and thermal escape of matter from comets is certainly of great importance for the loss of matter from comets. However, the loss in the plasma tail is no doubt due to plasma acceleration processes driven by the solar wind interaction as already postulated by Biermann (1951) – i.e. non–thermal escape processes. Thermal heating is necessary for building up the cometary atmosphere, the target for the ionizing solar UV and the solar wind interaction. Solar wind forcing, the energy and momentum transfer from the solar wind plasma, provides the high speed escape.

We present a simple model of the non–thermal escape based on the thermal expansion of a cometary atmosphere and ionosphere interacting with the solar wind at a given distance, r, from the Sun. The atmosphere and ionosphere are created as a result of solar electromagnetic and corpuscular radiation, solar heating being the prime reason for the expanding cometary coma during the approach of the Sun. Solar UV–radiation combined with electron impact by the solar wind are responsible for the formation of the cometary ionosphere. However, direct momentum exchange by the fast flowing solar wind with the cometary ionosphere is the prime reason for the fast outflow in the plasma tail. The extent of the plasma outflow is estimated by the present model.

The production of H_2O for comet 67P/C–G near perihelion as determined by Mäkinen (2004), is $\approx 10^{28} s^{-1}$, i.e. ≈ 600 kg/s. Since most of the mass flux should be generated on the sun–facing side, we assume gas production over the area $A = 2\pi r_c^2$, where r_c is the comet radius. Assuming a 4 km diameter of 67P/C–G we obtain $A \approx 2.5 \cdot 10^7 m^2$. If the gas move upward with a velocity 50 m/s we obtain a number density (n_0) at surface level $n_0 \approx 8.0 \cdot 10^{18}$ particles/m^3. Considering a steady–state atmospheric density distribution we may write:

$$n(z,r) = n_0(r)exp(-\tfrac{z}{H(r)}); \; H(r) = \tfrac{kT(r)}{g_c m}$$

where z is the altitude, g_c the acceleration of gravity, $T(r)$ the atmospheric temperature, m the mean molecular mass, $H(r)$ the scale–height and r the distance from the sun. An estimate of the surface pressure may be obtained from: $P_0(r) = n(r) \cdot k \cdot T(r)$, but our main interest is here to determine the atmospheric number density, $n(r)$, and the scale height, $H(r)$.

Assuming the following values for 67P/C–G: $g_c \approx 1.7 \cdot 10^{-3} m/s^2$, T(1.3 AU) ≈ 230 K, $m \approx 18 m_p$, we obtain a scale height, $H(1.3$ AU) $\approx 6.2 \cdot 10^7 m$, i.e. much higher than the scale height of the Earth–like planets. Neglecting absorption and sublimation within the atmosphere the number density of the expanding neutral gas from the comet then follows the relation:

$$n(z) = n_0(r) \cdot (r_c/z)^2$$

where r_c is the comet radius (≈ 2 km) and n_0 the surface level number density.

As for the production of water versus heliocentric distance we assume that the H_2O production is determined by the radiative input from the Sun, i.e. the water production scales as r^{-2}, where r is the heliocentric distance. In reality the water production involves sublimation ($Q_{tot} = Q_{subl} - Q_{heat}$) and heat loss that increases with distance such that the water production drops off faster than r^{-2} (see e.g. Benkhoff, 1999) The following tentative relation for the H_2O production rate is used:

$$S_{H_2O} \approx 10^{28}(r_p/r)^2 \cdot m_W,$$

where r_p is the perihelion distance and m_W is the molecular mass of water. The atmospheric density versus height, z, for 67P/C–G at a heliocentric distance r is then determined by:

$$n(z) = n_0(r) \cdot (r_c/z)^2$$
where

$$n_0(r) = n_0(1.3 AU)$$

Figure 4 gives model atmospheric densities for comet 67P/C–G at heliospheric distances 1.3 AU and 2.5 AU, assuming gas temperatures of 250 K and 170 K respectively. In the diagram we have also provided a dual temperature atmospheric density model for Venus and Mars:

Venus: T($<$100km)=350 K; T($>$100 km)=2000 K;

Mars: T($<$100km)=150 K; T($>$100 km)=1500 K.

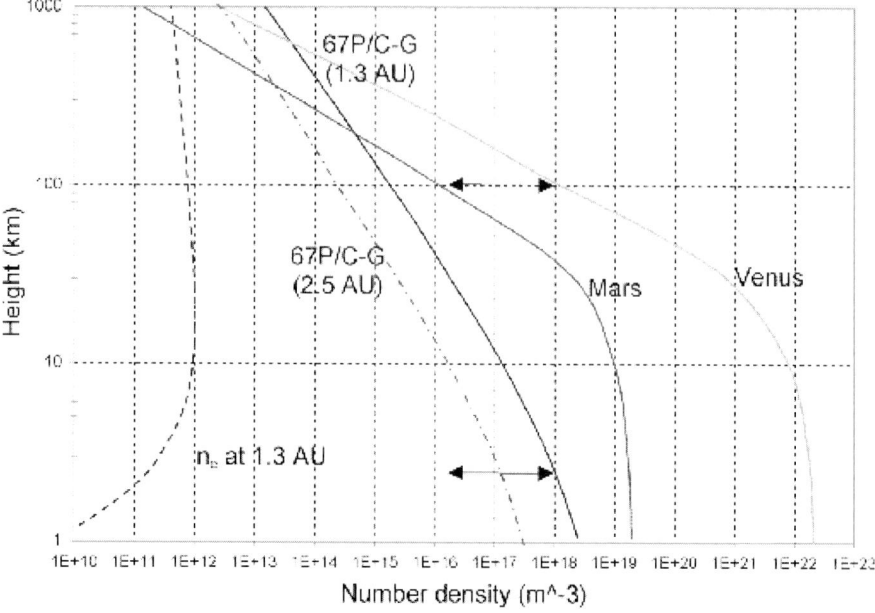

Figure 4. Model atmosphere density profiles for comet 67P/Churyumov–Gerasimenko at 1.3 AU and 2.5 AU, compared with the non–magnetized planets Venus and Mars. Arrows indicate the range of gas densities where ionization becomes important at Mars and Venus. Based on ionization rates similar to those for the Earth (and Mars and Venus) a model electron density profile is obtained for the comet near perihelion.

The atmosphere at Mars and Venus starts to ionize by solar UV and particle precipitation at an altitude of 90–100 km, developing an extensive ionosphere at even higher altitudes. A number density ratio between the neutral and the ionized gas of 10^{-7} is sufficient for electromagnetic forcing to become important for the gas ensemble. The same applies for comets. Assuming the same ionization rates as for the Earth (Venus and Mars) one gets an altitude profile for the ionosphere of comet 67P/C–G near perihelion as shown in Fig. 4. A more careful analysis of the ionosphere of comets by Körösmezey et al. (1987), indicates that the ionosphere electron density varies roughly as $1/z$ under both photochemical and transport controlled conditions as long as the transport velocity is constant. Notice that the low neutral gas density makes it possible for solar UV to maintain an ionosphere down to the surface of comet 67P/C–G. This implies that electromagnetic forcing is of significant importance for the dynamics of the atmosphere. Comparing with the Earth it seems likely that electromagnetic forcing is significant for the neutral gas of 67P/C–G above 10 km altitude. Plasma motions induced by solar wind electric fields will be transferred to the neutral gas by ion drag, leading to neutral winds (Rees, 1989). We therefore conclude that electromagnetic forcing and plasma effects should significantly influence the atmospheric dynamics of comet 67P/C–G.

The plasma tail is the visible result of plasma effects on a comet. The intrinsic filamentary structure of the tail is characteristic of plasma processes, the filamentation connected with electric currents and induced magnetic fields. In the same way as for Mars the plasma tail of a comet is expected to be dominated by energized plasma from the ionosphere (Balsiger et al., 1986; Korth et al., 1986; Johnstone et al., 1986), the energy and momentum in the energization process originating from the solar wind. We may then use a simple kinetic model for the solar wind momentum transfer (e.g. Pérez–de–Tejada, 1987, and Lundin and Dubinin, 1992) to describe the rate of energy and momentum transfer to a target ionosphere.

Fig. 5 illustrate how the total mass flux, S_i, incident on a target, A, of the solar wind mass flux, $\Phi_{SW} \cdot m_p$, can be obtained from (m_p is proton mass / average solar wind mass) :

$$S_i = A \cdot \Phi_{SW} \cdot m_p (kgs^{-1}) \qquad (1)$$

Notice that the target area is determined as a region where the solar wind impacts and interacts with ionospheric plasma. Using the analogy with Mars (Lundin and Dubinin, 1992), momentum exchange occurs predominantly inside the outer limit of heavy mass–loading, denoted the mass–loading boundary, MLB, and outside the magnetic pile–up boundary, MP. One may envisage at least two extremes, the area near the terminator, A, and an area taking into account the tail flaring of MP and MLB, A_{max}.

The energy and momentum exchange process can be described considering:

- the solar wind flux outside the interaction region ($\Phi_{SW} = n_{SW} \cdot v_{SW}$), where n denotes the number density and v the flow velocity),

- the solar wind flux inside the interaction region ($\Phi_{i,SW} = n_{i,SW} \cdot v_{i,SW}$), and

- the pick–up flux of ionospheric ions in the interaction region ($\Phi_C = n_C \cdot v_C$).

Under the assumption that the ionospheric ions are at rest before they interact with the solar wind ions in the interaction region, we obtain from the conservation of energy the following expression for the flux of ionospheric ions (same ion species):

$$\Phi_C = \frac{v_{SW} \cdot m_{SW}}{v_C \cdot m_C} \cdot \left(\Phi_{SW} - \frac{v_{i,SW}}{v_{SW}} \cdot \Phi_{i,SW} \right) \qquad (2)$$

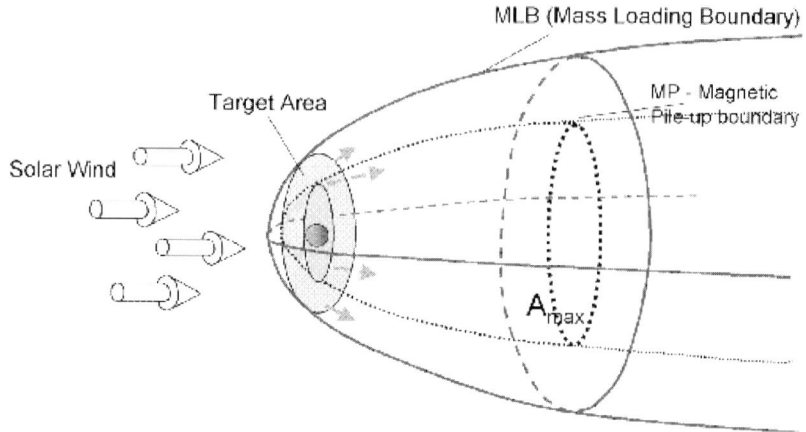

Figure 5.   Target areas for the solar wind energy and momentum transfer (see text)

where m_{SW} is the mass of the solar wind ions and m_C the mass of the iono-spheric ions.

A more accurate model involve an equivalent momentum exchange region thickness (δ_{SW} and δ_C) as discussed by Pérez–de–Tejada (1987). This leads to a correction term δ_{SW} / δ_C multiplied to the right–hand side of the above equation.

The net mass flux escape can now be obtained from:

$$S_C = A \cdot \Phi_C \cdot m_C (kg/s) \tag{3}$$

where m_C is the mass of the cometary ions. What remains to be determined is the distance to the boundaries (MP and MLB) determining the target area (A), the incident solar wind flux (mainly protons) and the pick-up flux of iono-spheric ions. We assume that there are no source problems, i.e. the loss in the target region is balanced by the upward transport of ionospheric plasma to the target region.

The target area in Fig. 4 is now given by $A = \pi(R_{MLB}^2 - R_{MP}^2)$.

R_{MP} and R_{MLB} may be estimated from the balance between the ionospheric and solar wind/sheath pressure (R_{MP}) and by scaling the pressure balance conditions of the MLB at Mars with that of comet 67P/C–G. This gives us $R_{MP} \approx 10^1 km$ and $R_{MBL} \approx 4 \cdot 10^3$ km. As for the comet ionosphere pressure it may be scaled as $(1/r_c)^2$ above 100 km altitude. This gives a total target area $A \approx 5 \cdot 10^{15} m^2$.

Using an average solar wind speed 400 km/s and a number density of $5 \cdot 10^6 (m^{-3})$ we may deduce the total solar wind mass flux impinging on the target area $A \approx 5 \cdot 10^{15} m^2$:

$$S_i = A \cdot \Phi_{SW} \cdot m_p \approx 33 kg/s$$

The production rate of water molecules ($m_W = 18 m_p$) from 67P/C–G near perihelion can be obtained from:

$$S_{H_2O} \approx 10^{28} \cdot m_W \approx 301 kg/s$$

which is ten times larger than the impacting solar wind mass flux on the comet target area. However, this is still sufficient in view of equation 2, leading to the model result of Fig 6. Moreover, target A corresponds to the terminator area. As Fig 5 shows the flaring angle of the MLB gives a larger target area further down the tail corresponding to a factor of 4 larger area (e.g. Lundin and Dubinin, 1992). This gives an effective target area of $A \approx 2 \cdot 10^{16} m^2$, i.e. an impinging solar wind flux corresponding to ≈ 130 kg/s. The upper curve of Fig. 6 shows the model for the tail effective area. Notice from Fig. 6 that the outflow, Φ_C, from the momentum exchange region depends on the average ion outflow velocity from the comet, v_C, the average ion outflow velocity ranging between escape velocity (≈ 2.6 m/s) and the solar wind velocity (400 km/s). Fig. 6 demonstrates an important aspect of the plasma momentum exchange process – the lower the outflow energy, the higher the mass loss rate. For instance, if the plasma outflow has about the same energy as the solar wind ($\approx$ 1 keV i.e. ≈ 100 km/s for O^+), the loss rate equals the production rate for the terminator area. For lower outflow energies solar wind erosion becomes very effective, i.e. the outflow may by far exceed the atmosphere production rate of a comet. This implies that it is the production rate of cometary ions that is the limiting factor rather than the solar wind momentum.

3. Discussion and Conclusion

The solar wind is important for the removal of volatiles from comets. The flyby of comet Halley (Korth et al., 1986; Johnstone et al., 1986) demonstrated that solar wind mass–loading results in the acceleration and subsequent removal of ions from the ionosphere of a comet. The removal process, non–thermal escape, comprises a multitude of heating and acceleration process energizing charged particles into the keV range. The energization is expected to be identical to that of a non–magnetized planet like Mars.
So far no spacecraft has provided adequate in–situ measurements to establish the rate of escape of ionospheric plasma from a comet. However, comparing with the ionospheric outflow from Mars (e.g. Lundin and Dubinin, 1992) we

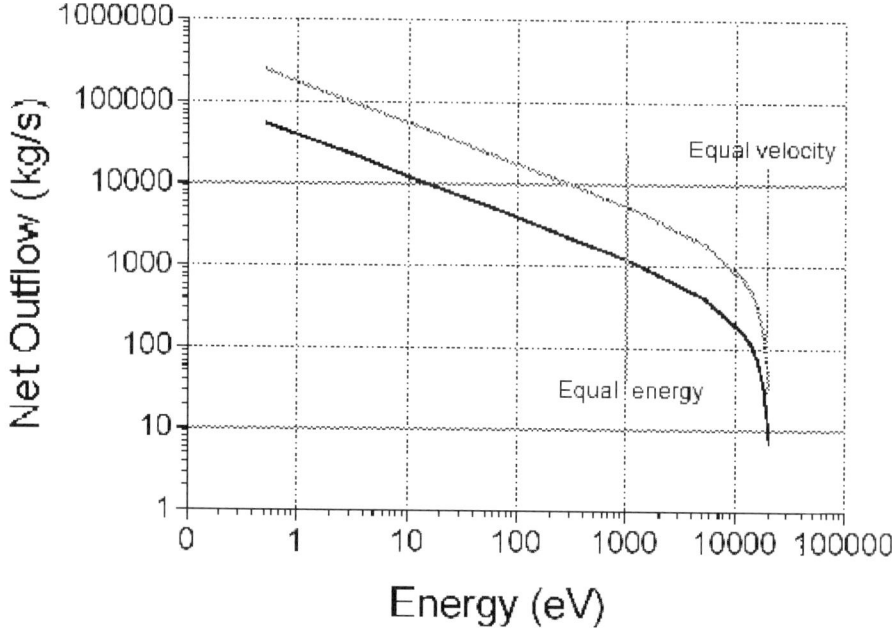

Figure 6. Theoretical mass loss from comet 67P/C–G due to the solar wind energy and momentum transfer process. Notice that the net outflow depends on the energy of the outflowing ion mass. Heavy mass–loading result in low energies (low velocities) and correspondingly higher outflow rates. The max and min outflow relates to the effective target area (Fig. 4).

here present a qualitative model of the escape from a comet such as 67P/C–G. Work on a refinement of the model is in progress, but it is not until real *in–situ* data exists that this or any other model can be confirmed or refuted. With Rosetta we expect to carry out detailed measurements of the ionospheric outflow from a comet.

Although the solar heat flux has a power about six orders of magnitude higher than that of the solar wind, we argue that non–thermal escape resulting from mass–loading of the solar wind by cometary ions is quite adequate for the removal of the volatiles produced by the solar wind heat flux on 67P/C–G. Moreover, the solar wind mass–loading process accelerate particles to high speeds, up to the solar wind speed (300–900 km/s) – i.e. an order of magnitude higher than the orbital velocity of a periodic comet and 3–4 orders of magnitude higher than the thermally induced outflow velocity of the neutral gas from a comet. The solar heat flux creates the atmosphere of a comet and the solar UV plus impact ionization of solar electrons are major reasons for establishing the ionosphere of a comet. In this way the erosion process, i.e. the removal of volatiles, depends on electromagnetic as well as corpuscular radiation from the Sun.

In summary, our study of the solar wind and solar radiation input to comet 67P/C-G suggests that:

- The solar wind is capable of accelerating and removing all water molecules produced by the solar radiation input on comet 67P/C–G.

- The atmosphere may be ionized by solar UV all the way down to the surface of comet 67P/C–G. This suggests that solar wind electric and magnetic forcing may strongly influence the dynamic of the comet atmosphere.

References

Alfvén, H. and Arrhenius G. (1974) Structure and Evolutionary History of the Solar System IV, *Astr. Space Sci.*, 29:63.

Balsiger H., Altwegg K., Bühler F, Geiss J., Ghielmetti A.G., Goldstein B.E., Goldstein R., Huntress W.T., Ip W.-H., Laxarus A.J., Meier A., Neugbauer M., Rettenmund U., Rosenbauer H., Schwenn R., Sharp R.D., Shelley E.G., Ungstrup E., Young D.T. (1986) Ion composition and dynamics at comet Halley, *Nature*, 321: 330–334.

Benkhoff, J. (1999). On the flux of water and minor volatiles from the surface of comet nuclei, *Space Science Review*, 90: 141–148.

Biermann, L.Z. (1951). Kometenschweife und solare Korpuskularstrahlung, Zeitschriftür Astrophysik, 29, 274.

Brace L.H., Kasprzak, W.T., Taylor, H.A., Theis, R.F., Russell, C.T., Barnes, A., Mihalov, J.D. and Hunten, D.M. (1987) The ionotail of Venus: Its configuration and evidence for ion escape. *J. Geophys. Res.*, 92: 15–26.

Johnstone, A. D., Coates A., Kellock S., Wilken B., Jockers K., Rosenbauer H., Stüdemann W., Weiss W., Formisano V., Amata E., Cerulli–Irelli R., Dubrowolny M., Terenzi R., Egidi A., Borg H., Hultqvist B., Winningham J., Gurgiolo C., Bryant D., Edwards T., Feldman W., Thomson M., Wallis M.K., Biermann L., Schmidt H., Lust R., Haerendel G., Paschmann G. (1986). Ion flow at comet Halley, *Nature*, 321: 344–347.

Korth A., Richter A.K., Loidl A., Anderson K.A., Carlson C.W., Curtis D.W., Lin R.P., Reme H., Sauvaud J.A., d'Uston C., Cotin F., Cros A., Mendis D.A. (1986). Mass spectra of heavy ions near comet Halley. *Nature*, 321: 335–336.

Körösmezey, A., Cravens T.E., Gombosi T.I., Nagy A.F., Mendis D.A., Szegö K., Gribov B.E., Sagdeev R.Z., Shapiro V.D., Shevchenko V.I. (1987). A new model of cometary ionospheres, *J. Geophys. Res.*, 92: 7331–7340.

Luhmann, J.G., Schwingenschuh K. (1990). A model of the energetic ion environment of Mars. *J.Geophys. Res.*, 95(A2): 939–945.

Lundin R., Norberg O., Dubinin E.M., Pisarenko N. and Koskinen H. (1991). On the momentum transfer of the solar wind to the Martian topside ionosphere. *Geophys. Res. Lett.*, 18: 1059–1062.

Lundin, R. and E.M. Dubinin (1992). Phobos-2 results on the ionospheric plasma escape from Mars. *Adv. Space Res.*, 12, 9: 255–263.

210

Mäkinen, T. (2004). Water production rate of comet 67P/Churyumov–Gerasimenko, *this volume*.

Pérez–de–Tejada, H. (1987). Plasma flow in the Mars magnetosphere, *J. Geophys. Res.*, 92: 4713–4718.

Rees, M.H. (1989). Physics and chemistry of the upper atmosphere, Cambridge University Press.

Trotignon, J.G., Boström R., Burch J.L., Glassmeier K.H., Lundin R., Norberg O., Balogh A., Szegö K., Musmann G., Coates A., Åhlén L., Carr C., Eriksson A., Gibson W., Kuhnke F., Lundin K., Michau J.L. and Szalai S. (1999). The Rosetta plasma consortium: Technical realization and scientific aims, *Adv. Space Res.*, 24: 1149–1158.

NON-RESONANT GRAVITATIONAL MOTION
AND EQUILIBRIUM STABILITY
OF ROSETTA

E. Mysen
Institute of Theoretical Astrophysics, University of Oslo
emysen@astro.uio.no

Abstract   First the mathematics of the two finite body problem is revisited, elaborated and used to derive the non-resonant gravitational component of Rosetta's motion around 67P/Churyumov-Gerasimenko. The classical expressions of pericenter and node precession are rediscovered, but now with a modifying excitation factor dependent on the complexity of the nucleus rotation state. Following non-resonant theory, high rotational excitation can lead to secularly stable orbits, but resonance and chaos are often encountered instead.

Above $6 - 7$ nucleus radii, neglecting probe-gas/dust interaction, radiation pressure is the main perturber of the probe's Keplerian motion at a rendezvous distance of $R = 3.6\,AU$. This type of motion has an equilibrium which is displaced by the comet's gravitational field close to the nucleus. Correcting for this perturbation does not in general decrease the probability of escape or impact for the cometary orbiter significantly, while using the true equilibrium values on the other hand, found by adding the short-periodic variations of the motion induced by radiation pressure, leads to such stability.

At last the success of the Rosetta gravity mapping campaign where the rotation state and the nucleus' detailed gravitational field is to be extrapolated, is predicted using non-resonant theory. Adopting a Gaussian shape for the comet, it is found that the amplitude of the second degree spherical harmonics in an expansion of the field are easily detectable over a time-scale similar to the comet rotation periods. However, only limited information is available on the third degree harmonics, and then only for close orbits ~ 5 radii stretching over typically a week. Chaotic and resonant trajectories are more sensitive to model parameters, and could therefore yield such low order information. On the other hand, Rosetta colliding with the nucleus is then possible.

1.   Introduction

One of the many scientific objectives of the Rosetta mission is the extraction of the detailed gravitational field and rotation state of the, most likely, irregularly shaped target through range- and Doppler- measurements on Rosetta

L. Colangeli et al. (eds.), The New ROSETTA Targets, 211–221.
© 2004 *Kluwer Academic Publishers. Printed in the Netherlands.*

(Pätzold et al., 2001). This is sought achieved by the Radio Science Investigations. The information will constrain the inner structuring of the comet and also reduce ambiguity in the interpretation of tracking data.

2. Model

The Keplerian motion of the probe in an orbit around a central body is disturbed by the body's finite dimension, leading to a perturbing acceleration which can be derived from an expansion of the body's gravitational field in normalized spherical harmonics $\bar{C}_{nm}$, $\bar{S}_{nm}$ (Heiskanen and Moritz, 1967)

$$U = \frac{\kappa^2 m_c}{r} \sum_{n=2}^{\infty} \sum_{m=0}^{n} \frac{1}{r^n} \left[\bar{c}_{nm} \bar{C}_{nm}(\lambda, \beta) + \bar{s}_{nm} \bar{S}_{nm}(\lambda, \beta) \right]. \qquad (1)$$

κ^2 is the gravitational constant, m_c the comet mass and r the probe's distance from the comet mass-center, while λ and β are the spacecraft's east longitude and north latitude in the comet nucleus' principal axis system, respectively. Another important perturber of the Keplerian spacecraft orbit is radiation pressure for which the secular motion has been analyzed in detail (Scheeres, 1999). As in Scheeres (1999), the area of Rosetta exposed to solar radiation is here assumed constant with a low mass to area ratio of $20 \, kg/m^2$. In addition, the spacecraft is affected by a gravitational pull from the Sun which is countered by an effective force equal in magnitude to the inertial acceleration of the nucleus, the net contribution also known as tidal acceleration.

To obtain specific values, the mass of the target comet 67P/C-G will be estimated. Using the approach of Marsden et al. (1972), the comet perturbing acceleration is decomposed into two components in the orbital plane $A_1 g(R)$ and $A_2 g(R)$ where the A_i's are solve-for parameters and $g(R)$ a sublimation function. If the post-perihelion water production rate $Q_{H_2O} \sim 1 \times 10^{28} \, mol./s$ is used together with the epoch 2002 perihelion distance $Q_c = 1.3 \, AU$ of 67P/C-G, its mass is given by

$$m_c = \frac{\zeta \times \eta \, v_{H_2O}(Q_c) \, Q_{H_2O} \, m_{H_2O}}{\sqrt{A_1^2 + A_2^2} \, g(Q_c)} = 1 \times 10^{13} \, kg \qquad (2)$$

Here the global $\zeta = 1$ and local $\eta = 0.5$ momentum transfer coefficients have been given a high value and a nominal value respectively. v_{H_2O} is the mean velocity of the escaping water vapour, molecular weight m_{H_2O}, at the nucleus surface evaluated at a typical temperature $T = 200 \, K$. For a nucleus with radius $r_c = 1.98 \, km$ the mass density is $\rho = 0.3 \, g/cm^3$, fairly at terms with other comet densities (Davidsson and Gutiérrez, 2004 and Rickman, 1989). A rotation period of $P_c = 12.3 \, h$ then equals the probe's Keplerian orbital period at a semi-major axis $a = 1.6 \, r_c$.

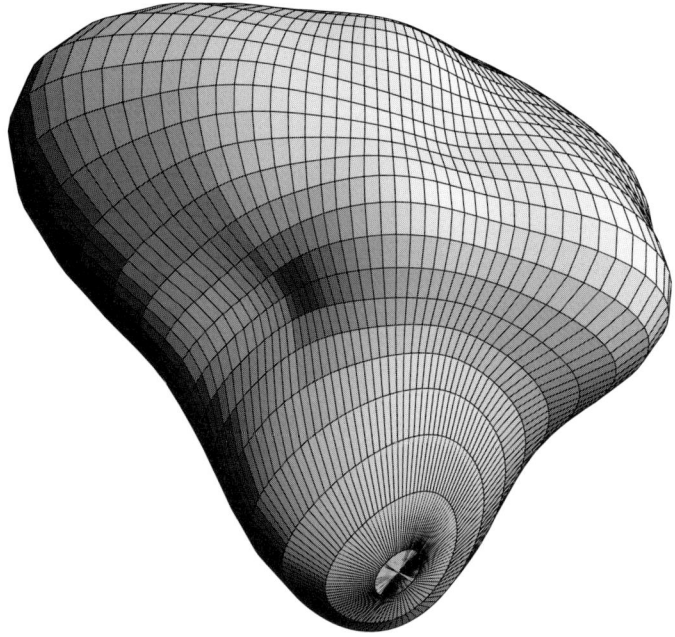

Figure 1. The Gaussian shape adopted as comet nucleus.

To generate a realistic irregular nucleus we use the Gaussian shape hypothesis (Muinonen, 1998): the surface of the body is given by a number of points for which the natural logarithm of the point radii follow multivariate normal statistics. A reasonable normalized radius standard deviation is $\sigma = 0.2$, opening for a $\sim 1 : 2$ axis ratio. Following Muinonen and Lagerros (1998) the surface irregularity parameter $\Gamma = 30°$ with a realization in Fig. 1, hereby adopted as the comet nucleus.

Now, assuming constant mass density and principal body axes, the normalized coefficients of the gravitational potential fail to converge, although the expansion itself converges outside the largest physical dimension of the body (Garmier and Barriot, 2001). The nucleus Fig. 1 can rotate in two different states, oblateness (short-axis mode, SAM) or prolateness (long-axis mode, LAM), each corresponding to a specific triaxiality parameter $e_c(A, B, C)$ (Kinoshita, 1972), a function of the comet's principal moments of inertia.

With SAM of Fig. 1, the first normalized harmonic coefficient is $\bar{c}_{20} = 0.046\, r_c^2$, and we can compare the accelerations of Rosetta not connected to outgassing, all in all giving the result of Fig. 2.

Clearly, neglecting the possibly strong gas/dust-probe interaction, the dynamics is generally a composite with radiation pressure as main perturber

214

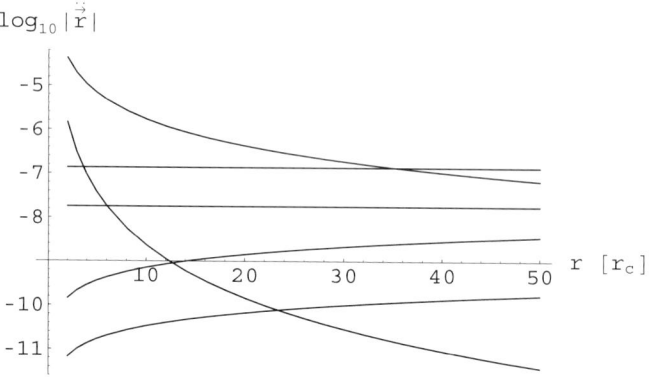

$\log_{10} |\ddot{\vec{r}}|$

$r\ [r_{\mathrm{c}}]$

Figure 2. Some of the accelerations $|\ddot{\vec{r}}|$ of Rosetta. The constant and increasing curves represent the radiation pressure and the tidal acceleration respectively. The upper curve for these effects are evaluated at $R = 1.3\,AU$, while the lower at $R = 3.6\,AU$. The gravitational keplerian and perturbative acceleration are given by the decreasing curves.

of the two-body problem for $r > 6 - 7\,r_c$ at a comet heliocentric distance $R = 3.6\,AU$. As can be seen, the tidal acceleration is weak, and only catches up with radiation pressure at about 2000 nucleus radii at $R = 1.3\,AU$.

3. Gravitational motion

To describe the nucleus and its rotation we use the canonical Andoyer variables, the impulses $p = (L_c, G_c, H_c)$ and coordinates $q = (l_c, g_c, h_c)$ (Kinoshita, 1972). The free rigid variation of (L_c, l_c, g_c), described by the Hamiltonian F_c, can be decomposed into two frequencies given by the time derivatives of the system's (action-) angle variables $\tilde{l}_c$ and $\tilde{g}_c$, namely $n_{\tilde{l}_c}$ and $n_{\tilde{g}_c}$. The Delauney variables are assumed to parameterize the orbit.

3.1 Averaging and regular dynamics

Consistent with first order perturbation theory (Hori, 1966), the secular motion of the Rosetta probe can be described by a set of averaged equations. The averaging is done in evaluating a new Hamiltonian $\bar{F}$

$$\bar{F} = \frac{1}{T} \int_0^T dt^* \, F \qquad (3)$$

from the total F, including its Keplerian part F_0 and all perturbations. t^* is Hori's pseudo-time defined by

$$\frac{dq}{dt^*} = \frac{\partial W}{\partial p}, \quad \frac{dp}{dt^*} = -\frac{\partial W}{\partial q}, \quad W = F_0 + F_c. \tag{4}$$

Assuming that the system is not in the vicinity of phase space regions where the free frequencies n, $n_{\tilde{g}_c}$ and $n_{\tilde{l}_c}$ for gravitational motion, and $n = \dot{l}$ and the derivative of the comet mean anomaly, N_c, for motion under radiation pressure, are close to certain rational multiples of each other (resonance), the integration (3) can be written

$$\bar{F} = \frac{1}{(2\pi)^3} \int_0^{2\pi} d\tilde{l}_c \int_0^{2\pi} dg_c \int_0^{2\pi} dl\, F. \tag{5}$$

The result is that the mean anomaly's (l) corresponding impulse, the orbital energy L, is conserved on average.

However, as can be verified by simulation of the full equations, the semi-major axis can change secularly both for motion under gravitational and radiative perturbation separately. These are so-called chaotic changes and occur in regions of phase space where the resonances mentioned above have fused, strongly violating the prerequisites for non-resonance.

If we for the time being restrict discussion to the gravitational potential, $F - F_0 - F_c = U$, its constituent parts given by the integrand of

$$\frac{1}{2\pi} \int_0^{2\pi} dg_c\, P_{nm}(\sin\beta) \exp[im\lambda]$$

$$= Q_n(u) P_{nm}(\cos J_c) \exp[-im(l_c - \pi/2)] \tag{6}$$

where the result can be shown from Kinoshita (1972) and Kinoshita et al. (1974). Here $\cos J_c = L_c/G_c$ and $u = g + f$ where f is the spacecraft's true anomaly, g the orbit's argument of pericenter and P_{nm} the associated Legendre function. The Q_n's are conveniently written $Q_n(u) =$

$$\sum_{m=0}^{n} P_{nm}(0) P_{nm}(\cos I) \frac{(n-m)!}{(n+m)!} (2-\delta_{0m}) \times \begin{cases} \cos(mu)\cos(m\pi/2), \\ \qquad\qquad n \text{ even} \\ \sin(mu)\sin(m\pi/2), \\ \qquad\qquad n \text{ odd} \end{cases} \tag{7}$$

if the reference-plane is defined to coincide with the spin-plane. Note that the orbit node h has already been removed, implying no secular drift of the orbital angular momentum component along the nucleus spin. I is the Rosetta orbit's inclination, $H \equiv G \cos I$.

Now, with the integrals of Byrd and Friedman (1971) and expressions for the principal axis spin components, it follows that

$$\int_0^{2\pi} d\tilde{l}_c \, P_{nm}(\cos J_c) \exp\left[-im(l_c - \pi/2)\right] \in \mathbf{R} \tag{8}$$

if m is even, otherwise the average is zero.

Writing

$$\frac{1}{2\pi} \int_0^{2\pi} d\tilde{l}_c \, P_{nm}(\cos J_c) \cos(m l_c) \equiv \psi_{nm} \tag{9}$$

we reach the simple result

$$\frac{1}{(2\pi)^2} \int_0^{2\pi} dg_c \int_0^{2\pi} d\tilde{l}_c \, U = \sum_{n=2}^{\infty} \frac{\kappa^2 \, m_c}{r^{n+1}} \, c_{n0} \, Q_n(u) \, \hat{\psi}_n \tag{10}$$

with the rotational excitation factor as

$$\hat{\psi}_n = \sum_{m=0}^{\leq n/2} (-1)^m \frac{c_{n(2m)}}{c_{n0}} \, \psi_{n(2m)}(k, e_c). \tag{11}$$

ψ_{20} and ψ_{22} can be inferred from Kinoshita (1972), while

$$\left\{ \begin{array}{c} \psi_{30} \\ \psi_{32} \end{array} \right\} = \sqrt{\frac{1 + \bar{b}e_c}{\bar{b}(1 + e_c)}} \frac{\pi}{K} \frac{1}{\bar{b}(-1 + e_c)(1 + e_c)} \left\{ \begin{array}{c} \frac{1}{4}(-5 + 3\bar{b} + 2\bar{b}e_c^2) \\ \frac{15\,e_c}{2}(1 - \bar{b}) \end{array} \right\}, \tag{12}$$

and so forth. K is the complete elliptic integral of the first kind and $\bar{b} = \bar{b}(F_c, G_c)$ (Kinoshita, 1972) is an excitation parameter, constrained by $1 \leq \bar{b} \leq 1/e_c$, which gives the elliptic modulus k

$$k^2 = \frac{2e_c}{1 - e_c} \frac{\bar{b} - 1}{1 + \bar{b}e_c}, \qquad \bar{b} = \frac{2e_c + k^2(1 - e_c)}{e_c \left[2 - k^2(1 - e_c)\right]}. \tag{13}$$

$k = 1$ means comet rotational motion on the separatrix separating LAM and SAM, while $\bar{b} = 1$ implies simple rotation around a principal axis.

It is easy to show that the Hamiltonian (10) with $\hat{\psi}_n = 1$ coincides with un-averaged perturbing functions for motion around principal axis rotating oblate bodies, motion where the period of rotation is of no relevance. In Fig. 3 the excitation factors for the low and high triaxiality case of Fig. 1 is plotted.

As can be shown, $\hat{\psi}_2$ is insensitive to other parameters than e_c and k, making the behaviour of Fig. 3 typical; the reduction in secular change of the orbit around highly triaxial shapes is only a fairly weak function of k, while at low triaxialities it is possible for the secular change to approach zero at high excitation. This property can, e.g., strengthen the stability of equilibrium points for other types of perturbation.

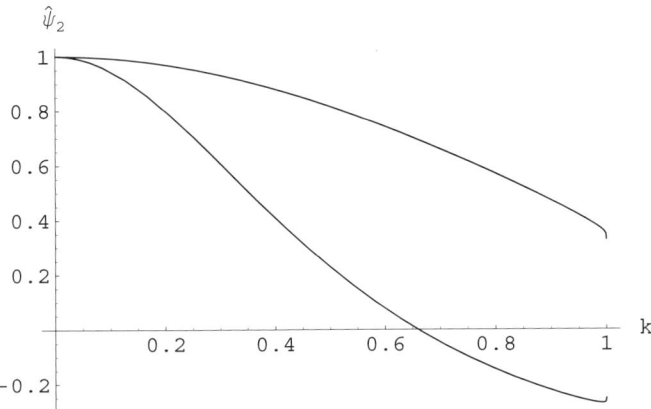

Figure 3. The second harmonic excitation factors $\hat{\psi}_2$ for Fig. 1 are plotted as functions of the rotational excitation. $e_c = 0.63$ for the upper curve, $e_c = 0.13$ for the lower.

Since g is removed from the second harmonic after averaging over the orbital true anomaly, the eccentricity is also conserved on average, but only approximately since g is contained in the averaged third harmonic.

It can easily be shown that gravitational resonance widths are zero for retrograde orbits around one-axis rotating comets. To what extent this is true for a complex rotating central body is not certain, but simulations indicate that retrograde resonances are narrow, even in the general case. Prograde orbits, on the other hand, rapidly deteriorate into resonance and chaos, large orbital energy alterations over short time-scales, with increasing excitation. Chaotic motion is highly sensitive to model parameters, and can therefore be used as a bold strategy for extrapolating the low order harmonics of the gravitational field. However, probe impact on 67P/C-G is then possible.

4. Radiation pressure equilibrium

According to Scheeres (1999), motion under radiation pressure has two equilibria (which are relevant here); orbits with mean angular momentum $\vec{G}$ pointing towards or away from the Sun. It can be shown that although correcting their positions, under the influence of non-resonant gravitational perturbation near the comet, stabilizes for instance the probe inclination and argument of pericenter, such corrections do not in general reduce the risk of probe escape or impact significantly. However, far from the nucleus, motion under radiation pressure has a large short-periodic amplitude, representing the deviation of the probe's actual (true) elements from its mean orbital elements. Since the equilibria's coordinates have only been given previously in mean values, their true values will be derived below so that the probe can be placed nearer to these stable orbits.

The non-resonant short-periodic components of the probe's Hill variables, $q = (r, u, h)$, $p = (\dot{r}, G, H)$ (Hill, 1913), can be found following Hori (1966). Let the reference plane now coincide with the comet's orbital plane so that that the X-axis points in the anti-solar direction at the comet's perihelion. We can then define the relative node $\lambda = h - \nu$ where ν is the cometocentric true anomaly of the anti-Sun direction. First we must find the generating function S

$$S = -\int df \, \frac{r^2}{G} \, \frac{\alpha}{R^2} \left[\cos \lambda (r \cos u + \frac{3}{2} ae \cos g) \right.$$
$$\left. - \sin \lambda \cos I (r \sin u + \frac{3}{2} ae \sin g) \right] \tag{14}$$

which is easily obtained if we expand r in true anomaly to first order in eccentricity, giving

$$\int df \, \frac{r^2}{G} \left(r \exp [iu] + \frac{3}{2} ae \exp [ig] \right)$$
$$= -\frac{i}{G} a^3 \left(\exp [iu] - \frac{3}{4} e \exp [i(u + f)] \right). \tag{15}$$

$\alpha = \frac{L_\odot}{4\pi c \chi}$ where $L_\odot$ is the solar luminosity, c the velocity of light and χ the mass to area factor previously mentioned. Expansions to first order are maintained due to the loss of an order in taking the partial derivatives

$$\delta q = \frac{\partial}{\partial p} S, \quad \delta p = -\frac{\partial}{\partial q} S \tag{16}$$

where

$$\frac{\partial f}{\partial(\dot{r}, r, G)}, \quad \frac{\partial e}{\partial(\dot{r}, r, G)} \tag{17}$$

are found in Aksnes (1970) yielding for $e = 0$

$$\delta \dot{r} = \frac{3}{4} a^{3/2} \left(\cos I \cos u \sin \lambda + \cos \lambda \sin u \right)$$
$$\delta G = a^{5/2} \left(\cos \lambda \cos u - \cos I \sin \lambda \sin u \right)$$
$$\delta H = a^{5/2} \left(\cos I \cos \lambda \cos u - \sin \lambda \sin u \right)$$
$$\delta r = \frac{3}{4} a^3 \left(\cos \lambda \cos u - \cos I \sin \lambda \sin u \right)$$
$$\delta u = \frac{1}{2} a^2 \left(7 \cos I \cos u \sin \lambda + 5 \cos \lambda \sin u \right)$$
$$\delta h = -a^2 \cos u \sin \lambda \tag{18}$$

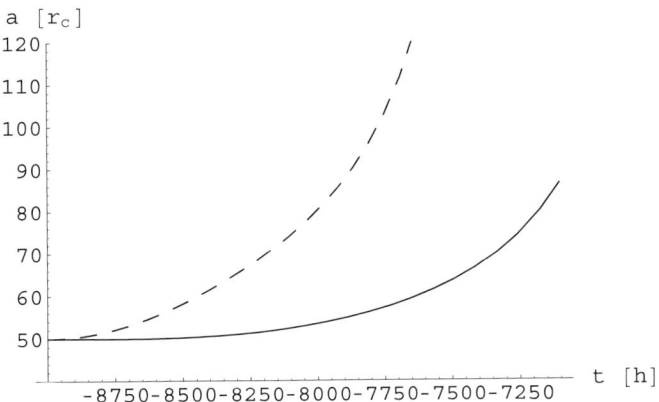

Figure 4. Motion from $R = 3.6\,AU$ under radiation pressure using true (solid) and mean (dashed) equilibrium. $t = 0$ at perihelion.

where each expression above is to be multiplied with α/R^2. The frozen orbits are likely to have low eccentricity, implying that the expressions above are quite representative.

So, if the mean true anomaly is assumed equal to zero, the stable orbit's true inclination is found by adding $\delta H = \alpha\, a^{5/2}/R^2$ to $\pi/2$. Using the true equilibria can lead to superior orbital stability as illustrated in Fig. 4, a simulation of the (chaotic) semi-major axis variation under radiation pressure is plotted. The stability of the orbit eccentricity is likewise improved. With the values in this work, the difference between mean and true equilibrium is $16°$ at $a = 50\,r_c$.

5. The Rosetta gravity mapping campaign

To evaluate the non-resonant variations in the probe velocity induced by the second harmonics, with frequency similar to those given by the nucleus rotation periods, first assume that we have the probe velocity V as a function of its cometocentric orbital elements. The mean amplitude δV of the velocity variations connected to the second harmonics can then be estimated with the use of basic perturbation theory ($\kappa^2 m_c \equiv 1$ and $r_c \equiv 1$)

$$\delta V \sim \frac{\partial V}{\partial u}\,\delta u \sim \frac{1}{a^{1/2}}\frac{\partial}{\partial G}\int df \frac{r^2}{G}\frac{\bar{c}_{20}}{a^3} \sim \frac{1}{a^{5/2}}\bar{c}_{20} = 0.087\,mm/s \quad (19)$$

at $a = 10\,r_c$, where here and below the values from the oblate nucleus Fig. 1 are used. The value is well above the instrumental Doppler noise $\sim 0.01\,mm/s$ at $1000\,s$ integration time for X-band measurements (Pätzold et al., 2001).

Similarly for the third harmonic

$$\delta V \sim \frac{1}{a^{7/2}} \, \bar{c}_{30} = 0.015 \, mm/s \tag{20}$$

at $a = 5 \, r_c$. So, to extrapolate the third harmonic one would have to rely more on its secular effects on the orbit. Assuming that the orbit is at least slightly eccentric, the signature of the third harmonic induced pericenter drift on the Doppler observable is measurable after a time Δt

$$\Delta V = \frac{\partial V}{\partial g} \, \dot{g} \, \Delta t \sim \frac{1}{a^5} \, \bar{c}_{30} \hat{\psi}_3 \, \Delta t \tag{21}$$

which equals the noise at $1000 \, s$ integration time after $\sim 10 \, h$ at $a = 5 \, r_c$ and after $\sim 230 \, h$ at $a = 10 \, r_c$. Of course the knowledge obtained in this way is of the product $\bar{c}_{30} \, \hat{\psi}_3$ only.

Acknowledgments

This work was prepared as part of project no. 153382/431, The Research Council of Norway. The author thanks Kaare Aksnes for his support.

References

Pätzold, M. et al. (2001). Gravity field determination of a Comet Nucleus: Rosetta at P/Wirtanen. *A&A*, 375:651-660.

Heiskanen, W.A. and Moritz, H. (1967). Physical Geodesy. W.H. Freeman and Company.

Scheeres, D.J. (1999). Satellite Dynamics about Small Bodies: Average Solar Radiation Pressure Effects. *The Journal of the Astronautical Sciences*, 47 (1-2):25-46.

Marsden, B.G., Sekanina, Z. and Yeomans, D.K. (1972). Comets and nongravitational forces. V. *AJ*, 78 (2):211-225.

Davidsson, B.J.R. and Gutiérrez, P., (2004). Estimating of the Nucleus Density of Comet 19P/ Borelly. *Icarus*, 168:392-408.

Rickman, H., (1989). The Nucleus of Comet Halley: Surface Structure, Mean Density and Dust Production. *Advances in Space Research*, 9 (3):59-71.

Muinonen, K., (1998). Introducing the Gaussian shape hypothesis for asteroids and comets. *A&A*, 332:1087-1098.

Muinonen, K. and Lagerros, J.S.V., (1998). Inversion of shape statistics for small solar system bodies. *A&A*, 333:753-761.

Garmier, R. and Barriot, J.-P., (2001). Ellipsoidal harmonic expansions of the gravitational potential: theory and application. *Celest. Mech.& Dyn. Astr.*, 79:235-275.

Kinoshita, H., (1972). First-Order Perturbations of the Two Finite Body Problem. *Publ. Astron. Soc. Japan*, 24:423-457.

Hori, G., (1966). Theory of General Perturbations with Unspecified Canonical Variables. *Publ. Astron. Soc. Japan*, 18 (4):287-296.

Kinoshita, H., Hori, G. and Nakai, H., (1974). Modified Jacobi polynomials and its applications to expansions of disturbing functions. *Annals of the Tokyo Astronomical Observatory*, Second series, Vol. XIV N.1.

Byrd, P.F. and Friedman, M.D., (1971). Handbook of Elliptic Integrals for Engineers and Scientists. Springer-Verlag.

Hill, G.W., (1913). Motion of a System of material Points under the Action of Gravitation. *AJ*, 27:171-182.

Aksnes, K., (1970). A Second-Order Artificial Satellite Theory Based on an Intermediate Orbit. *NASA Tech. Report*, 32-1507.

VIRTIS EXPERIMENT AT CHURYUMOV – GERASIMENKO COMET, NEW ROSETTA TARGET

A. Coradini, F. Capaccioni, G. Filacchione, G. Magni, E. Ammannito, M. T. Capria, G. Piccioni
Istitituto di Fisica dello Spazio Interplanetario – INAF
ARTOV, Roma, Italy
coradini@rm.iasf.cnr.it

Pierre Drossart
Observatorie de Paris, Meudon, Paris, France
pierre.drossart@obspm.fr

Gabriele Arnold
DLR, Institut fur Planetenerkundung, Berlin, Germany
gabriele.arnold@dlr.de

Abstract   The targets of the Rosetta mission are the most primitive solar system bodies: comets and asteroids. The previous Rosetta mission was essentially devoted to study in detail a comet nucleus and to fly by one or two asteroids. The new mission Rosetta is set to rendezvous with Comet 67P / Churyumov – Gerasimenko, orbiting around it and making observations as it journeys towards the Sun. En route to the comet it will flyby at least one asteroid. The study of small bodies is crucial to understand the solar system formation. In fact it is believed that comets and, to a lesser extent, asteroids underwent a moderate evolution so that they preserve some pristine solar system material. The scientific payload of Rosetta includes a Visual InfraRed Thermal Imaging Spectrometer (VIRTIS) among the instruments on board the spacecraft orbiting around the comet. This instrument is fundamental to detect and study the evolution of specific finger-prints – such as the typical spectral bands of minerals and molecules – arising from surface components and from materials dispersed in the coma. Their iden-tification is a primary goal of the Rosetta mission as it will allow us to identify the nature of the main constituents of the comets. Moreover, the surface thermal evolution during comet approach to Sun will provide essential information on the thermal and structural evolution of the comet nucleus. VIRTIS data will per-mit to identify as well the chemical evolution of the surface. The VIRTIS design

L. Colangeli et al. (eds.), The New ROSETTA Targets, 223–236.

and its detailed science goals will be discussed, as well as the new observation strategy and the on–ground calibrations.

Introduction

The delay in the launch of the Rosetta mission has obliged the Rosetta project to change the selected target. Comet 67P/Churyumov – Gerasimenko (hereafter C–G) is the current selected target for the international ROSETTA mission. Given the great variability of comets, the new target is profoundly different from the previous one. This has pushed the Rosetta experiment responsible persons (Principal Investigators, hereafter PIs) to re–evaluate the foreseen activity, and to re–plan the near comet observation strategy. Unfortunately little is known about C–G but some preliminary evaluation of the comet physical parameters, such as surface temperature, percentage of active surface, intensity of gas and dust fluxes has been already done on the basis of ground based observations as well as of theoretical models (Capria et al., 2004). Comet C–G was discovered on 1969 September 11, and has been observed at 6 apparitions, of which the latest was in 2002 – 2003. C–G is larger than 46P/Wirtanen, the initial target for which the ROSETTA mission was designed. Like P/Wirtanen, C–G is a Jupiter family comet with a perihelion distance of 1.29 AU and a period of 6.57 years; Tancredi et al., 2000 estimated a radius of 2.5 km for C–G, one of the largest sizes of Jupiter family comets. Radar observations (Kamoun et al., 1998) seem to indicate a even larger radius (upper limit of 3.7 km) consistent with the estimate of Tancredi et al., 2000. Similarly, the observations by Mueller et al., 1992 give a radius of 3.2 km with an albedo of 0.03. The estimated minimum axis ratio is 1.7, in good agreement with the previously cited values. Like P/ Wirtanen, C–G underwent a recent orbital change, as noted by Baliaev et al., 1986 and Carusi et al., 1985. Both papers agree about the fact that C–G should have had close encounters with Jupiter in 1840 and 1959 that reduced the perihelion distance significantly. Usually these events tend to rapidly increase the comet's activity, as shown by Coradini et al., 1997. The main characteristics of C–G is to be a dusty comet (Kiselev et al., 1998) and combining the estimated radius of the nucleus (Tancredi et al., 2000) with the production rates (Osip et al., 1992) it seems that the active area of the nucleus of C–G should be about 6% (Kidger et al., 2004), that is an effective active area of ≈ 2 km^2. The nucleus rotation period should have a value of 12.3 h. The peak water production is $\approx 10^{28}$ mol/s.

1.   VIRTIS description and scientific objectives

The main scientific objectives of the VIRTIS instrument will be to characterize the nucleus of C–G, it surface composition and detect early the set on of its activity. Recent images of comet Wild, collected by the Stardust mission have shown a completely cratered surface (Tsou et al., 2003), different from the Borrelly (Buratti et al., 2004a). These data show how important is to combine good image quality with high and medium resolution spectroscopy. VIRTIS will achieve these goals through a combination of the high spatial resolution channels (–M) and a high spectral resolution channel (–H). In particular, our expectation is: 1) to determine the nature of the solids on the nucleus surface (composition of ices, dust and characterization of organic compounds); 2) to identify the gaseous species; 3) to monitor the gaseous activity, and its spatial distribution; 4) to characterize the physical conditions of the coma, to measure the temperature of the nucleus; 5) to help the selection of the landing sites and to give support for other instruments.

2.   VIRTIS technical description

The instrument is made of 4 modules: the Optics Module – which houses the two –M and –H optical heads and the Stirling cycle cryocoolers used to cool the IR detectors to 70 K –, the two Proximity Electronics Modules (PEM) required to drive the two optical heads, the Main Electronics Module – which contains the Data Handling and Support Unit, for the data storage and processing, the power supply and control electronics of the cryocoolers and the power supply for the overall instrument. A detailed description of the experiment is given in Coradini et al., 1998; in what follows we will briefly summarize the main features of the instrument along with a description of the most recent activities carried out. The *VIRTIS–M* optical head perfectly matches a Shafer telescope to an Offner grating spectrometer to disperse a line image across two FPAs. The Shafer telescope produces an anastigmatic image, while coma is elimi- nated by putting the aperture stop near the center of curvature of the primary mirror and thus making the telescope monocentric. The result is a telescope system that relies only on spherical mirrors and yet remains diffraction limited over an appreciable spectrum and field: at $\pm$ 1.8 degrees the spot diameters are less than 6 microns in diameter, which is 7 times smaller than the slit width. The Offner grating spectrometer allows to cover the visible and IR ranges by realizing, on a single grating substrate, two concentric separate regions hav- ing different groove densities: the central one, approximately covering 30% of the grating area is devoted to the visible spectrum, while the external region is used for the IR range. The IR region has a larger area as the reflected infrared solar irradiance is quite low and is not adequately compensated by the infrared emissions of the cold comet. The visible region of the grating is laminar with rectangular grooves profile, and the groove density is 268 grooves/mm. More- over, to compensate for the low solar energy and low CCD quantum efficiency in the ultra–violet and near infrared regions, two different groove depths have been used to modify the spectral efficiency of the grating. The resulting efficiency improves the instrument's dynamic range by increasing the S/N at the extreme wavelengths and preventing saturation in the central wavelengths. Since the infrared channel does not require as high a resolution as the visible channel, the lower MTF caused by the visible zone's obscuration of the infrared pupil is acceptable; the groove density is 54 grooves/mm. In any case, the spot diagrams for all visible and infrared wavelengths at all field positions are within the dimension of a 40 microns pixel. For the infrared zones, a blazed groove profile is used that results in a peak efficiency at 5 μm to compensate for the low signal levels expected at this wavelength. In *VIRTIS–H* the light is collected by an off–axis parabola and then is collimated by another off–axis parabola before entering a cross–dispersing prism made of Lithium Fluoride. After exiting the prism the light is diffracted by a flat reflection grating which disperses in a direction perpendicular to the prism dispersion. The prism allows the low groove density grating, which is the echelle element of the spectrometer, to achieve very high spectral resolution by separating orders 9 through 13 across a two–dimensional detector array: the spectral resolution varies in each order between 1200 and 3500. Since the –H is not an imaging channel, it is only required to achieve good optical performance at the zero field position. The focal length of the objective is set by the required IFOV and the number of pixels allowed for summing. While the telescope is f/1.6, the objective is f/1.67 and requires five pixels to be summed in the spatial direction to achieve a 1 mrad^2 IFOV (5 x 0.45 mrad x 0.45 mrad).

3.   Observation strategy

After the long journey to the comet, the spacecraft will start to observe the comet from the beginning of its Sun–driven activity, then will follow and orbit it until at least the perihelion. At a distance from the Sun of 3 AU the probe will land on the surface of the nucleus. We have developed a simulation program to help in defining what is the best observation strategy for VIRTIS during the different Rosetta mission phases. Our goal is to achieve the maximum coverage and the minimum redundancy optimizing the VIRTIS sequences, selecting them among the several modes that can be implemented on VIRTIS. Through this simulation algorithm it is possible to compute the relative position of the probe in respect to the surface, the illumination condition of the surface, and the dwell time of a surface element in each VIRTIS IFOV. The simulation algorithm permits also to simulate bodies with irregular shapes and large surface markers, such as reliefs and craters. The simulation program allows to compute the orbital motion of the probe under the action of a central body gravity, also when the body is extremely irregular. The peculiarity of the program consists in the possibility to work in the coordinate system that is more appropriate in order to express the physical simulation that shall be performed. We have three main coordinate systems. As starting point the central body is simulated as a tri–axial ellipsoid: this ellipsoid can be considered as the reference point of the first coordinate system (S1), that is coincident with the three axes of the ellipsoid. The second coordinate system (S2) has the z–axis coincident with the rotation axis of the body, and rigidly rotates with the body itself. The third reference system (S3) is the inertial one, and is centered on the central body orbit, with the y–axis directed toward the sun, and the xy–plane coincident with the orbital plane of the body. In this reference system the motion of the probe and its orbital evolution are represented. The surface of the comet can be suitably modified and deformed by means of a specific algorithm, in order to describe several different morphologies, like mountains, fractures and craters. The coordinates transformations allowing to move from one system to the other are included in the algorithm. In this way it is possible to evaluate the probe motion in the coordinate system more appropriate to represent what it is necessary. The orbital evolution of the probe is computed in the inertial reference system, while all the representations of the observation are computed in the rotating system. In each time step of the evolution, through the appropriate coordinates transformations, it is possible to know the relative position of the imaged body in respect to the Sun and to the probe. It is also possible to know what is the phase angle of the observed area on the surface of the body and the position of the terminator on the body. For C–G the observation sequences in the different mission phase can be summarized as follows: 1) the comet should be acquired for the first time at a distance of 70000 – 100000 km; 2) the spectral phase curve will be measured; 3) closer to the comet, the –H channel will begin to search for early coma activity; 4) close approach and transition to global mapping phases; 5) the early mapping of nucleus surface and its observation under different illumination and phase angle conditions will be performed (–M channel); 6) the monitoring of a possible weak coma activity will continue (–H channel). In order to describe the monitoring phase, a simulation program has been developed to help in the definition of observation strategy. The relative position of the probe with respect to the surface of an irregular body, and the projection of the slit on the surface, can be computed. In Figure 1 the simulated shape of C–G and the projection of the spacecraft orbit on the plane perpendicular to the orbit of the comet are shown. The parameters used for the simulation are spin period = 12.3 h, mass = $2.8 \cdot 10^{13}$ kg, b/a = 0.76, c/a = 0.61, A = 2.4 km.

The spacecraft orbit is polar and retrograde, with a radius of 31.7 km and a period of 9.46 days. The duration of the simulation is 7.4 days with a ground resolution of the order of 20 m; in Figure 2 the comet coverage is shown. If we assume 14 hours of acquisition and 10 hours of downlink and that we work in –M nominal mode, then the data volume results of about 2 Gybits. The

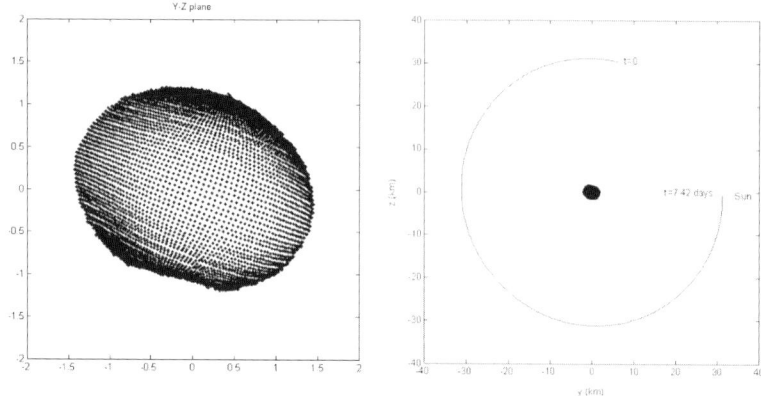

Figure 1. Left: simulated shape of comet C–G. Right: simulated orbit of Rosetta mission around C–G.

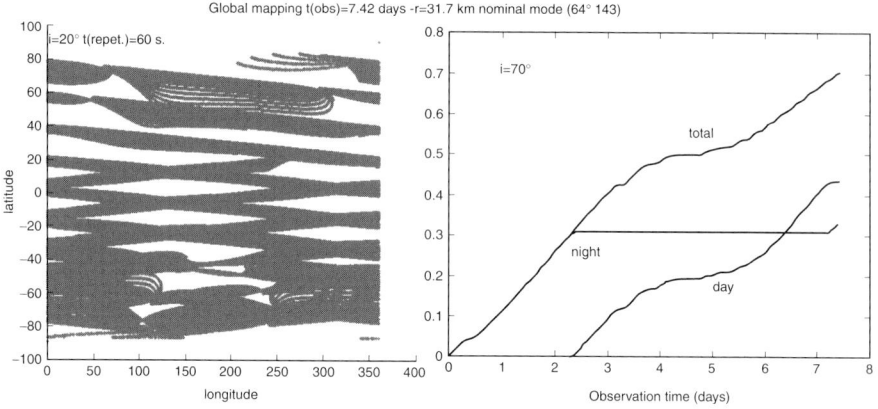

Figure 2. Left: coverage of C–G by means VIRTIS in nominal mode. Right: fraction of the C–G surface covered in about one week of continuous VIRTIS observations in nominal mode

coverage, globally, results of about 70% of the surface considering also night observations, and reaches 40% of the illuminated surface (see Figure 2).

During the close observation phase we assume that the orbit has apocenter of 10 km, pericenter of 5 km, period of 1.1 days and observation time of 8 days. The repetition time is assumed to be 5 s. It result that, in high resolution mode, the coverage is 1%, the data volume is 2.4 Gb, the mean dwell time is 11 s and the mean resolution is 2.5 m. With respect to the precedent target, the comet P/Wirtanen, we foresee a smaller surface coverage, a smaller redundancy in the coverage of the surface and a smaller surface resolution.

4.   VIRTIS–M on ground calibrations

Given the previously described observation strategy it is now possible to plan the observation sequences needed to optimize the VIRTIS scientific return. However, another component is also necessary to accurately perform such a planning, namely an accurate knowledge of the geometric and spectral characteristics of VIRTIS and its radiometric transfer function. This is necessary in order to plan the right integration time. A complete calibration campaign was performed at channel level (–M in Florence, Galileo Avionica, GA; –H in Meudon, LESIA) and at instrument level, in the large thermo-vacuum chambers in Orsay, IAS. The two channels, integrated on the VIRTIS–FM, including the Main Electronic (DHSU), were tested at IAS/Orsay. The objective of performing separate calibrations was to characterize, through specific setups, the single components, while the final activity was devoted essentially to the optical co–alignments, data handling and thermal stability. During GA tests, specific setups were realized regarding the geometrical, spectral, radiometric and internal performances of VIRTIS–M: the geometrical tests allow us to evaluate the overall instrumental imaging quality (IFOV, FOV and scan unit); the spectral characterization concerns with the instrument capability to recognize different spectral wavelengths (spectral range and resolution); the radiometric calibrations allows to convert the raw DN in physical units of spectral radiance.

4.1   Experimental setup

The basic experimental setup consisted of the following devices: VIRTIS–M optical head placed inside the thermovacuum chamber at operative temperature: the focal planes are cooled thanks to a dedicated cryocooler while the input beam enters through a CaF_2 window; PEM–M (Proximity Electronics Module); Unit Tester, simulating the Main Electronics module, with a master PC: this unit is used to send commands to the instrument and to receive back both telemetries and data packets; OGSE–M (Optical Ground Support Equipment) with a slave PC: this unit is used to command the different optical devices (monochromator, folding mirrors actuators, VIS and IR lamps–sources, blackbodies, reference photodiodes) used to stimulate the instrument with calibrated signals. Operating on the different elements is possible to modify the optical bench configuration to prepare a specific measurement setup: for the previously listed calibrations in fact we need to use different OGSE configurations. The acquired data are saved as telemetry packets and then converted in standard PDS (Planetary Data Format) by the Unit Tester.

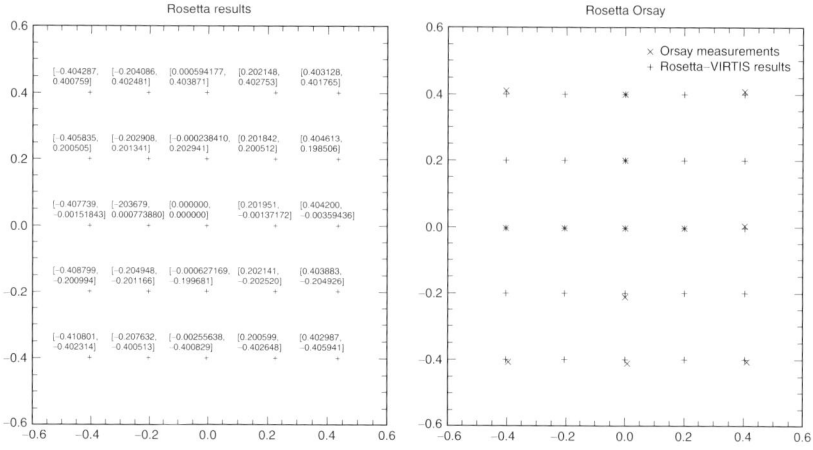

Figure 3. VIRTIS–M geometrical calibration results.

4.2 Geometrical calibrations

The geometric calibrations were realized through different spatially definite targets placed at the collimator focal plane; several scans over pinholes, slits (oriented parallel or orthogonal to the VIRTIS–M slit) or fixed patterns of micro lamps were acquired to evaluate the instrument geometrical performances; from these measurements resulted that the instrument geometrical response meets the proposed requirements: the IFOV is equal to 250 μrad/pixel, while the FOV is 64 mrad. An overall view of the instrumental imaging capabilities is given by the scan over a pattern of equi–distanced micro lamps (tungsten filament, bulb diameter 3 mm) disposed over 5 row for 5 columns with inter–distances of 1.5 cm. This target, with the micro lamps switched on, was placed at the collimator focal plane and acquired by moving the VIRTIS–M scan mirror. The resulting data cube, corrected for background and dark current allows to study the overall geometrical performances along the scan direction. By means of the target, it was possible to identify the presence of a limited spectral tilt. This effect has been removed by a translation of the columns along the slit direction. The resulting mean VIS spectral tilt along the sample direction is $< \Delta\_sample_{vis} >= 8.00817 \pm 0.17$ while along line direction is $< \Delta\_line_{vis} >= 1.01202 \pm 0.41$ The resulting mean IR spectral tilt along the samples is $< \Delta\_sample_{ir} >= -0.723183 \pm 0.23$ and along the lines is $< \Delta\_line_{ir} >= 1.31969 \pm 0.48$. During the VIRTIS integrated calibration campaign in Orsay the absolute position of 13 micro lamps (on 25) was measured by means of a theodolite. These measurements can be uses to estimate the micro lamps relative positions (see Figure 3) and to evaluate the geometrical distortions on the field of view.

	VIS	IR
λ_0 (nm)	221.853± 0.036	989.96± 0.58
$\Delta\lambda$ (nm)	1.88393 ± 0.00016	9.439± 0.036

Table 1.   VIRTIS–M VIS and IR spectral calibration results.

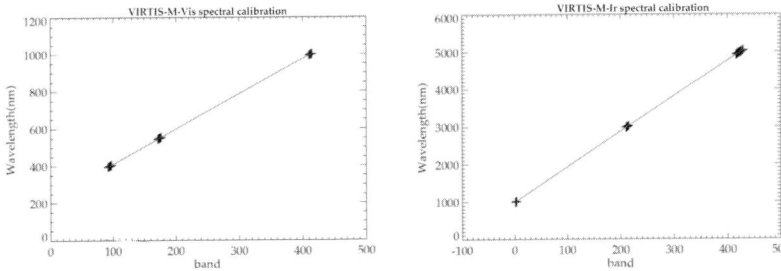

Figure 4.   VIRTIS–M spectral calibration results: Vis channel (left); IR channel (right).

4.3   Spectral calibrations

The spectral calibrations consist in the determination of the correspondence between the spectral bands and the input wavelengths. The test beam coming from the monochromator slit shall pass through a slit aligned in the collimator focal plane which works as a spatial filter. The system shall be adjusted in order to illuminate uniformly both the VIRTIS–M pupil and grating. This calibration consisted in the acquisition of some spectral scans realized through a monochromator placed along the VIRTIS–M optical boresight. From these measurements it is possible to evaluate the sample central wavelength, λ_c (nm), corresponding to the wavelength of the centroid of its spectral response function; the full spectral range corresponding to the wavelength interval between the minimum and the maximum wavelength at which the instrument is sensitive; the spectral width, SW(nm), corresponding to the FWHM of the spectral response. A linear fit applied over these experimental points gave the initial wavelength (corresponding to b=0), λ_0, and the mean spectral sampling, $\Delta\lambda$. At the n–th band will correspond the wavelength $\lambda(n) = \lambda_0 + \Delta\lambda \cdot n$. The linear fit results over the experimental data points are in Table 1, while in Figure 4 the best–fit straight lines over the points are plotted.

4.4   Visible radiometric calibration

The radiometric calibration, based on the evaluation of the ITF, Instrumental Transfer Function, allows us to convert the raw DN in physical units of spectral radiance (W cm^{-2} nm^{-1} sterad^{-1}) through the relation $DN(b,s,l) = ITF(b,s)Rad(b,s)t_{exp}$ where (b,s,l) are the pixel coordinates in the data–

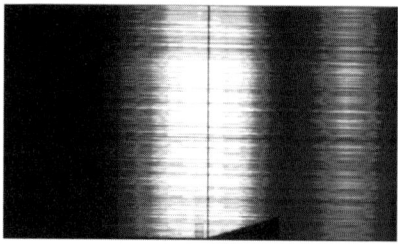

Figure 5. Monochromator scan signal interpolated over the VIS focal plane.

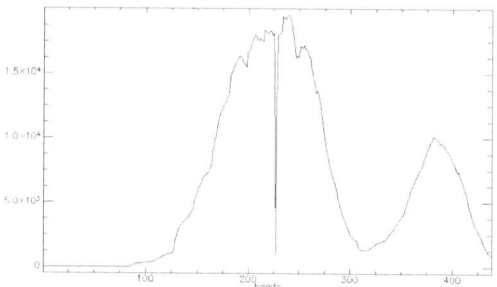

Figure 6. Monochromator scan signal interpolated over the VIS focal plane and plotted at slit center (sample = 128).

cube (band, sample, line), $Rad(b, s)$ is the calibrated input radiance and t_{exp} is the exposure time. The radiometric calibration represents the last and more important phase as includes all the previously discussed characterizations. VIRTIS–M observed at the output beam of a calibrated monochromator scanning the entire VIS range from 380 to 1060 nm at steps of 2 nm; the VIRTIS–M slit was completely illuminated. The light source was a QTH lamp placed at the entrance slit of the monochromator. Each acquisition is normalized for the exposure time, is corrected for the corresponding background (dark current and ambient light) and is re–arranged in order to avoid the overlapping between two adjacent sessions. As the monochromator step (2 nm) is larger than the spectral resolution of VIRTIS–M (1.88393 nm) we need to resample the input signal over the spectrometer focal plane. As the monochromator signal changes at steps greater than the instrumental spectral resolution, the generic pixel (b, s) will be illuminated with a gaussian–like temporal profile during the scan. The associated signal is estimated as the area included below the fitted gaussian curve centred over the spectral pixel (b).

As the monochromator scan starts at 380.0 nm the first 84 bands of the interpolated data–cube will not contain any signal. In the Figures 5 and 6 is shown the interpolated signal over the entire VIS focal plane (b along x di-

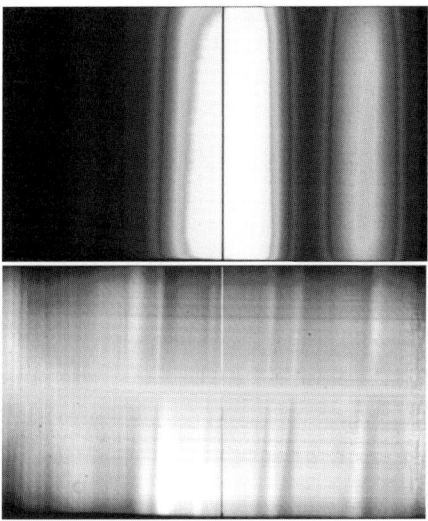

Figure 7. Top: mean VIS flat signal. Down: visible flat field coefficients frame.

rection, s along y) from which it is possible to detect the following effects: in the UV–blue range ($0 < b < 84$) there isn't signal as the monochromator scan started at 380.0 nm; at $b = 226$ the order sorting filter is present; the disuniformities along the spatial direction (sample, s) are introduced by the monochromator slit. To take into account the monochromator signal disuniformities along the slit direction, we preferred to consider only the signal at slit center (sample = 128) to calibrate; the ITF was extended to the remaining pixels of the frame with a flat field. The flat–field coefficient matrix was evaluated from an acquisition of a spatially flat white target (coated with spectralon paint) illuminated by a QTH lamp stabilized in current; the data cube dimensions (b, s, l) are (438, 256, 257); the flat–field frame, obtained as the mean over 257 frames and normalized respect to the $s = 128$ spectrum, is given as $Flat(b, s) = \overline{DN(b, s)}/\overline{DN(b, s = 128)}$.

In Figure 7 the mean signal over 257 frames of the flat target and the flat–field coefficient matrix are shown. The flat–field, shown in the figure allows us to remove the disuniformities introduced by the optical design (telescope, spectrometer, grating), by the spectrometer's slit and by the different responsivity of the pixels; a statistical analysis over the flat gives min = 0.004938, max = 1.125128, mean = 0.951109 and $\sigma = 0.112827$. The input irradiance, $irr_{PD}(b)$, is measured with a photodiode at steps of 40 nm during the monochromator scan. This quantity is fundamental to evaluate the instrumental transfer function as represent the input stimulus to the instrument. Finally the ITF can be written as:

234

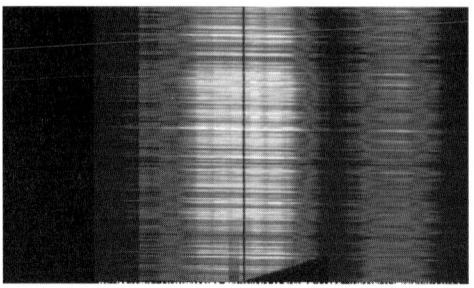

Figure 8.   ITF for VIRTIS–M–VIS.

$$ITF(b, s) = \frac{DN(b, s) \cdot \Delta\lambda \cdot \Omega}{irr_{PD}(b)t_{exp}} \tag{1}$$

where $\Delta\lambda$ = 1.88393 nm is the spectral sampling, $irr_{PD}(b)$ is the input irradiance measured by the photodiode and interpolated over the VIRTIS–M bands (b); t_{exp} = 50000 msec is the exposure time; $\Omega = (\pi/4)(D_T/F_{coll})^2 = 3.01907 \cdot 10^{-3}$ sterad is the solid angle subtended by the diffusive target placed at the collimator focal plane as viewed by the photodiode (D_T = 62 mm is the diffusive target diameter and F_{coll} = 1000 mm is the collimator focal length). The ITF–VIS matrix is shown in Figure 8.

4.5   Infrared radiometric calibration

The method used to evaluate the infrared radiometric calibration is the same discussed for the visible: nevertheless some additional difficulties were introduced by some limitation in the experimental setup. In particular the main problems were: the impossibility to have a black body with an aperture large enough to cover the entire VIRTIS–M slit length; this imposed a different approach to flat field measurements; the lack of a spectral source with enough signal in the entire IR range (1.0 – 5.0 μm): blackbodies at different temperatures (ranging from 323 to 690 K) must be used to calibrate radiometrically. For the previously discussed reasons, the flat–field IR was computed using both GA and Orsay data: for the range $0 < b < 150$ we used a GA acquisition with a setup similar to that used for the VIS flat field: in this spectral range the QTH lamp has enough signal. For the range $151 < b < 438$ we used a 50 °C blackbody scan along slit acquired at Orsay. The resulting IR flat field, normalized with respect to $s = 128$, is shown in Figure 9. With respect to the Vis flat field, this is clearly more noisy, especially at mid wavelengths where the blackbody signal is too low.

The ITF was evaluated by using different blackbodies acquisitions at variable temperatures (323 to 690 K). The signal used to calibrate was typically

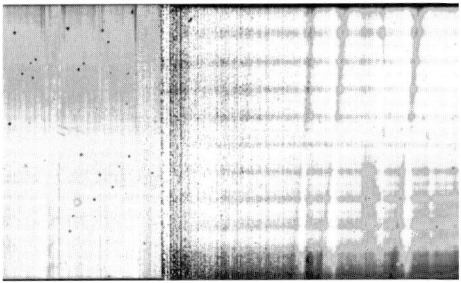

Figure 9.   IR flat field normalized respect to slit centre (sample=128).

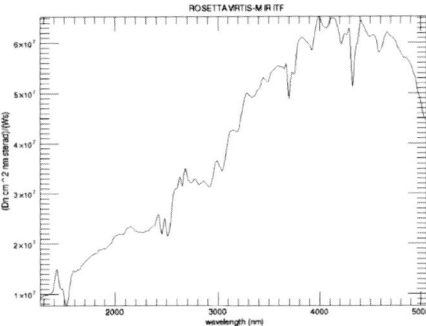

Figure 10.   IR ITF (DN cm^2 nm sterad W^{-1} s^{-1}) evaluated at slit centre (sample=128) and corrected for the odd–even effect.

comprised over the slope of the blackbody emission, between the noise region (short wavelengths) and the saturation (long wavelengths). The calibrated input radiance is given by the Planck's law for a fixed temperature of the blackbody; the solid angle of the BB (D_{BB} = 40 mm in diameter) at the collimator focal plane (F_{coll} = 4000 mm) was estimated as $\Omega = (\pi D_{BB})/(4F_{coll}) = 7.85 \cdot 10^{-5}$ This ITF presents some peculiarities: the 4 filters are correctly placed at their nominal spectral positions while some odd–even noise is still present over the function. This effect is introduced by the FPA electronics read circuits and can be removed by applying a de–noise algorithm operating on the separate odd and even channels signals. The resulting ITF is shown in Figure 10 and expressed in physical units (DN/s)(cm^2 nm sterad/W).

References

Beliaev, N. A., Kresak, L., and Pittich, E. M. (1986). in *Catalogue of short-period comets* (Bratislava: Slovak Academy of Sciences, Astronomical Institute).

Bockelée-Morvan, D., Biver, N., Colom, P., Crovisier, J., Henry, F., Lecacheux, A., Davies, J. K., Dent, W. R. F., and Weaver, H. A. (2004). The outgassing and composition of Comet 19P/Borrelly from radio observations. *Icarus*, 167:113-128.

Buratti, B. J., Hicks, M. D., Soderblom, L. A., Britt, D., Oberst, J., and Hillier, J. K. (2004a). Deep Space 1 photometry of the nucleus of Comet 19P/Borrelly. *Icarus*, 167:16-29.

Buratti, B. J., Britt, D. T., Soderblom, L. A., Hicks, M. D., Boice, D. C., Brown, R. H., Meier, R., Nelson, R. M., Oberst, J., Owen, T. C., Rivkin, A. S., Sandel, B. R., Stern, S. A., Thomas, N., and Yelle, R. V. (2004b). 9969 Braille: Deep Space 1 infrared spectroscopy, geometric albedo, and classification. *Icarus*, 167:129-135.

Capria, M. T., Coradini, A., De Sanctis, M. C., and Fulle, M. (2004). This volume.

Carusi, A., Kresak, L., Perozzi, E., and Valsecchi, G. B. (1985). Long-term evolution of short-period comets. Bristol (England and Accord, MA, Adam Hilger, Ltd.).

Clark, R. N. (1981). Water frost and ice - The near-infrared spectral reflectance 0.65-2.5 microns. *J. Geophys Res.*, 86-B4:3087-3096.

Coradini, A., Capaccioni, F., Capria, M. T., De Sanctis, M. C., Espinasse, S., Orosei, R., Salomone, M., and Federico, C. (1997). Transition Elements between Comets and Asteroids. *Icarus* 129:337-347.

Coradini, A. et al. (1998). Virtis: an imaging spectrometer for the rosetta mission. *Plan. Space Sci.*, 46:1291-1304.

Crovisier, J. (1992). Radio spectroscopy of comets: Recent results and future prospects. in *Asteroids, Comets, Meteors*, p. 137-140.

Kamoun, P., Campbell, D., Pettengill, G., and Shapiro, I. (1998). Radar observations of three comets and detection of echoes from one: P/Grigg-Skjellerup. *Planet. and Space Sci.*, 47:23-28.

Keller, H. U., Kram, R., and Thomas, N. (1988). Surface features on the nucleus of Comet Halley *Nature*, 331:227-231.

Kidger, M. R. and Mart'yn-Luis, F. (2004). in preparation.

Kiselev, N. N. and Velichko, F. P. (1998). Polarimetry and Photometry of Comet C/1996 B2 Hyakutake. *Icarus*, 133:286-292.

Moroz, L. V., Arnold, G., Korochantsev, A. V., and Wasch, R. (1998). Natural Solid Bitumens as Possible Analogs for Cometary and Asteroid Organics. *Icarus*, 134:253-268.

Mueller, B. E. A. (1992). CCD-photometry of comets at large heliocentric distances. in *Asteroids, Comets, Meteors 1991* , p. 425-428.

Osip, D., Schleicher, D. G., and Millis, R.L. (1992). Comets - Groundbased observations of spacecraft mission candidates. *Icarus*, 98:115-124.

Tancredi, G., Fern'andez, J. A., Rickman, H. and Licandro J. (2000). A catalog of observed nuclear magnitudes of Jupiter family comets. *Astron. and Astroph. S.*, 146:73-90.

Tsou, P., Brownlee, D. E., Sandford S. A., Hörz, F., and, Zolensky, M. E. (2003). Wild 2 and interstellar sample collection and Earth return. *J. Geophys Res.*, 108:3-1.

CONSERT EXPERIMENT: DESCRIPTION AND PERFORMANCES IN VIEW OF THE NEW TARGET

W. Kofman, A. Herique
Laboratoire de Planétologie de Grenoble, UJF/CNRS, Bâtiment D de Physique, B.P. 53, 38041 Grenoble Cedex 9, France

J–P. Goutail
Service d'Aéronomie, B.P. 3, 91371 Verrières–le–Buisson, France

CONSERT team *

Abstract   We summarize the scientific objectives of the CONSERT experiment. We point up the practical realisation of the instrument describing the main points necessary to understand how the instrument is working. The transponder structure and the principle of the temporal synchronization are explained. The calibration of the instrument is discussed and its performances as results of tests are shown. We also discuss future operations which have to be planned so as to prepare the in flight operations and to maintain the instrument knowledge for 11 years. Finally, the CONSERT ability to explore the 67P/Churyumov–Gerasimenko comet is discussed.

Introduction

The objective of the ROSETTA mission is to study the physical properties, the chemical composition and the structure of comets in order to be able to deduce information on their formation and history and from this, information on the formation and history of the Solar System. For this it is crucial to get some

*List of CONSERT team: E. Nielsen, Max Planck Institut für Aeronomie, Germany; T. Hagfors, Max Planck Institut für Aeronomie, Germany; A–C. Levasseur–Regourd, Service d'Aéronomie, France; J–P. Barriot, Observatoire Midi–Pyrenees, France; Y. Barbin, LSET, France; P. Edenhofer, Institute for High–Frequency Technique, University of Bochum, Germany; D. Plettemeier, Institute for High–Frequency Technique, University of Bochum, Germany; G. Picardi, INFOCOM, Italy; R. Seu, INFOCOM, Italy; C. Elachi, JPL, United States of America; P. Weissman, JPL, United States of America; H. Svedhem, ESA/ESTEC, The Netherlands; S. Hamran, PFM, Norway; I. P. Williams, Queen Mary College, United Kingdom.

L. Colangeli et al. (eds.), The New ROSETTA Targets, 237–256.
© 2004 *Kluwer Academic Publishers. Printed in the Netherlands.*

knowledge on the distribution of matter within the comet nucleus. Indeed, the distribution of voids and the presence or absence of big substructures (boulders, cometesimals) could be a diagnostic of the formation of the body. The CONSERT experiment is the only experiment on board the ROSETTA mission which will provide information about the deep interior of the comet. In this experiment, an electromagnetic signal is transmitted between the lander and the orbiter. The transmitted signal will be measured as a function of time and as a function of the relative position of the orbiter and the lander for a number of orbits. According to our current understanding of the composition of cometary materials, a simulation of electromagnetic wave propagation through assumed cometary materials showed that it is possible to propagate signals between the satellite orbiting the comet and the lander located on the comet surface and led to the choice of the carrier frequency (90 MHz) (Kofman et al., 1998). If a sufficient number of orbits were available, one would be able to obtain many cuts of the interior of the comet and therefore to build up a tomographic image of the interior. Doing this as part of a space mission introduces many constraints on the instruments. These will inevitably limit the amount of detail that can be generated in our image reconstruction; nevertheless, the information that will be obtained will significantly increase our understanding of cometary physics.

During the initial planning stages, it was assumed that two landers, Champollion and Roland, would be available. Unfortunately, the Champollion experiment was later cancelled, significantly diminishing the amount of data that we might expect and also restricting our ability to build up a complete image of the cometary interior.

As it was not possible to get sufficiently stable clocks within the given constraints of mass and power consumption to use a one–way transmission of a radio wave from the orbiter to the lander, the transponder technique was adopted. This transponder system makes the experiment scientifically richer. The wide bandwidth signal will allow the measurements of a signal propagating through main and secondary paths. These measurements will allow not only the same scientific objectives as was proposed in the original proposal, through the measurements of the phase delay and amplitude of the main path, but also the distribution of the secondary path for a deeper description of the comet interior and better space coverage. The main constraints on the experiment introduced by incorporating it in a space mission arise through a strict limit on the total mass available (3 kg) and on the total power consumption (3 W). As the lander will be deployed near the beginning of the ROSETTA mission, when the spacecraft (and the comet) are far from the Sun, the amount of power available through the solar panels will be low and this is especially true for the lander. Hence, the amount of electrical power available to us to transmit between the lander and the orbiter at this stage will correspondingly be very low. A further problem is that the number of independent orbits to be performed by the orbiter is not precisely known at present, but is unlikely to be very large and will restrict our ability to reconstruct a detailed image of the interior. Irrespective of the amount of details that we will reconstruct, we will be able to determine some of the fundamental properties of the nucleus, for example whether it is homogenous or layered or an accumulation of several blocks. The internal structure is ultimately tied to the formation process and so knowing one is a vital step in understanding the other.

The change from 46P/Wirtanen (hereafter 46P/W) to 67P/Churyumov–Gerasimenko (hereafter 67P/CG) implies the new limitation of CONSERT experiment which we define as a probability to explore the given fraction of the comet.

CONSERT is a new type of experiment and to maintain the knowledge of the experiment for the next 11 years we plan to have field experiments in canyons, glaciers and eventually in Antarctica.

1.   Scientific objectives

The scientific objectives of the CONSERT experiment (Kofman et al., 1998) are the determination of the main dielectric properties and, through modelling, to set constraints on the cometary composition (materials, porosity, etc.), to detect large–size structures (several tens of meters) and stratification, to detect and characterise small–scale irregularities within the nucleus. A detailed analysis of the radio–waves which have passed through all or parts of the nucleus will put real constraints on the materials and on inhomogeneities and will help to identify blocks, gaps or voids. From this information, we attempt to answer some fundamental questions of cometary physics. How is the nucleus built up? Is it homogeneous, layered or composed of accreted blocks (cometesimals, boulders)? What is the nature of the refractory component? Is it chondritic as generally expected or does it contain inclusions of unexpected electromagnetic properties? With the answer to these questions, it should also be possible to provide answers to the basic question of the formation of the comet. Did it form directly from unprocessed interstellar grain–mantle particles or from grains condensed in the pre–solar nebula? Did the accretion take place in a multi–step process leading first to the formation of cometesimals which then collided to form a kilometre size body?

In more detail, the purpose of CONSERT experiment is to measure the following quantities:

- The mean permittivity of the comet nucleus is derived from the group delay of the main path introduced when the comet is inserted into the propagation path. The permittivity enables to identify the electrical properties of the material found in the comet nucleus.

- The mean absorption of the comet nucleus is derived from the radiowave path loss as the signal propagates through the comet nucleus. The absorption identifies the class of materials found in the comet nucleus.

- The structure of the received signal, the number of different paths and their variation with the propagation path are related to the size of the cometesimals and to the reflection coefficient at internal interfaces.

- The correlation length of the measured signal as a function of the orbit position is related to the size of the irregularities or small structures inside the comet.

- The volume scattering coefficient is derived from the nature of the observed signal. The volume scattering coefficient measures the homogeneity of the interior of the comet nucleus.

2. Basic principle of the CONSERT experiment

The basic principle of the experiment consists in using the electromagnetic propagation through the cometary interior. An electromagnetic wave–front propagates through the cometary nucleus at a smaller velocity than in free space and loses energy in the process. Both the change in velocity and the energy loss depend on the complex permittivity of the cometary materials. They also depend on the ratio of the wavelength used to the size of any inhomogeneities present. Thus, any signal that has propagated through the medium contains information concerning this medium. The change in velocity of the electromagnetic wave induced by propagation through the cometary material is calculable from the time taken by the wave to travel between the orbiter and the lander, while the loss of energy is deducible from the change in signal amplitude.

The orbiter will send a signal which will be picked up by the lander. As the orbiter moves along its orbit, the path between it and the lander will vary and so pass through differing parts of the comet. In addition, the rotation of the comet nucleus will also change the relative position of the lander and the orbiter. Hence, over several orbits, many different paths will have been obtained. A schematic representation of the situation is shown in Fig. 1.

To extract the time information, the instrument has to measure the propagation delay between the lander and the orbiter signals with a very high accuracy. In addition, the precise knowledge of the orbit relatively to the surface of the comet with an accuracy of 10 m in three directions is necessary. This information will be provided by the cameras of the ROSETTA mission. The CONSERT measurements should be stable over the whole orbit in order to be able to process the signal covering the orbit in a coherent way. This is necessary to build the image of the interior of the comet (Kofman et al., 1998).

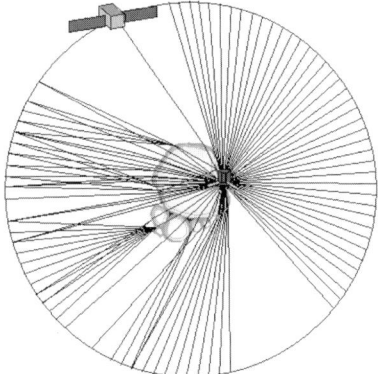

Figure 1. The ray propagation throughout the comet nucleus between orbiter and lander for different orbiter positions on the orbit.

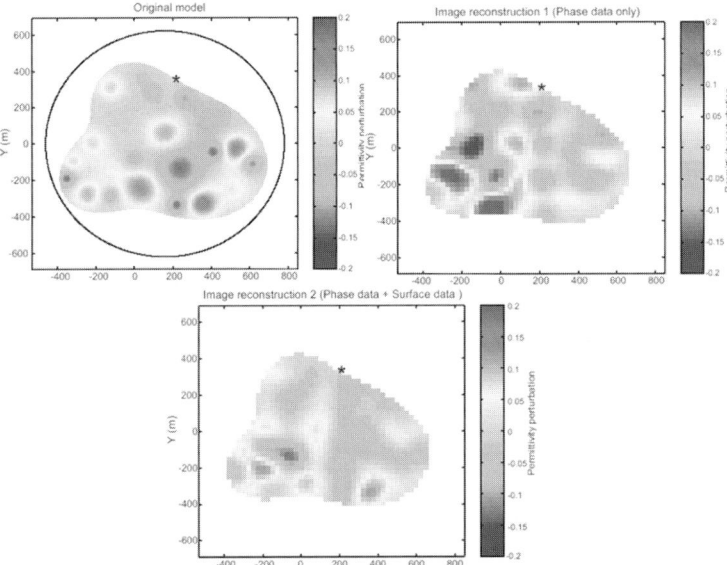

Figure 2. Simulation of the inversion for the CONSERT experiment. The upper left panel shows the used model of the cometary interior, the upper right panel shows the reconstruction using the propagated signals and the lower panel is the reconstruction using in addition the electromagnetic properties of the surface (Benna, 2002)

Two inversion philosophies have been developed for CONSERT tomographic data. The deterministic inversion (Barriot et al., 1999; Benna et al., 2002; Benna, 2002) derives a low resolution image of the interior of the comet nucleus from the measurements. This method is close to that used in the ray tomography. Fig. 2 shows the quality of the inversion in two dimensions. Even the use of direct rays only allows an acceptable inversion. One can clearly distinguish the large structures inside the nucleus. This approach requires a few orbits with a complete coverage.

On the other hand, the statistical approach (Herique et al., 1999) is more robust in terms of lacks of data and requires "a low a priori knowledge". From the measurements of the time delay, one will be able to deduce the average properties of the cometary materials, starting with their dielectric constant. This inversion extracts statistical parameters of the comet nucleus and follows their variations along an orbit. These statistical parameters give the average complex permittivity along the propagation path and their variation give the size of the nucleus irregularities at different scales. The statistical approach gives information on the internal large scale structure of the nucleus and it allows the inversion of a stratified nucleus. The estimate of the permittivity gives good results without too many artefacts and biases.

The next step will be to model the cometary materials, constrained by the measurements of CONSERT and other experiments and by laboratory data. The permittivity is a function of the composition of the nucleus. The measurement of the permittivity introduces a constraint in a composition model and reduces by one the number of free parameters. For example, assuming a mean composition of the material part, one can infer the internal porosity and density of the nucleus. In order to introduce this constraint, we have two possible strategies: laboratory measurement or theoretical modelling of a mixture permittivity. For a given mixture composition and a given porosity range, the effective permittivity can be directly measured in the 100MHz frequency range. On the other hand, the permittivity can also be theoretically modelled by the Maxwell–Garnett equation (Sihvola & Kong, 1988), already used in the direct problem. This requires that we know the permittivities of components in the right range of frequencies and temperature: some of them are already well known in the literature (Campbell & Ulrichs, 1969; Stevels, 1980; Evans & Smith, 1969), while others may have to be characterised by laboratory measurements.

3.   System constraints, signal definition and operations

The system design is constrained on the one hand by the space mission conditions, mass, power and temperature limitations, and on the other hand by the ignorance of the nucleus structures which determine the propagation of the signal. In order to process the signal globally, i.e., to combine the measurements corresponding to the different positions on the orbit, the signals on the orbiter and on the lander must be coherent with the stable relative phase and have an adequate signal–to–noise ratio (SNR). These conditions put constraints on:

- system structure;

- clocks stability;

- antennas;

- operations.

The points enumerated above are related one to another.

The required time accuracy and range resolution define the necessary clock stability and the total signal bandwidth (Kofman et al., 1998). The bandwidth determines the total noise power. Our calculations show that the expected SNR will be high and this will allow the use of a large bandwidth. The total attenuation range as a function of the nucleus radius and the sounding frequency for the two types of cometary materials, ice and silicates and ice and chondrites, has been studied in Kofman et al. (1998). This study obtained the ranges of attenuation for which it is possible and for which it is impossible to build the instrument because of the technical limitations.

The bandwidth of the signal and the time resolution are closely related. The time resolution is given by the desired discrimination between the various paths and, in this sense, it is dictated by the scientific requirement. However, the better the time resolution, the larger the bandwidth and the more difficult is to build the system. A large bandwidth demands a higher sampling rate and a higher ratio between the bandwidth and the carrier frequency, and so more data to process and transfer. The antenna should be more complex; the dipoles should be replaced by a surface antenna or by loaded antennas, which are less efficient. This would increase the mass and the power consumption. For all these reasons we had to accept a compromise between the science requirements and the technique.

As a result of a trade–off between signal penetration, spatial resolution, antenna design, electronic speed, mass and power, we adopted a carrier frequency equal to 90 MHz, a bandwidth of 10 MHz and a sampling rate of 20 MHz. This bandwidth gives a resolution of 30 m in free space and about 20 m in the cometary material, which is a good enough compromise between the scientific requirement and the technology. For this bandwidth the estimated SNR after the signal processing will be large for a comet with a 1 km radius.

The major constraint, which is the stability of the clocks, comes from the desired accuracy of the dielectric constant measurements. As it was said, the average dielectric constant will be obtained by the measurements of the time of propagation of the signal between the orbiter and the lander. This time delay depends on the propagation in the free space and in the comet and on the stability of the clocks.

In one–way experiments, in order to measure the absolute time difference between the transmitted and received signals with the needed time accuracy

better than 100 ns during the whole occultation period (this gives a space resolution of the order of 20 m), one should use an ultra–stable oscillator on
each side, with a long–term stability of the orbiter and lander clocks of 10^{-12},
which is clearly impossible with the mass and power assigned to CONSERT
by the ROSETTA project. To by pass this constraint, we adopted the time–
transponder technique. The method is described in Barbin et al. (1999). This
allows a relaxation of the constraint on the clocks stability to $\Delta f / f = 10^{-7}$
during a 30 hour period. The fulfilment of this constraint does not protect us
from ageing and a large variation between the lander and the orbiter clocks
due to various environmental conditions. In the system we have to include the
tuning circuits, PLL loop, which will tune and synchronise the master clock of
the orbiter to the lander clock. This operation has to be done at the beginning
of the occultation phase (measurements phase) when the orbiter and the lander are in line of sight. This condition is necessary in order to have a signal
propagating in the free space, directly between two spacecrafts. This tuning
has to be accurate in frequency (better than $\Delta f / f = 10^{-7}$) and the two
clocks have to be synchronised in time with an accuracy over 10 ms. This time
synchronisation is necessary to give simultaneous measurement cycles on the
two vehicles.

To improve the SNR, we transmit the periodic signal for the coherent integration of the received signal. Taking into account the relative radial velocity
between the orbiter and the lander, and the offset between two clocks, one can
accumulate the sounding signal over a 26 ms period. This coherent integration creates a 54 dB increase in the SNR, as compared to a single pulse of the
same power. The coherent integration is made in two ways: the first is the
transmission of the phase shift coded signal, called pseudo–random code and
the second is the use of the periodic repetition of this coded signal (Barbin et
al., 1999). The final resource budget is 3 kg, 2 liters and 3 W average power
for the orbiter and 2.3 kg, 2.5 l and 3 W for the lander. The main system
characteristics are given in Table 1.

In Fig. 3, the block description of the CONSERT experiment is shown.
The coded signal is transmitted from the orbiter through the comet and is received on the lander. This signal is coherently integrated and compressed,
which means that the cross–correlation between the model of the coded signal and the received integrated signal is calculated, and the time delay of the
strongest cross–correlation peak is measured by the instrument with a $\pm$ 50 ns
accuracy. Then the coded signal is transmitted from the lander back to the orbiter with a delay equal to the one just measured. The signal is received on the
orbiter where it is coherently integrated, stored in the memory and will be sent
to the Earth. On the lander, the measured time delay, a part of the calculated
cross–correlation, and the gain values are saved and will be sent to Earth with
CONSERT telemetry.

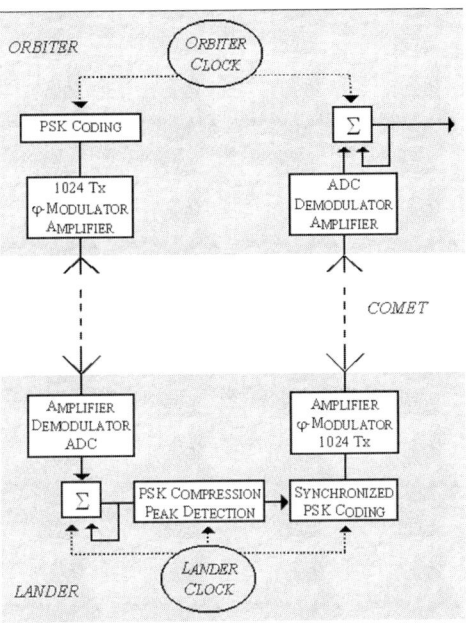

Figure 3.   Block diagram of the instrument.

Table 1.   Main characteristics of the instrument important to the achievement of the science goals

	Lander	Orbiter
Transmitted power	0.2 W	0.4 W
Instrument noise	4000 K	4000 K
Carrier frequency	90 MHz	90 MHz
Transmitted bandwidth	10 MHz	10 MHz
Received bandwidth	8 MHz	8 MHz
Sampling rate (on I and Q components)	10 MHz	10 MHz
Linearity of the receivers	good	good
Total harmonic distortion	< -50 dB	-50 dB
Gain control linearity	good	good
Clock stability	10^{-7}	10^{-7}
Ratio of main correlation peak to secondary peaks	< 30 dB	< 30 dB
Radiometric accuracy	< 1 dB	< 1 dB
Interference immunity	good	good

Fig. 4 shows the time sketch in more detail: the code transmitted by the lander is delayed by the lander synchronization delay, that is equal to the orbiter to lander propagation time of the strongest signal. The lander to orbiter propagation induces its own propagation delay, that is equal to the previous

246

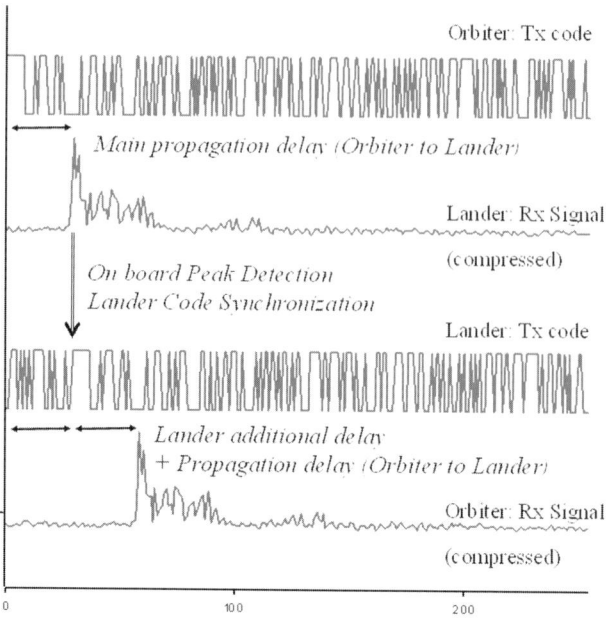

Figure 4. Lander synchronization: effect on the measured signal taking into account the periodicity of the calculated signal.

one. On the science signal, the main propagation time corresponds to twice the real propagation delay while, in comparison to the main peak, the secondary peaks have the right relative position corresponding in one way propagation. So, the first propagation provides the synchronization from the orbiter to the lander, while the second way corresponds to the signal of interest.

In Fig. 1, we display a cartoon of measurement sequence over one CONSERT orbit. For the line of sight of visibility between the lander and the orbiter, the following operations will take place: a warming of the instrument clocks in order to reach the frequency stability, a tuning phase to put the master clocks at the same frequency and synchronise them in time. The last operation is the start of the measurement phase. This phase will last during the whole occultation period, for many hours. During this phase, the system will take several thousands periodic measurements, sampling the waveform of the signal which propagated through the comet. The amount of data needed to analyse the signal correctly is fixed by the sampling theorem. On the orbiter, one has to take samples (to receive the propagating signal, whose duration will be 25.5 μs) every Δx meters, where Δx is given by:

$$\Delta x = \lambda R_o / 2 R_c$$

with R_o the orbit radius, R_c the cometary radius and λ the wavelength.

This sampling step is calculated for a circular orbit and assumed a circular comet. The amount of measurements can be obtained from the above and gives about 3000 measurements to be carried out, for the nominal model of the 46P/W comet, and about 9000 measurements, for 67P/CG comet.

Each measurement, which will be transmitted to Earth, consists of 255 complex samples of the signal which propagated from the lander to the orbiter and a 21 correlation function around the maximum of the signal which propagated from the orbiter to the lander. This measurement phase will terminate at the end of the occultation phase with the new line of sight data acquisition which will allow the control of the signal quality and the time delay measurements, as the time delay in the free space will be easily estimated knowing the orbit parameters.

4.   Antennas

The CONSERT experiment has a transmit/receive antenna on both the orbiter and the lander. A fundamental requirement on the antenna is that it must have a broad main antenna lobe centred on the antenna normal to ensure that the whole comet is 'illuminated' by the main lobe when the antenna normal is pointing towards the comet. No steering of the antenna is required in order to explore the whole comet. A further requirement for the antenna is a bandwidth of 10 % of the centre frequency. In order not to impose restraints on the relative orientation of the orbiter and lander antennas, it was decided to use circular polarisation. These 'electrical' requirements must now be considered in view of 'mechanical' limitations on the antenna (especially limit on the mass), which are different for the orbiter and the lander. This leads to two different solutions for the antenna on the orbiter and lander (Nielsen at al., 2001).

A dipole has a broad radiation pattern as required. A half–wave dipole has a large real component of the impedance, as needed for a wide bandwidth antenna. Two dipoles oriented 90 degrees to each other and fed 90 degrees out of phase form a circular antenna. The orbiter antenna is formed of two crossed dipoles and two crossed dipole reflectors. To minimise the interaction with other experiments on the spacecraft, the dipoles and reflectors are mounted on a mast that deploys the antenna out to the side of the instrument panel of the spacecraft. The linear scale length of the antenna elements is about half a wavelength at the centre frequency of 90 MHz, i.e. about 1.5 m. Clearly such a large antenna has to be folded during launch and then deployed on command once the spacecraft is in space. The antenna is therefore constructed of basically 10 linear elements: 2 masts (one to support the dipoles and reflectors, and one to deploy the antenna away from the spacecraft), while the remaining 8 elements form pairwise the dipoles and the reflectors. In the folded state, the 10 elements are placed parallel to each other, and are lashed together with a steel wire, which can be cut by a pyro cable cutter. The elements are connected by springs such that when released the spring forces deploy the antenna. The folded and deployed antenna is shown in Fig. 5.

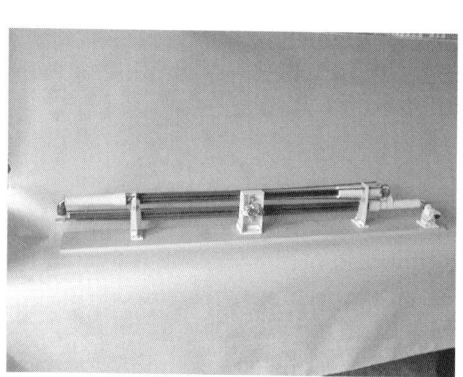 

Figure 5. The CONSERT orbiter antenna folded in launch position (left) and deployed (right).

On the lander there are severe restrictions on mass and on the location of the antenna. In order to ensure a good coupling of the energy radiated by the antenna into the comet, it is desirable to place the antenna as close to the comet surface as possible, and possibly at a height less than 1/10 of a wavelength. In order not to interfere with possible later movements of the lander the antenna could not be placed on the feet or legs of the lander. Using the lander body as ground plane for monopoles, the antenna elements could be placed on top of the body to yield good gain in the nadir hemisphere, but that was mechanically not possible, and electrically the antenna was far from the surface. In the end, two monopoles were placed about 1/10 of a wavelength over the surface on the lander body base plate. A monopole has a broad antenna diagram and a relatively large real component of the impedance. Two monopoles oriented 90 degrees to each other and fed 90 degrees out of phase form a circular antenna. These monopoles form the lander antenna.

The linear scale of the monopoles is a quarter of a wavelength, or 0.8 m. During launch, the monopoles are folded back along the base plate and fixed in place by the collapsed lander legs. As the legs deploy, spring forces acting at the foot of the monopoles deploy the antenna.

5.   CONSERT instrument performances

During the instrument design and development, the performance specifications have been derived from the knowledge of the comet nucleus structure and composition in order to be able to characterize the nucleus and have been translated into technical specifications (Kofman et al., 1998). Considering the challenge of the instrument design within the allocated budgets, several instrument performances have not been specified like the radiometric accuracy and stability. This incomplete set of instrument specification is not directly useful in the data inversion.

The calibration process has required a specific study to define a relevant set of associated instrument performances. Indeed for all the space–borne radar instruments, a general agreement exists concerning this set of instrument performances resulting from a long experience, including science and technical aspects. For CONSERT, the classical background is not directly transposable due to an original configuration measurement (transponder and propagation) and specific scientific objectives.

Our instrument consists in three functions:

- a chronometer to measure the propagation delay, which is the main function;

- an imager to separate the multipath propagation;

- a radiometer to estimate the wave attenuation which is more a secondary function.

Each function has been separately characterized and a specific attention has been applied for the statistical properties (accuracy and stability) in order to allow the inversion of the signal fluctuation along an orbit by the study of the correlation length. The complete definition of the instrument performances parameter is given in Herique & Kofman (2001).

Imager

All the CONSERT models have been calibrated on the ground and fulfil the specifications. The time resolution is about 150 ns. The ratio of principal to secondary peaks is larger than 30 dB. This limitation comes essentially from the electronics non linearity and illustrates the sensitivity of the used transmitted signal to this kind of distortions.

Radiometer

The instrument factor of merit is equal to 192 dB, while the total dynamic range is 130 dB. The radiometric stability is better than 0.3 dB at short term and 1.5 dB at long term, including the thermal effects.

Chronometer

After pre–processing, the time measurement accuracy is limited by the lander resynchronization: the on–board peak detection degrades the time accuracy to 100 ns (10 MHz sampling frequency). On the measured propagation delay, this degradation could be described as an additive noise with a 100 ns large uniform density of probability. This noise is correlated along an orbit since the frequency drift of the lander master clock versus orbiter is slow and risks to be interpreted as nucleus internal structures.

This accuracy can be improved by ground–processing: indeed for each sounding the signal received by the lander is partially down–loaded: 21 signal samples around the on–board detected peak–position. The principle is to interpolate this short signal in order to estimate the lander synchronization error that means the difference between the on–board calculated peak position used for the transponder resynchronization and the on–ground interpolated peak position. The same interpolation scheme is applied to the orbiter full–length signal in order to measure the twice propagation delay with sufficient accuracy and the lander error estimate is subtracted to improve the synchronization accuracy.

A preliminary study shows a short–term stability of 10 ns which is good by regards to the 8 MHz signal bandwidth. The long–term stability is 50 ns for the whole lander temperature range (- 40 / +50 °C): this variation could be explained by the clock frequency drift versus temperature. These values give the expected chronometric accuracy with a large SNR.

6.   Future operations

Due to mass constraints, there is no specific calibration channel for CONSERT but the transponder structure allows an end–to–end calibration of each way (Orbiter Tx / Lander Rx and vice et versa). This "internal" in flight calibration will provide the electronics characteristics in terms of link budget, radar impulse response and the relative clock drift. This sequence of measurement will be repeated every year in order to follow the instrument evolution during the travel period.

Nevertheless, the in flight calibration configuration is drastically different than the science operation: the lander antenna is folded; the lander / orbiter coupling is expected without direct visibility by the spacecraft structures and the solar panels. The measurement geometry will change during the flight. Indeed, the spacecraft attitude is determined during the cruise phase by the three constraints: the main antenna dish orientation in Earth acquisition; the solar panel orientation in Sun acquisition and the lander thermal constraints out of Sun illumination. So there is no remaining freedom degree in the Rosetta attitude and it is impossible to operate CONSERT with the same solar panel position for all the operation slots during the cruise phase. It is also difficult to

estimate impact of this geometry change to the CONSERT impulse response
and link budget by ground measurement or by simulation due to the 3–meter
wavelength.

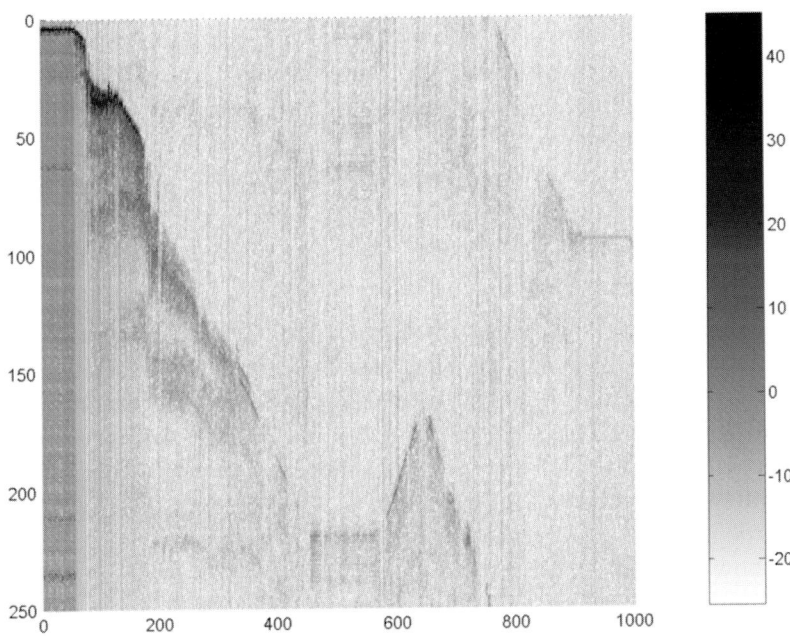

Figure 6.   CONSERT signal acquired in Grand Goulet (French Alps). Power in dB as a
function of propagation time between the lander and orbiter (vertical) and receiver position
(horizontal line) which corresponds to a few kilometres of lander displacement relatively to
orbiter. The strongest first echoes (black traces), are the first reflections from the cliff, the
secondary reflections arriving later are seen with a weaker power

This impact will have to be quantified in detail during the instrument com-
missioning. A specific sequence will be dedicated to the solar panel influence
during which we calibrate all the possible solar panel orientations and acquire
a reference set of CONSERT data in order to be able to separate the Solar
Panel coupling effect to the aging effect in the cruise operation signals. In
parallel to the in–flight activities, we are developing on–ground operations. In
order to develop and validate the inversion methods, we have to have a set
of realistic signals presenting a complexity similar to future signals acquired
on the comet. The novelty of the tomography in propagation and also of the
transponder structure imposes specific on–ground operation to acquire these
signals. The first natural field we tested is canyons in the French Alps, near to
Grenoble (Fig. 6). The multipath propagation in the canyon with reflections on
cliffs provides a complex signals while the cliffs roughness introduces a prop-

agation delay variability which is expected to be quite similar to the variability introduced by the comet nucleus blobs. The inversion is oriented on the statistical study of the cliffs by the estimation of the roughness standard deviation (perpendicular to the cliff) and its correlation length (parallel to the cliffs).

Other experiment fields are envisaged in the future with propagation throughout material media: mountain glaciers, sand dunes and also Antarctica ice sheet. In particular, the tomography of large tabular icebergs could provide an experimental configuration very similar to the Rosetta one. For this kind of media, the use of Ground Penetrating Radar could provide profiles of internal structures in order to validate our inversions (Herique & Kofman, 1997).

7. CONSERT performances and its ability to explore interior of comets

The modification of the Rosetta target corresponds in an increase of the expected nucleus radius. The estimated radius of the 46P/W comet is 650 ± 100 m (Lamy, 1996: 620 ± 20 m; Boehnhardt, 1999: 590 ± 50 m or 510 ± 45 m in function of albedo) and for 67P/CG is about 2000 m. This evolution has an impact on the CONSERT link budget. Could it reduce the CONSERT ability to explore interior of comets ? To estimate the performances of CONSERT to explore the interior of the comet we adopted the probabilistic approach.

We know the CONSERT performances: the minimum detectable signal is of about -150 dBW which gives the signal to noise ratio at the input of the CONSERT receiver of $\sim$ SNR$_{input}$ = -30 dB and after the code compression and coherent addition of $\sim$ SNR$_{output}$ = 24 dB. This result corresponds to the comfortable signal to noise ratio allowing easy signal detection.

On the other side, our a priori knowledge of the cometary material properties is low. Assuming that material is principally a porous ice with some silicate and olivine additions, the propagation losses vary between 0 and 2 dB / 100 m (α is 0 to 4.6×10^{-3} m^{-1}). In the following, we will suppose the losses normally distributed around the mean value of 1 dB / 100 m. We will estimate the attenuation as function of cometary diameter which we take variable up to 5000 m.

To calculate losses we suppose that Rosetta is on an orbit of radius 5 to 10 cometary radii. The lander and orbiter antennas gains are both equal to 1. These assumptions compensate the fact that the gain of the orbiter antenna is about 5 dB and that of lander is minus few dB, strongly depending on the lander position and nature of the ground. The attenuation is than calculated using the following formula:

$$attenuation = (\lambda^2/4\pi)G_rG_e exp(-\alpha D)/(4\pi R^2)$$

where R is the distance between lander and orbiter, D the diameter of the comet, antennas gain G_e and $G_r = 1$ and $\lambda = 3.3$ m. We add the 10 dB loss due to the 0.4 W transmitted power and internal system loss.

We calculate the probability that the signal is stronger than -150 dBW (attenuation lower that this value) in function of the cometary diameter. This estimation is done using the 30,000 random draws of loss coefficient for each diameter. This probability is shown in Fig. 7. For comets with diameter smaller than 2200 m (like 46P/W) one has 100 % of probability to go through the cometary interior. This also means that the probability to traverse the whole comet is 100 %. For larger comets like 67P/CG this probability is only of the order of 60 %.

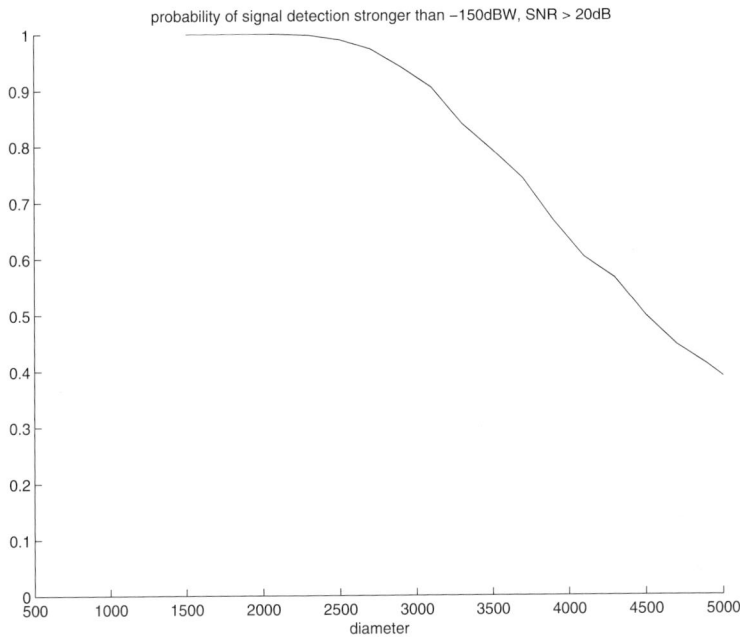

Figure 7.   Probability of signal detection stronger than -150 dBW.

We can also calculate the probability of the exploration of the given fraction of the volume of the 67P/CG comet assuming 4 km diameter. The spherical and camembert like form of the comet were assumed. In Fig. 8, this probability is shown for the spherical comet with 2 km radius. The probability to explore the whole comet is $\sim$ 65 %.

If the comet is close to a cylinder, like camembert (3 km $\times$ 1.5 km) and if the landing site is in the center of the long dimension, the whole comet will be explored. For a cylindrical comet, like camembert (4 km $\times$ 2 km) and if the

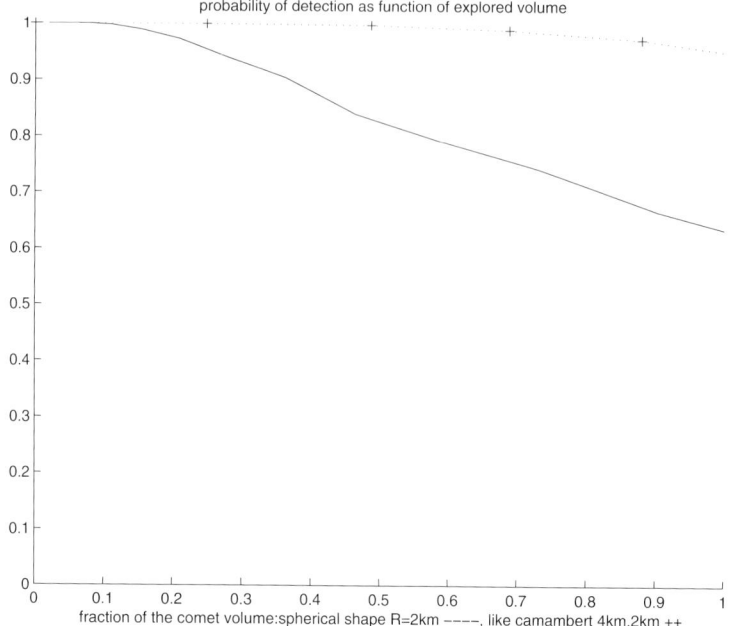

Figure 8. Probability of detection as function of fraction of explored cometary volume. Continuous line: 67P/CG comet assuming a sphere with 2 km radius; crosses: 67P/CG comet assuming the camembert–like shape 4 km × 2 km.

landing site is in the center of the long dimension, the probability to explore the whole comet is 90 % (Fig. 8).

8. Conclusions

CONSERT instrument is ready for launch. Technical realization of CONSERT experiment flight models fulfills the scientific requirements which where defined in the initial project. The probabilistic approach used in this article shows that we have a large possibility of exploring the comet. Even if we only explore a fraction of the comet, the important scientific objectives of the experiment as determination of: value of permittivity, structure of irregularities and partial imaging, will be fulfilled. However, the global image of the interior will not be accessible. Of course this conclusion depends on the cometary materials largely unknown today, on landing site and orbits which should be defined after the first observation period of the comet.

Acknowledgments

The CONSERT experiment is the effort of an international consortium with experienced scientists and engineers from many countries. The principle of the instrument was established in the Laboratoire de Planetologie de Grenoble and in Service d'Aeronomie. The management of the experiment, the design and the construction of the electronic part are made by Service d'Aeronomie in Paris and the design and the construction of the antenna by MPAE in Lindau.

The development of the CONSERT experiment is supported by CNRS, CNES, MPAE and DLR. In the first development stage it was also supported by ESTEC Space Science Department and ASI.

References

Barbin, Y., Kofman, W., Nielsen, E., Hagfors, T., Seu, R., Picardi, G. and Svedhem, H. (1999). THE CONSERT INSTRUMENT for the ROSETTA mission. *Advances in Space Research*, 24:1115–1126.

Barriot, J–P., Kofman, W., Herique, A., Leblanc, S. and Portal, A. (1999). A two dimensional simulation of the CONSERT experiment (Radio tomography of comet Wirtanen). *Advances in Space Research*, 24:1127–1138.

Benna, M. (2002). Generation et inversion de donnees de propagation d'ondes radio a travers un nouyau cometaire (Simulation de l'experience Consert). PhD Thesis, Universite Toulouse III Paul Sabatier.

Benna, A. Piot, Barriot, J.–P. and Kofman, W. (2002). Data set generation and inversion simulation of radio waves propagating through a two–dimensional comet nucleus (CONSERT experiment). *Radio Science*, 37:3-1.

Boehnhardt, H., Delahodde, C., Sekiguchi, T., Hainaut, O., West, R., Spyromilio, J., Tarenghi, M., Schulz, R., Schwehm, G. (1999) First Science Observations from the ESO VLT Kueyen Telescope: Comet 46P/Wirtanen, a Tiny Iceball for ROSETTA. *American Astronomical Society*, DPS meeting n. 31, n. 27.02.

Campbell, M. J. and Ulrichs, J. (1969). Electrical properties of rocks and their significance for Lunar radar observations. *Journal of Geophysical Research*, 74:5867–5881.

Evans, S. and Smith, B. M. E. (1969). A radio equipment for depth sounding in polar ice sheets. *Journal of Scientific Instruments*, 2:131–136.

Herique, A., Kofman, W., Hagfors, T., Caudal, G. and Ayanides, J.–P. (1999). A characterization of a comet nucleus interior: Inversion of simulated radio frequency data. *Planetary and Space Research*, 47:885–904.

Herique, A. and Kofman, W. (2001). Definition of the CONSERT / Rosetta radar performances. *Committee on Earth Observation Satellites (CEOS) SAR Workshop*. Tokyo, 2–5 April 2001, CEOS–SAR01–006:275.

Herique, A. and Kofman, W. (1997). Determination of the ice dielectric permittivity using the data of the test in Antarctica of the Ground Penetrating Radar for Mars'96. *IEEE Geoscience and Remote Sensing*, 35,N. 5:1338–1349.

Kofman, W. et al. (1998). Comet Nucleus Sounding Experiment by Radiowave Transmission. *Advances in Space Research*, 21:1589–1598.

Lamy, P. (1996). Comet 46P/Wirtanen *IAU Circ.* 6478, 2.

Nielsen, E., Engelhardt, W., Chares, B., Bemmann, L., Richards, M., Backwinkel, F., Plettemeier, D., Edenhofer, P., Barbin, Y., Goutail, J.P., Kofman, W., Svedhem, L.H. (2001). Antennas for sounding of a cometary nucleus in the Rosetta mission. *Eleventh International Conference on Antennas and Propagation (IEE Conf. Publ. No. 480)*, IEE, London, UK; pp. 436–441.

Sihvola, A. H. and Kong J. A. (1988). Effective permittivity of dielectric mixtures. *IEEE Transactions on Geoscience and Remote Sensing*, 26:420–429 (correction 27:101–102).

Stevels, J. M. (1980). Local motions in vitreous systems, *Journal of non crystalline solids*, 40:69–82.

ROSINA'S SCIENTIFIC PERSPECTIVE AT COMET CHURYUMOV-GERASIMENKO

Kathrin Altwegg, Annette Jäckel and Hans Balsiger
Physikalisches Institut, Universität Bern, CH-3012 Bern, Switzerland

E. Arijs
Belgisch Instituut voor Ruimte-Aeronomie, B-1180 Brussel, Belgium

J. J. Berthelier
Institute Pierre Simon Laplace, F-94107 Saint-Maur-des-Fossés, France

S. Fuselier
Lockheed Martin Advanced Technology Center, Palo Alto, CA 94304, USA

F. Gliem
Technische Universität Braunschweig, D-38106 Braunschweig, Germany

T. Gombosi
University of Michigan, Space Physics Research Laboratory, Ann Arbor, MI 48109, USA

A. Korth
Max-Planck-Institut für Aeronomie, D-37191 Katlenburg-Lindau, Germany

H. Rème
Centre d'Etude Spatiale des Rayonnements, F-31028 Toulouse Cedex 4, France

Abstract   The Rosetta Orbiter Spectrometer for Ion and Neutral Analysis (ROSINA) on board Rosetta is designed to analyze the volatile material of comet 67P/Churyumov-Gerasimenko. We show in this paper that the scientific requirements related to the volatile material of comets

257

L. Colangeli et al. (eds.), The New ROSETTA Targets, 257–269.
© 2004 *Kluwer Academic Publishers. Printed in the Netherlands.*

in general and specifically to the present Rosetta target can be met by the three ROSINA sensors. ROSINA will answer scientifically important questions concerning the elemental, isotopic and molecular composition of Churyumov-Gerasimenko.

Keywords: COPS, DFMS, RTOF, dynamic range, mass resolution, mass range

Introduction

Comets present a reservoir of well-preserved material from the time of the creation of the solar system, containing important clues to the origin of the solar system material and to the processes that led from the solar nebula to the formation of planets. Some of the material present in comets may even be traced back to the dark molecular cloud from which our solar system emerged. Several interesting questions on the history of the solar system materials can therefore be answered by studying comets, and in particular by studying the composition of the volatile material which is the main goal of the ROSINA instrument.

ROSINA (Rosetta Orbiter Spectrometer for Ion and Neutral Analysis) will answer outstanding questions related to the main objectives of the Rosetta mission. To accomplish the very demanding goals, ROSINA will have unprecedented capabilities, including very wide mass range from 1 amu to > 300 amu, very high mass resolution ($m/\Delta m \approx 3000$, i.e. the ability to resolve CO from N_2 and ^{13}C from ^{12}CH), very wide dynamic range and high sensitivity as well as the ability to determine cometary gas velocities and temperatures.

So far four comets have been visited at close distance by a spacecraft, namely the comets Halley, Grigg-Skjellerup, Borrelly and Wild 2. But only in the case of the Giotto Mission to comet Halley data on the elemental, isotopic and molecular composition of the volatile material have been collected, although with limited mass resolution, mass range and sensitivity (Balsiger et al., 1986; Krankowsky et al., 1986). Additional data on comet composition have been obtained by remote sensing, especially from the two big comets during the last decade, comets Hale-Bopp and Hyakutake (e.g. Crovisier and Bockelée-Morvan, 1999). Both these comets as well as Halley belong to the Oort cloud family. With remote sensing only sufficiently abundant molecules, which have strong absorption lines within the observable frequency ranges, can be detected. This makes it hard to detect minor molecules in Kuiper belt comets by remote sensing. No other mission to a comet with the capability to analyze the volatile material is foreseen in the near future except Rosetta. ROSINA will therefore play a key role in analyzing, for the first time, the elemental, isotopic and molecular composition of the volatile material of a Kuiper belt comet.

Some open questions to be answered by ROSINA

Below are listed the most important of the many topics ROSINA will address once measurements will start near the comet.

- The outgassing rate of comet Churyumov-Gerasimenko (hereafter C-G) varies over many orders of magnitude during its journey around the sun (Tab. 1). It is expected that the composition of the coma will vary with the heliocentric distance, far out releasing the highly volatile material (e.g. CO, CO_2), later on near the sun also less volatile molecules (e.g. water, organics), thus giving evidence of the differentiation in the outer layers of the comet.

Table 1. Expected water production rate and corresponding pressure at 2 km from the nucleus for comet C-G (after Schleicher and Millis, 2003).

Heliocentric Distance	$Q(H_2O)$ $[s^{-1}]$	H_2O density $[cm^{-3}]$ @ 2 km	Pressure [mbar]
1.3 AU/Perihelion	4.1 x 10^{27}	2.0 x 10^{11}	6.0 x 10^{-6}
Peak activity	1.0 x 10^{28}	8.0 x 10^{11}	2.5 x 10^{-5}
3 AU	1.0 x 10^{23}	1.0 x 10^{7}	3.0 x 10^{-10}

In order to cope with these different outgassing rates ROSINA needs a very large sensitivity and a large dynamic range to be able to measure minor components of the volatile material all the way from 3 AU to perihelion.

- From measurements at Halley it is known that the composition of Oort cloud comets is almost solar with respect to most elements except, of course, hydrogen and, more surprisingly, nitrogen (Geiss et al., 1999). It is not known if there is any N_2 in the cometary ice at all. In the outer planets there is a significant enrichment of nitrogen as well as of carbon compared to solar values (Tab. 2).

Because the origin of Oort cloud and Kuiper belt comets is thought to be located at different radial distances in the solar nebula, the amount of carbon and nitrogen may give an indication on the variation of the solar nebula with radial distance or in the subnebulae around the big planets.

Table 2. Enrichment of N and C in the outer solar system relative to the solar abundance.

	N in the form of N_2 or NH_3	C in the form of CH_4, CO or CO_2
Jupiter	3.6 (Atreya et al., 1999)	2.9 (Atreya et al., 1999)
Saturn	2.0 (Briggs & Sackett, 1989)	2.85 (Kerola et al., 1997)
Uranus, Neptune	?	45 (Gauthier et al., 1995)
Comet Halley	0.2 (Geiss et al., 1999)	≈ 1 (Geiss et al., 1999)

N_2, in particular can hardly be detected by remote sensing. In order to determine the N_2 abundance by mass spectrometry the mass resolution of the instrument has to be sufficient to separate N_2 from the abundant CO, which requires a mass resolution of $m/\Delta m > 2500$ at the 1 % level of the CO peak.

- Radio astronomy has allowed the detection of a surprisingly high number of organic molecules in dark molecular clouds. The heaviest molecule observed is Coronene with a mass of 300 amu (Tab. 3). In order to test the hypothesis that such material was incorporated in comets almost unchanged a mass spectrometer has to be able to cover the whole mass range of these molecules with a sufficiently high mass resolution and sensitivity.

Table 3. Examples of molecules within a broad mass range observed in dark molecular clouds (Irvine, 1999).

Molecule	Mass [amu]
$(CH_3)_2CO$	58
C_5O	76
CH_3C_5N	89
HC_7N	99
HC_9N	123
$HC_{11}N$	147
$C_{24}H_{12}$/Coronene	300

- The importance of isotopic measurements has since long been recognized. The high deuterium abundance in the water of comet Halley (e.g. Balsiger et al., 1995) or in HCN (Meier and Owen, 1999) in comet Hale-Bopp gives an important insight into the origin of cometary material and into the physical and chemical conditions under which comets formed. It will be of utmost importance to compare the D/H ratio of a Kuiper belt comet in various molecules with the previously obtained values from Oort cloud comets. New measurements by means of optical determinations of the $^{14}N/^{15}N$ ratio in the CN molecule in comets Hale-Bopp and LINEAR indicate an enrichment of a factor of 2 in ^{15}N (Arpigny et al., 2003) compared to the value on earth. The authors conclude "that the HCN-molecule cannot be the only "parent" of the CN-molecule; the latter must also be produced by some as yet unknown parent(s) in which the ^{15}N isotope is even more abundant". The only way such a molecule can be identified is by in situ mass spectrometry.

- Comet C-G shows very interesting features like seasonal effects, probably a high obliquity of the rotational axis, outbursts far away from the sun, etc. It will be very important to assess the difference in the composition of the bulk with respect to the composition of vents. From these measurements clues on the inhomogeneity of the nucleus can be drawn. ROSINA will be able to measure the composition with a narrow field of view and to determine the dynamic parameters of the neutral gas.

ROSINA Characteristics

The instrument package ROSINA consists of two mass spectrometers and one pressure sensor. The mass spectrometers are the Double Focusing Mass Spectrometer (DFMS) and the Reflectron Time-Of-Flight mass spectrometer (RTOF) that are designed to analyze the cometary neutral gases and cometary ions. The third sensor, the COmetary Pressure Sensor (COPS), consists of a pressure gauge assembly. The digital fully redundant Data Processing Unit (DPU), which weights 2.2 kg and has a mean power consumption of 5 W, controls all three sensors.

The DFMS is a high resolution mass spectrometer with a wide dynamic range and a good sensitivity. It covers a mass range of 12 - 140 amu/e and has a mass resolution of $m/\Delta m > 3000$ at the 1% peak height which corresponds to a resolution of $m/\Delta m > 7000$ at the 50% level.

The RTOF complements DFMS with an extended mass range from 1 to > 300 amu/e and a higher sensitivity. An advantage of the RTOF sensor is that a full mass spectrum of the entire mass range, that is only limited by the signal accumulation memory, is recorded within 100 μs. The mass resolution in the triple reflection mode is $m/\Delta m > 4500$ at the 50% peak height.

The COPS consists of two gauges in order to determine the gas dynamics of the comet. It measures the total density as well as the ram pressure of the comet. Combining the results from both gauges and the known spacecraft orientation relative to the comet nucleus the gas velocity can be calculated. In addition, this sensor serves as a safety instrument for Rosetta.

A special Calibration system termed CASYMIR (CAlibration SYstem for the Mass spectrometer Instrument ROSINA) (Graf et al., 2004; Westermann et al., 2001) has been designed to establish the key performances and key parameters like the mass resolution, the dynamic range and the sensitivity of the sensors. Isotopic ratios and fragmentation patterns for the calibration gas, which will also be used for in-flight calibration, have been determined. All results presented in this publication were obtained at this calibration facility.

Instrument Performance

DFMS The DFMS (Fig. 1) is a very compact state of the art high-resolution double focusing mass spectrometer (Mattauch and Herzog, 1934) realized in the Nier-Johnson configuration (Johnson and Nier, 1953).

Figure 1. ROSINA DFMS in flight configuration without the multi layer insulation. The dimension of the electronic box is 40 x 23 x 18 cm.

Figure 1 shows the fully assembled DFMS model without the surrounding Multi Layer Insulation (MLI). The sensor weights 16 kg and the power consumption averages 22 W. DFMS covers a mass range from 12 amu/e (C^+ ions) to about 140 amu/e (Xe isotopes) and has a high dynamic range as well as a good sensitivity. The mass resolution of $m/\Delta m$ is greater than 3000 (at 1% peak height) and allows to separate, e.g. ^{13}C and ^{12}CH. For a more detailed description of the DFMS sensor see Balsiger et al. (2003).

Fig. 2 demonstrates the dynamic range of DFMS and substantiates the ability of this sensor to measure in the specified mass range from 12 to 140 amu/e. This figure represents a calibration gas mixture spectrum (CO_2, Ne and Xe) in the low resolution mode taken at 5×10^{-7} mbar with the Channel Electron Multiplier (CEM) detector. With an integration time of typically one second the recording of the whole spectrum took approximately two hours. Whereas the larger panel displays mainly the mass peaks of the fragments of CO_2 (m/q = 12, 16, 22, 28, and 44), Ne (m/q = 20 and 22) and the residual gas, the inner panel shows the higher mass peaks of the single and doubly charged Xe ions. In this spectrum DFMS covers a dynamic range (ratio of highest to lowest peak in one measurement) of nearly five orders of magnitude. Even at high pressures (5×10^{-7} mbar) DFMS is able to measure minor species, e.g. m/q = 32 thus covering the calculated pressure ranges at comet C-G (Tab. 1).

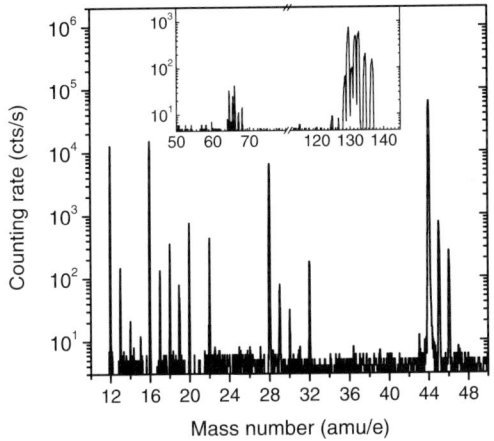

Figure 2. Mass spectrum from 12 to 140 amu/e of the calibration gas (CO_2, Ne and Xe) taken at 5×10^{-7} mbar with the CEM detector.

Figure 3 shows that the mass resolution of DFMS is high enough to measure interesting isotopic ratios of e.g., the two nitrogen isotopes ($^{14}N^+$, $^{15}N^+$). This is of great importance in order to determine and explain the anomalous nitrogen isotope ratios in comets (Tab. 2).

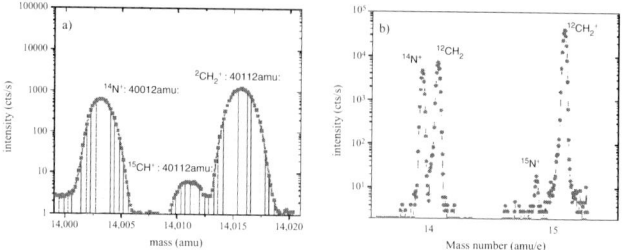

Figure 3. Mass spectrum of 14 and 15 amu/e. The three different peaks at 14 amu/e (a) corresponding to $^{14}N^+$, $^{13}CH^+$, and $^{12}CH^{2+}$ can be separated as well as the two peaks at 15 amu/e ($^{15}N^+$ and $^{12}CH^{3+}$)(b).

RTOF The RTOF (Fig. 4) is the Reflectron Time-Of-Flight mass spectrometer and complements DFMS. RTOF is characterized by a very high mass range and sensitivity in order to identify organic material, e.g. polyaromatic hydrocarbons. RTOF also has a high dynamic range. The sensor weights 15 kg and needs about 30 W. For more details of the RTOF sensor characteristics see Balsiger et al. (2003).

Figure 4. ROSINA RTOF in flight configuration without the surrounding HV protection foil. In this image the instrument is fixed on a baseplate due to tranport reasons. The length of the black electronic box is 68 cm.

Fig. 5 shows the high sensitivity of the RTOF sensor with respect to the pressure range expected at C-G (Tab. 1). This spectrum represents a residual gas spectrum, taken at 1×10^{-9} mbar with an acquisition time of 100 seconds. The time of flight displayed reaches from 10 μs to 20 μs, representing only a part of the whole spectrum. The Kr isotope group at 16.8 μs can clearly be identified. These peaks correspond to $^{78}Kr^+$, $^{80}Kr^+$, $^{82}Kr^+$, $^{83}Kr^+$, $^{84}Kr^+$ and $^{86}Kr^+$. The water group at around 12 μs is observable, as well as the CO_2 peak and the doubly charged

Kr isotope group. Comparing this spectrum to the water production rates of C-G it is obvious that the RTOF sensor will be able to analyze cometary gases, e.g. CO_2 far away from the sun already at the onset of activities of the new Rosetta target.

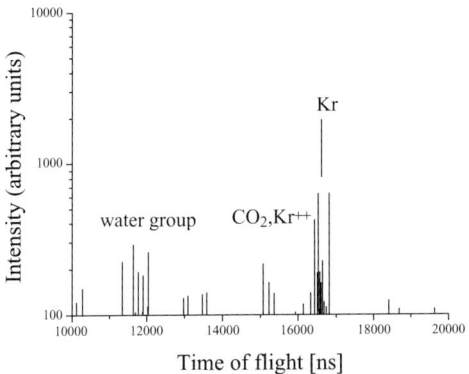

Figure 5. Detail of a residual gas spectrum 10 to 20 μs taken at 1×10^{-9} mbar with an acquisition time of 100 seconds. The water group at around 12 μs is observable, as well as the CO_2 peak and the doubly charged Kr isotope group. The peaks at 16.8 μs belong to the six single charged Krypton isotopes.

In order to cover the mass range of the heavy molecules listed in Table 3 in the environment of C-G, an extended mass range up to at least 300 amu/e in combination with a high sensitivity is required.

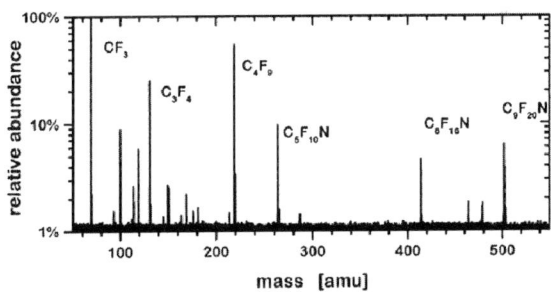

Figure 6. High mass range spectra from 50 amu/e to 550 amu/e was taken by recording a spectrum of the mass compound $(CF_3(CF_2)_3)_3N$ (Heptacosafluorotributylamine) with the mass of the parent molecule at 671 amu/e.

The spectrum in Fig. 6 demonstrates the ability of the RTOF sensor to record high mass range spectra up to the heaviest masses known from remote sensing in dark molecular clouds. This spectrum was taken

266

by using the mass compound $(CF_3(CF_2)_3)_3N$ (Heptacosafluorotributy-lamine) with the mass of the parent molecule at 671 amu. The spectrum shows a characteristic fragment pattern for electron impact ionization.

COPS Fig. 7 shows the COPS instrument in flight configuration without the multi layer insulation. This sensor weights 1.7 kg and needs 7 W. The left gauge is the Nude Gauge, a hot filament extractor type Bayard-Alpert ionisation gauge (Redhead, 1966). This gauge measures the total particle density with a nitrogen sensitivity of about 20 mbar^{-1} at 100 μA electron emission current. The RAM gauge is attached on top of the electronic box, perpendicular to the Nude Gauge. This gauge, with its opening facing towards the comet, measures the molecular flow from the comet. Resulting from these measurements, the velocity and density of the cometary gas can be calculated. More detailed information about COPS and about calibration results can be found in Balsiger et al. (2003) and Graf et al. (2004).

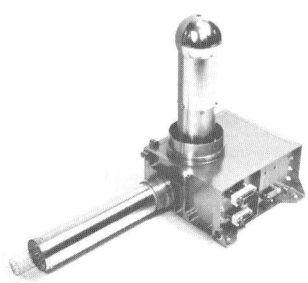

Figure 7. ROSINA COPS in flight configuration without the multi layer insulation. The Nude gauge points to the left side, the RAM gauge is mounted on top of the electronic box that measures 17 x 16 cm.

Fig. 8 represents results from the calibration of the COPS-FS Nude gauge. Pure Argon gas in the pressure range from 5×10^{-10} mbar, corresponding to the residual gas pressure, to 1×10^{-6} mbar was used. The pressure is proportional to the ratio between the ion current and the emission current. In the high pressure region above 1×10^{-7} mbar a low ion current range as well as a medium ion current range have been calibrated.

In Fig. 9 the velocity of the measured test gas Neon is calculated by the pressure difference in the RAM gauge flux. The reference value was measured independently in the calibration system CASYMIR. Combining the ability to determine the gas velocity with COPS and the narrow

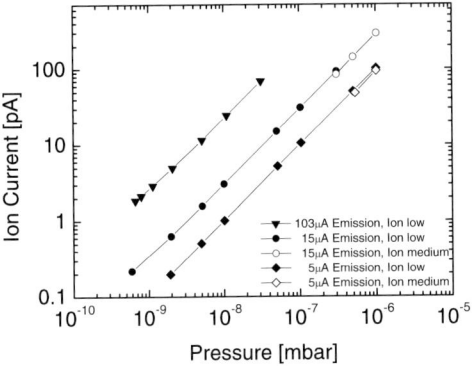

Figure 8. ROSINA COPS-FS Nude gauge calibration (left filament) in the static mode with pure Argon in the pressure range from 5×10^{-10} to 1×10^{-6} mbar. The different emission currents used are 103, 15 and 5 μA.

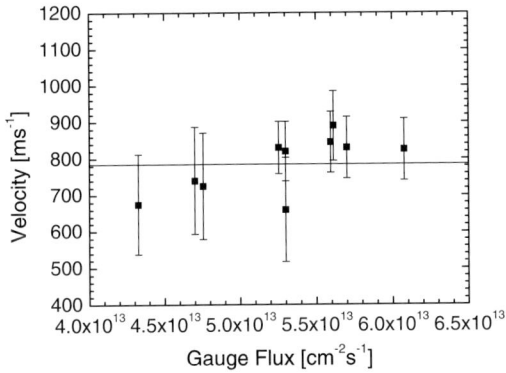

Figure 9. Calculated velocity results by pressure differences in the RAM gauge with Neon test gas at CASYMIR. The solid line represents the reference value that was measured independently.

field of view ($2° \times 2°$) of DFMS the dynamic of vents at comet C-G can be studied.

Conclusions

By comparing the scientific requirements with the characteristics of the ROSINA sensors (Tab. 4) it can be stated that ROSINA is well adapted to the new target of the Rosetta mission.

Table 4. Characteristics of the ROSINA sensors compared to the scientific requirements at comet C-G.

	Scientific Requirements	*RTOF*	*DFMS*
Mass range [amu/e]	1 to > 300	1 to > 300	12 to 140
Mass resolution	> 2500 at 1% level	> 500 at 1% level	> 3000 at 1% level
Pressure range [mbar]	$<10^{-10}$ to $>10^{-6}$	$<10^{-11}$ to $>10^{-6}$	$<10^{-10}$ to $>10^{-6}$
Dynamic range	10^{-5}	$>10^{-5}$	$>10^{-5}$

In particular, ROSINA will:

- measure relevant elemental, isotopic and molecular abundances from the onset of activity through perihelion,

- cope with the activity of C-G at 4 AU as well as at perihelion,

- analyze the composition of the volatile material over a large mass range with a large dynamic range,

- help to understand the dynamics of a comet.

Comet C-G will, due to its different features (Kuiper belt comet, seasonal effects, high activity, etc.), be a challenging but extremely interesting target for the ROSINA instrument.

Acknowledgments

The authors would like to express their gratitude to the whole ROSINA team (e.g. Balsiger et al., 2003) for building this wonderful set of instruments.

References

Arpigny, C., Jehin, E., Manfroid, J., Hutsemékers, D., Schulz, R., Stüwe, J.A., Zucconi, J.-M., and Ilyin, I. (2003). Anomalous Nitrogen Isotope Ratio in Comets. *Science*, 301: 1522–1524.

Atreya, S.K., Wong, M.H., Owen T.C., Mahaffy, P.R., Niemann, H.B., dePater, I., Drossart P., and Encrenaz, T. (1999). A comparison of the atmospheres of Jupiter and Saturn: deep atmospheric composition, cloud structure, vertical mixing, and origin. *Planet. Space Sci.*, 47: 1243–1262.

Balsiger, H., Altwegg, K., Bühler, F., Fischer, J., Geiss, J., Meier, A., Rettenmund, U., Rosenbauer, H., Schwenn, R., Benson, J., Hemmerich, P., Säger, K., Kulzer, G., Neugebauer, M., Goldstein, B.E., Goldstein, R., Shelley, E.G., Sanders, T., Simpson, D., Lazarus, A.J., and Young, D.T. (1986). The Giotto Ion Mass Spectrometer. *ESA SP–1077*, 129–148.

Balsiger, H., Altwegg, K., and Geiss, J. (1995). D/H and $^{18}O/^{16}O$ – ratio in the hydronium ion and in neutral water from in situ ion measurements in comet Halley. *J. Geophys.Res.*, 100: 5827–5834.

Balsiger, H., Altwegg, K., Arijs, E., Bertaux, J.-L., Berthelier, J.-J., Block, B., Bochsler, P., Carignan, G.R., Duvet, L., Eberhardt, P., Fiethe, B., Fischer, J., Fisk, L.A., Fuselier, S.A., Ghielmetti, A.G., Gliem, F., Gombosi, T.I., Illiano, M., Koch, T., Kopp, E., Korth, A., Lange, K., Lauche, H., Livi, S., Loose, A., Magoncelli, T., Mazelle, C., Mildner, M., Neefs, E., Nevejans, D., Rème, H., Sauvaud, J.A., Scherer, S., Schönemann, A., Shelley, E.G., Waite, J.H., Westermann, C., Wilken, B., Woch, J., Wollnik, H., Wurz, P., and Young, D.T. (2003). ROSINA: Rosetta Orbiter Spectrometer for Ion and Neutral Analysis. *ESA SP–1165, in press.*

Briggs, F.H and Sackett, P.D. (1989). Radio Observations of Saturn as a probe of its atmosphere and cloud structure. *Icarus*, 80: 77–103.

Crovisier, J. and Bockelée-Morvan, D. (1999). Remote Observations of the Composition of Cometary Volatiles. *Space Sci. Rev.*, 90: 19–32.

Gauthier, D., Hersant, F., Mousis, O., and Lunine, J.I. (1995). Enrichments in Volatiles in Jupiter: A New Interpretation of the Galileo Measurements. *ApJ*, 550: L227; Erratum, *ApJ*, 559: L183.

Geiss, J., Altwegg, K., Balsiger, H., and Graf, S. (1999). Rare atoms, molecules and radicals in the coma of P/Halley. *Space Sci. Rev.*, 90: 253–268.

Graf, S., Altwegg, K., Balsiger, H., Jäckel, A., Kopp, E., Langer, U., Luithardt, W., Westermann, C., and Wurz, P. (2004). A cometary Neutral Gas Simulator for Gas Dynamic Sensor and Mass Spectrometer Calibration. *J. Geophys. Res.. in press*

Irvine, W.M. (1999). The composition of interstellar molecular clouds. *Space Sci. Rev.*, 90: 203–218.

Johnson, E.G. and Nier, A.O. (1953). Angular aberrations in sector shaped electromagnetic lenses for focusing beams of charged particles. *Phys. Rev.*, 91: 10–17.

Kerola, D.X., Larson, H.P., and Tomasko, M.G. (1997). Analysis of the Near-IR Spectrum of Saturn: A Comprehensive Radiative Transfer Model of Its Middle and Upper Troposphere. *Icarus*, 127: 190–212.

Krankowsky, D., Lämmerzahl,P., Dörflinger, D., Herrwerth, I., Stubbemann, U., Woweries, J., Eberhardt, P., Dolder, U., Fischer, J., Herrmann, I.U., Jungck, M., Meier, F.O., Schulte, W., Berthelier, J.J., Illiano, J.M., Godefroy, M., Gogly, G., Thévenet, P., Hoffman, J.H., Hodges, R.R., and Wright, W.W. (1986). The Giotto Neutral Mass Spectrometer. *ESA SP–1077*, 109–118.

Mattauch, J. and Herzog, R. (1934). Über einen neuen Massenspektrographen. *Z. Physik*, 89, 786.

Meier, R. and Owen, T. (1999). Cometary Deuterium. *Space Sci. Rev.*, 90: 33–43.

Redhead, R.A. (1966). New Hot-Filament Ionization Gauge with Low Residual Current. *J. Vac. Sci. Technol.*, 13: 173–180.

Schleicher, D.G. and Millis, R.L. (2003). Results from Narrowband Photometry of ROSETTA's New Target Comet 67P/Churyumov-Gerasimenko. *DPS 35th Meeting*, oral presentation, 30.06.

Westermann, C., Luithardt, W., Kopp, E., Koch, T., Liniger, R., Hofstetter, H., Fischer, J., Altwegg, K., and Balsiger H. (2001). A high precision calibration system for the simulation of cometary gas environments. *Meas. Sci. Technol.*, 12: 1594–1603.

THE GIADA EXPERIMENT FOR THE ROSETTA MISSION

L. Colangeli, V. Della Corte, F. Esposito, E. Mazzotta Epifani and E. Palomba
INAF – Osservatorio Astronomico di Capodimonte, Napoli, Italy,

colangeli@na.astro.it

J.J. Lopez–Moreno, J. Rodriguez, R. Morales, A. Lopez–Jimenez, M. Herranz and F. Moreno
Instituto de Astrofisica de Andalucia, Granada, Spain

P. Palumbo and A. Rotundi
Università "Parthenope", Napoli, Italy

M. Cosi
Galileo Avionica, Firenze, Italy

The International GIADA Consortium
I, E, UK, F, D, USA

Abstract   The Grain Impact Analyser and Dust Accumulator (GIADA) instrument, on board the ESA Rosetta mission, shall analyse the physical and dynamical properties of grains ejected by the target comet and monitor the coma evolution in terms of dust flux and spatial distribution vs. time. The mission, formerly planned to visit comet 46P/Wirtanen, is now targeted to a rendezvous with comet 67P/Churyumov–Gerasimenko. The present operative mission plan foresees that Rosetta will follow the comet from about 4 AU pre–perihelion to about 2 AU post–perihelion. This will allow us to study, for the first time, the onset and evolution of activity of a comet nucleus and its environment. GIADA is composed by different sub–systems designed to measure mass, momentum and speed of single grains larger than about 30 μm in size and to monitor the cumulative flux of smaller grains coming from different directions. GIADA technical characteristics and scientific performances will guarantee a full monitoring of the

L. Colangeli et al. (eds.), The New ROSETTA Targets, 271–280.

dust environment and the achievement of unprecedented scientific results about cometary dust physics.

Keywords: Rosetta, GIADA, comet dust environment

Introduction

The ESA Rosetta space mission shall perform a rendezvous with the short period comet 67P/Churyumov–Gerasimenko (67P/CG in the following). The mission was originally planned to visit 46P/Wirtanen (46P/W in the following) comet, with a launch in January 2003. Due to problems with the Ariane launcher, the mission had to be postponed for a launch in February 2004 towards the new target.

Among the various scientific objectives of the Rosetta mission, a key aspect is the study of physical – chemical characteristics of cometary grains and their dynamic evolution in the coma. In fact, according to tail and coma models (Crifo & Rodionov, 1997, Fulle et al., 1999), dust plays a relevant role in the coma evolution, during comet approach to the Sun. Despite this interest, the dust–gas links and overall interactions in the inner coma are poorly known. Actually, previous in situ measurements were performed during the Giotto fly–by of comet 1P/Halley (Reinhard, 1986). However, the relative velocity was too high (about 70 km s^{-1}) to retrieve any time resolved information about dust evolution in the coma (McDonnell et al., 1991).

The GIADA (Grain Impact Analyser and Dust Accumulator) experiment, onboard the Rosetta orbiter, is aimed at performing in situ dust measurements for a long period of time. The general aims are to study the dust flux from different directions and to measure the dust mass, size and velocity distribution while the comet approaches the Sun. This information is essential to correlate the coma evolution with the nucleus characteristics in terms of dust emission properties (e.g., active areas or jets) and to determine the dust–to–gas ratio (in combination with gas measurements).

Moreover, real time information on dust environment is crucial to interpret effectively data provided by other instruments onboard Rosetta (e.g., imaging spectrometer, camera, microscope, dust chemical analyser) and to generate alerts whenever dust flux could increase to levels so critical to produce degradation of experiments and/or of the spacecraft (e.g., optical elements, radiators, solar panels).

In the following sections we will describe the GIADA scientific goals, technical characteristics and performances based on pre–flight calibrations. Finally, we will discuss the implications of the change of target comet on the expected GIADA behaviour.

1.   Scientific tasks of GIADA

The GIADA experiment will provide unique scientific data on dust fluxes and dynamic properties, which are a fundamental source of information on dust emission processes, on dust–gas relations in the inner coma environment and on overall evolution of cometary dust in the coma.

Grains ejected by the comet nucleus are sensitive to solar radiation pressure. Therefore, two types of populations exist in the coma: *direct* grains come from the nucleus, while *reflected* particles come from the sun direction, under the action of the solar radiation pressure. The two populations have different dynamic evolution and are characterised by different ejection times. No ground–based or space measurements have allowed us to study the two components separately, so far. In the case of Rosetta, the low spacecraft velocity with respect to the nucleus will allow to measure the contribution of the two dust populations. This is a fundamental task to be achieved in order to determine the dust size distribution at the nucleus (Fulle et al., 1995) and, thus, the dust mass loss rate of the nucleus.

The ejection velocity of grains depends both on their size and on time; a wide dust velocity distribution must be expected. Models applied to the interpretation of ground based dust tail observations have provided global properties on long periods (days) about the time and size dependence of the dust velocity. The models are completely insensitive to the velocity dispersions. With Rosetta it will be possible to measure momentum and velocity of single *direct* grains.

If velocity and momentum will be measured for single grains, their mass will, then, be derived, so providing the functional dependence dust velocity vs. mass, as well as the velocity dispersion.

All the previous measurements will be possible for a long period of time, nominally when the comet is between about 4 AU pre–perihelion and 2 AU post– perihelion. Therefore, the flux of grains vs. size/mass will be measured during different phases of the comet nucleus evolution. The retrieved information will allow us to study the coma and nucleus changes vs. heliocentric distance, to identify any asymmetric behaviour with respect to perihelion and to look for correlation between coma anisotropy (e.g., jets) and surface inhomogeneity (e.g., active areas). One of the crucial parameters characterising the comet nucleus, the dust to gas ratio, will be determined.

GIADA is designed to accomplish all the previous scientific tasks. In particular, the instrument will measure: flux and fluence of dust, momentum, scalar velocity and mass of single grains. Some information about the optical properties (hence, on morphology and composition) of single grains will also be obtained, as described in the following section.

274

Information provided by GIADA about dust abundance and physical–dynamic properties is also essential in a synergistic effort with other Rosetta experiments. In fact, GIADA results are complementary to data about chemical properties, morphology and structure of grains coming from, e.g., COSIMA and MIDAS experiments. The results coming from dust monitoring will be also useful to estimate coma opacity, needed to interpret at best optical and infrared images. Moreover, the GIADA data will be essential to characterise the dust population and emission rate of different surface areas, since early mission phases: a key information for the proper selection of the site where to deliver the lander. Last but not least, estimates of deposition rates of solid materials vs. time on the spacecraft and/or critical surfaces (e.g., optical elements, radiators, solar panels) will be critical to guarantee performances of dust sensitive parts.

2. GIADA instrument description

The GIADA instrument (Figure 1) includes three main modules: GIADA–1, 2 and 3. GIADA–1 is aimed at measuring the momentum and the scalar velocity of single grains. It is oriented towards the nucleus so to detect *direct* particles, mainly. The underlying GIADA–2 module is the electronic box and contains the DPU, which is interfaced with sub–systems and spacecraft. It controls the acquisition of data and operation of the whole instrument. The GIADA–3 module is devoted to the measurement of the cumulative dust deposition in time, and thus of the flux, by five quartz micro–balances. They point towards different directions: the nucleus and other four, to cover the widest possible solid angle. A multi–shot cover protects GIADA–1 and GIADA–3 sensors and is maintained closed during the cruise to the comet.

The GIADA–1 sub-system has a sensitive area of 10×10 cm^2 and an acceptance angle of about 40 deg. It includes two measurement devices placed in cascade. The first stage, called *Grain Detection System* (GDS), is devoted to detect single grains entering the instrument, to measure their velocity and to obtain information about their morphology and, possibly, composition. The working principle of the GDS is based on optical detection of grains crossing a light curtain (3 mm thick) generated on the XY plane by an illumination – collimator system. The light sources are four laser diodes (emission at 915 nm). The radiation scattered at 90 $^\circ$ by the crossing grain is monitored by two series of receivers, which measure also the time of flight across the curtain, so that the grain speed is derived. Since the detected signal is related to the geometric cross–section and the scattering efficiency of the grain, hints about particle size, composition and aggregation status can be retrieved. The optical detection guarantees that the dynamical properties of the grain are not altered by the detection (see Mazzotta Epifani et al., 2002 for more technical details on the

Figure 1. The Flight Model of the GIADA instrument during calibration. GIADA is shown with the cover open. The 5 MBS's and the entrance to the GDS + IS system are clearly visible (courtesy Galileo Avionica, Firenze).

GDS). The bottom *Impact Sensor* (IS) stage is formed by a square aluminium diaphragm equipped with five piezoelectric transducers (PZT's), placed below the centre and each corner of the plate. Each grain impact event generates acoustic waves which propagate along the plate. When a wave reaches a sensor, this begins to vibrate at its resonant frequency and generates a voltage signal proportional to the incident grain momentum. The proportionality factor is $(1 + e)$, with e = coefficient of restitution. The maximum peak of the principal wave (the first wave packet) is proportional to the normal component of the grain momentum (see Esposito et al., 2002 for more technical details on the IS). A clock is started when the GDS detects a grain, that is stopped when the grain hits the IS plate. The time–of–flight between the two stages is derived and, thus, the grain velocity (knowing the GDS–IS distance). Then, the coupled GDS – IS system will provide for each grain entering the instrument the momentum, p, and the speed v, from which the mass, m, is obtained. The measurement of m can be converted in size, when the density of grains is measured by other experiments on board Rosetta. Counting of single grain events will be used to determine dust flux vs. size and time.

The GDS – IS system is mainly sensitive to the population of grains larger than about 30 μm (see the next section), thus an alternative system is to be used for monitoring micron / sub–micron grains. In the latter case, single grain detection is rather complex, also considering the increasing abundance of grains expected vs. size decrease (e.g., Mueller & Gruen, 1998), and cumulative mass detection is more suitable. *Micro–Balances* (MBS's) provide an output frequency signal proportional to the mass deposited on the sensor. According to the working principle, the measured physical quantity is the shift of the resonance frequency of a quartz oscillator. The shift is due to the variation of its mass, as a result of material accretion. By using specially cut crystals, whose frequency has an extremely small temperature dependence, a high sensitivity can be achieved. An improvement of the detection system is obtained by mixing the signal from the sensing crystal with that of a second quartz crystal, used as reference, and measuring the beat frequency of the mixed signals. Each of the five MBS's used in GIADA consists of a matched pair of quartz crystals, resonating at frequencies of ~ 15 MHz. The sensing crystal (surface ~ 0.1 cm^2) is exposed to the environment for dust accumulation and is displaced in frequency approximately 1 KHz below the reference crystal. The output of a mixer circuit gives a signal which is linearly related to the mass deposition in a frequency range up to about 1 % of the resonating frequency.

3.   GIADA calibration and scientific performances

In order to test actual GIADA scientific performances extensive calibration campaigns have been carried on the various versions of GIADA experiment, including the Flight Model. In this section we give some hints about the methodology applied for determining the key parameters needed to convert engineering readings (e.g., voltages) in actual physical quantities.

For the MBS's the calibration has been performed on (commercial–type) sensors identical to the space qualified devices used on–board GIADA. In fact, cleanliness considerations prevent the possibility to perform actual dust deposition tests on the flight units without affecting their characteristics in flight. Among others, scientific tests performed on the MBS laboratory units have been targeted to demonstrate the capability of these systems to detect deposition of dust with different sizes. The results of these measurements have been already reported in a previous paper (Palomba et al., 2002). They demonstrate that the sensitivity to mass deposition of grains smaller that ~ 5 μm is close to the nominal value (about $\sim 10^{-10}$ g), while the saturation limit is 10^{-4} g.

For the GDS and IS systems, calibrations have been performed by using actual grains, selected with different sizes (classes: $50 - 100$, $100 - 200$ and $200 - 500$ μm), speed (between 1 and 70 m s^{-1}) and composition (e.g., silicates, such as andesite and nontronite, and amorphous carbon grains). Due to the difficulties in handling such small grains and in order to reduce contamination risks, the use of real grains has been limited to tests in some reference points of the GDS and IS sensitive areas.

To have an information, complementary to the one obtained with real grains, about the sensitivity behaviour over the whole GDS and IS sensitive areas, a *relative* calibration has been also performed. A sensitivity map for the two (left and right) GDS detectors has been measured by applying a fixed stimulus (generated by moving a 13 μm chromel wire) in a raster of points through the whole sensitive area. Similarly, for each PZT sensor of the IS, a sensitivity map has been determined by applying a fixed stimulus to a raster of points on the aluminium plate. The stimulus has been generated by using a metallic tip, free moving perpendicularly to the plate and stimulated by another PZT piloted by a 5 V square wave. An example of IS sensitivity map for the central PZT is shown in Figure 2.

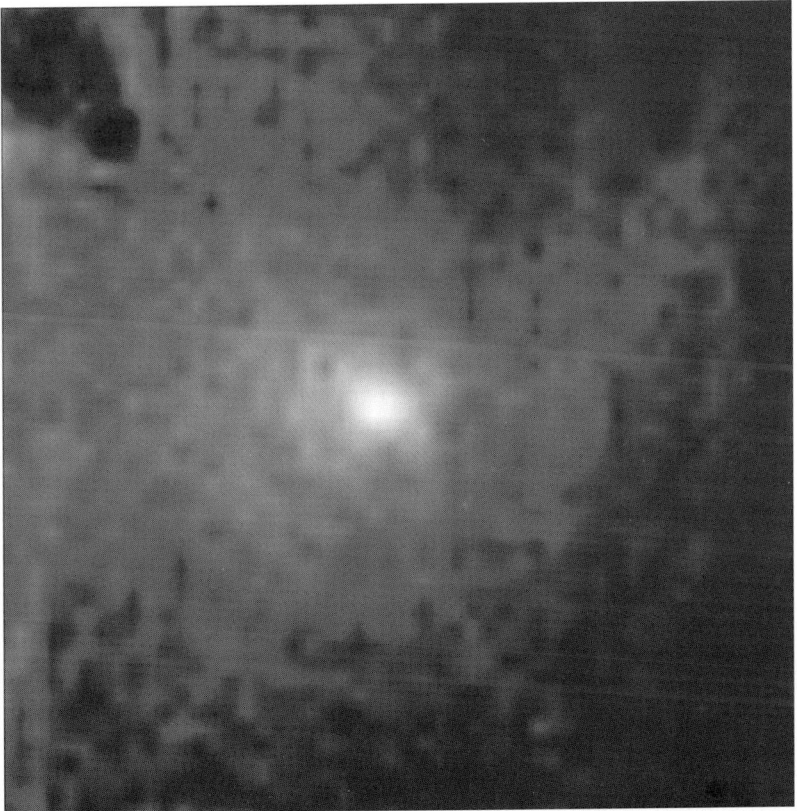

Figure 2.   Sensitivity map for the central PZT sensor over the whole 10×10 cm^2 IS area. According to the colour coding, darker areas are less sensistive. As expected, the maximum sensitivity is attained at the (central) PZT position.

The calibration program has allowed us to demonstrate that the sensitivity limit of the GDS corresponds to grain radius $= 15$ μm (for andesite grains) and that saturation occurs at $\sim 500 - 1000$ μm. Particles larger than the saturation limit are anyway detected, although the signal is saturated. Moreover, calibration curves shown in Figure 3 demonstrate the capability of GDS to discriminate particles of different grain composition (i.e., scattering efficiency). It is useful to stress that this result is an add–on with respect to the actual instrument requirements and can be used to derive some information about the nature of cometary grains tested by GIADA, to be compared with the results of other Rosetta experiments dedicated to grain chemical analysis (e.g., COSIMA).

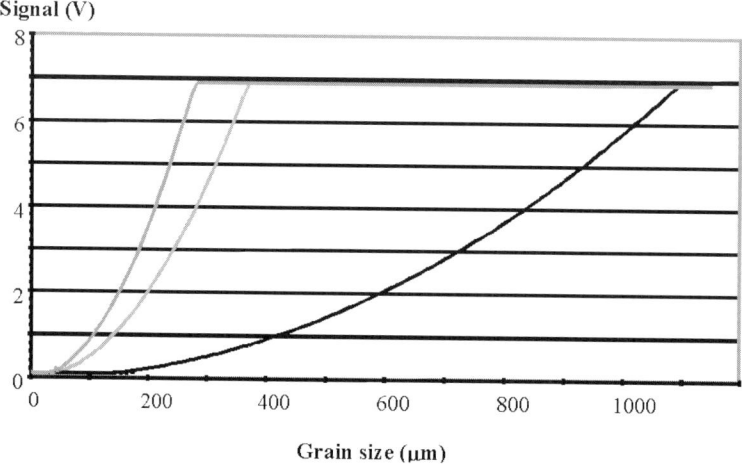

Figure 3. Calibration curves of GDS. The three curves refer to andesite, nontronite and carbon grains from left to right, respectively.

For the IS, the calibration campaign with grains has allowed us to determine a sensitivity limit of 6.5×10^{-10} N s and a saturation level of 4.0×10^{-4} N s. Moreover, the sensitivity maps of the IS diaphragm for each of the five PZT's allow to derive the exact momentum released by a particle impinging on any point of the sensing plate. Of course, to this aim the impact position must be determined by analysing the time delays of signal detection at each of the sensors. It is useful to note that the momentum detection limit of the IS roughly corresponds to a grain of 20 μm at 20 m s^{-1}. This is nicely complementary to the GDS size detection limit and confirms that the two systems shall monitor the same class of grains.

4.   GIADA vs. comet 67P/Churyumov–Gerasimenko

The results reported in the previous section demonstrate that GIADA is able to detect grains from sub–micron to mm sizes in a wide speed range.

According to its working principle, GIADA is an *event–driven* experiment. This means that to collect an amount of information statistically meaningful, GIADA will have to be switched on for long periods during the various mission phases at the comet. Of course, the actual results will depend on the encountered comet dust environment. Therefore, the change of Rosetta mission target had to be carefully considered to check that the wide ranges of GIADA detection were compatible with model predictions. Data about 46P/W dust environment were derived, based on various observation campaigns (see, e.g., Mueller & Gruen, 1998; Fulle et al., 1999). Predictions for the dust environment of 67P/CG have been recently reported by Fulle et al., 2004. Between 8.5×10^5 and 3.0×10^6 events (depending on model assumptions) were expected from 46P/W for the whole mission (from 3.2 to 1.0 AU) for grains with radius > 15 μm, i.e. detectable by the GIADA GDS + IS system. In the case of 67P/CG the expected events for the same class of grains is between 4.3×10^5 and 5.9×10^6, from about 4.0 AU pre–perihelion to 2.0 AU post– perihelion. In the size range (radius < 2 μm) of MBS's sensitivity, the comparison is in terms of integrated mass: 2.4×10^{-7} g $\div$ 6.5×10^{-8} g for 46P/W and 1.1×10^{-4} g $\div$ 1.6×10^{-5} g for 67P/CG. All these evaluations are based on the (simplified) assumption of a constant distance of the Rosetta spacecraft from the comet of 50 km along the whole mission. The comparison demonstrates that the total events and cumulative mass are expected more abundant for the new target, but still fully compatible with the sensitivity and saturation limits of the GIADA sensors.

In conclusion, GIADA characteristics and performances are fully compatible with the expected comet environment, despite the change of mission target. We confirm that GIADA will produce data never obtained so far and capable to provide a new and deeper knowledge of the comet coma environment and its relations with nucleus evolution.

Acknowledgments

The GIADA instrument has been funded by the Italian (ASI) and the Spanish space agencies.

References

Crifo, J.F. and Rodionov, A.V. (1997). The dependence of the circumnuclear coma structure on the properties of the nucleus. *Icarus*, 127:319–353.

Esposito, F., Colangeli, L., Della Corte, V., Palumbo, P., & the International GIADA Team (2002). Physical aspect of an impact sensor for the detection of cometary dust momentum onboard the Rosetta space mission. *Adv. Space Res.*, 29:1159–1163.

Fulle, M., Colangeli, L., Mennella, V., Rotundi, A. and Bussoletti, E. (1995). The sensitivity of the size distribution to the grain dynamics: simulation of the dust flux measured by GIOTTO at P/Halley. *Astronomy and Astrophysics*, 304:622–630.

Fulle, M., Crifo, J.F. and Rodionov, A.V. (1999). Numerical simulation of the dust flux on a spacecraft in orbit around an aspherical cometary nucleus - I. *Astronomy and Astrophysics*, 347:1009–1028.

Fulle, M., Barbieri, C., Cremonese, G., Rauer, H., Weiler, M., Milani, G. and Ligustri, R. (2004). The dust environment of comet 67P/Churyumov-Gerasimenko. *This volume*.

Mazzotta Epifani, E., Bussoletti, E., Colangeli, L., Palumbo, P., Rotundi, A., Vergara, S., Perrin, J.M., Lopez Moreno, J.J. and Olivares, I. (2002). The grain detection system for the GIADA instrument: design and expected performances. *Adv. Space Res.*, 29:1165–1169.

McDonnell, J.A.M., Lamy, P.L. and Pankiewicz, G.S. (1991). Physical properties of cometary dust, in Comets in the post Halley era, eds. R.L. Jr. Newburn, M. Neugebauer, J. Rahe (Dordrecht, Kluwer), 1043–1073.

Müller, M. and Grün, E. (1998). ESA document RO-ESC-TA-5501.

Palomba, E., Colangeli, L., Palumbo, P., Rotundi, A., Perrin, J.M. and Bussoletti, E. (2002). Performance of micro-balances for dust flux measurement. *Adv. Space Res.* 29:1155–1158.

Reinhard, R. (1986). The Giotto Encounter with Comet Halley, *Nature*, 321:313–318.

IMPLICATIONS OF THE NEW TARGET COMET ON SCIENCE OPERATIONS FOR THE ROSETTA LANDER

J. Biele, S. Ulamec
DLR, Institut für Raumsimulation, 51170 Köln, Germany
jens.biele@dlr.de

Abstract   The change of target comet from 46P/Wirtanen to 67P/Churyumov-Gerasimenko will have consequences on the Rosetta Lander mission, because the nucleus of 67P/Churyumov-Gerasimenko is expected to be considerably larger ($r \approx$ 2km instead of $r \approx$ 600m) than that of Wirtanen and its dust production seems to be far greater. A mass of about thirty times larger than that of the original target is currently estimated, making nominal landing increasingly risky in case of a high nucleus density (i.e., $>1g/cm^3$). Science operations, however, seems only to be affected by a probably (much) shorter descent time. Operations on the surface can basically follow the plans as prepared for the Wirtanen case with some exceptions.

1.   Introduction

The original Rosetta mission foresaw a launch in January 2003 to reach comet 47P/Wirtanen in 2011 at a heliocentric distance of about 3 AU. Due to some uncertainties regarding the reliability of the Ariane 5 launcher (after a catastrophic failure in December 2002) the Rosetta launch had been postponed and a new mission was studied. Both, spacecraft constraints (e.g. no swing-by at Venus due to thermal reasons) and given launcher performance strongly limited the number of possible alternatives. After careful investigation a decision was taken for a mission to comet 67/P Churyumov-Gerasimenko (see Fig. 1) with a launch date in February 2004. This new mission scenario does have major consequences on the Rosetta Lander mission because the nucleus of Churyumov-Gerasimenko is expected to be considerably larger than that of Wirtanen. The current best estimate (Lamy & Schulz, 2003) assumes a radius of about 2.0 km and thus a mass of about two orders of magnitude larger than that of the original target. This impacts strongly on the Lander separation, descent and landing scenario. Analyzes on the increased landing risk

L. Colangeli et al. (eds.), The New ROSETTA Targets, 281–288.
© 2004 *Kluwer Academic Publishers. Printed in the Netherlands.*

Figure 1.   New target comet, courtesy Rita Schulz

on Churyumov-Gerasimenko have lead to modifications of the landing gear in order to cope with the higher impact velocities.

2.   Scientific objectives and payload as is

It is the general task of the scientific investigations carried out by the Rosetta Lander to get a first in-situ analysis of primitive material from the early solar system, to study the structure of a cometary nucleus, reflecting growth processes in the early solar system and to provide ground truth for Rosetta Orbiter instruments (see also Biele et al., 2002). The scientific objectives of the Lander comprise:

- The determination of the composition of cometary surface matter: bulk elemental abundances, isotopes, minerals, ices, carbonaceous compounds, organic volatiles - in dependence on time and insolation.

- The investigation of the structure, physical, chemical and mineralogical properties of the cometary surface: topography, texture, roughness, mechanical, electrical optical and thermal properties.

- The investigation of the local depth structure (stratigraphy), and the global internal structure.

- Investigation of plasma environment.

These scientific objectives are independent of the target comet and unchanged for the new mission!

The payload of the Rosetta Lander, which consists of 10 individual instruments, is listed as follows:

Table 1. Rosetta Lander scientific instruments

Instrument	Principal investigator	Responsible (PI-)institute
APX-spectrometer	R. Rieder	Max Planck I. f. Chemistry (D)
COSAC (evolved gas	H. Rosenbauer	MPI for Aeronomy (D)
MODULUS analyzers)	I. Wright / C. Pillinger	Open University (UK)
ÇIVA (imaging system)	J.P. Bibring	IAS (F)
ROLIS (imaging system)	S. Mottola	DLR (D
ROMAP (magnetometer/ plasma monitor)	U. Auster, I. Apathy (merge of 2 instrument proposals)	MPI for extraterr. Physics, TU Braunschweig(D), KFKI (H)
SESAME (acoustic properties, dust impact analyzer, permittivity probe: merge of 3 instrument proposals)	K. Seidensticker W. Schmidt, I. Apathy D. Möhlmann	DLR (D), FMI (SF), KFKI (H) DLR (D)
MUPUS (temperature, physical properties)	T. Spohn	University of Münster (D)
CONSERT (radio wave experiment)	W. Kofman	LPG (F)
SD2 (drill and sampler)	A. Ercoli-Finzi	Politecnico Milano (I)

The Lander is shown in figs. 2a, 2b and 3.

3. The new mission and the Lander

Possibilities for alternative missions for Rosetta have been worked out by ESOC, considering restrictions like launcher availability (Ariane 5G+ or Proton-DM, respectively), storage time on ground (< 2.5 years) cruise time (<10 years) and, most importantly, scientific interest of the target comet. Only three scenarios turned out be considered as realistic alternatives:

284

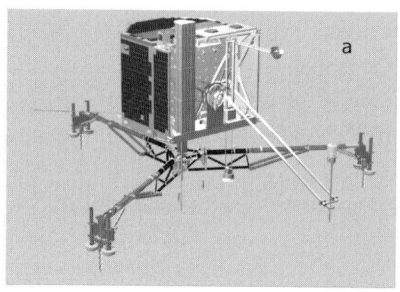
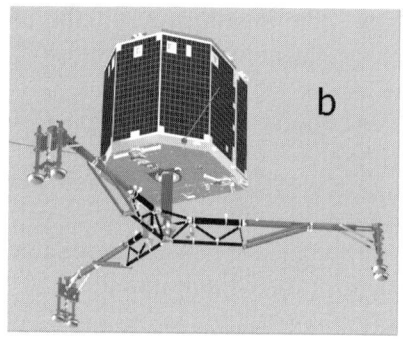

Figure 2a. View of the Lander as operating on the comet (CAD)

Figure 2b. View of the Lander just before touch-down

Figure 3. Rosetta Lander FM mounted to Orbiter PFM

- Mission to 46/P Wirtanen; launch January 2004, requiring modified Proton-DM

- Mission to 67/P Churyumov-Gerasimenko; launch end February 2004, Ariane 5 G+

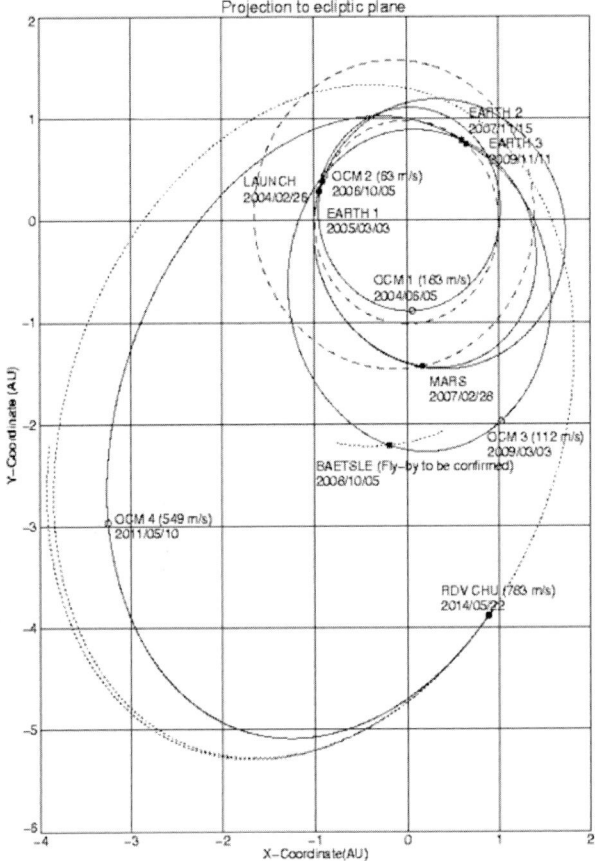

Figure 4. Rosetta trajectory (courtesy ESOC)

- Backup mission to 67/P Churyumov-Gerasimenko, launch 2005, Proton-DM or Ariane 5 ECA

The (Proton) mission to Wirtanen, although preferred for scientific reasons as well as in thought of minimal modification of the descent-landing strategy, was too difficult to achieve due to the necessary adaptations of the Proton launcher within the relatively short time frame.

Figure 4 shows the chosen trajectory (Sánchez Pérez & Rodríguez Canabal, 2003). After swing-by manoeuvres at Earth (March 2005, November 2007 and November 2009) and Mars (February 2007), and up to two asteroid flybys the target comet, 69P/ Churyumov-Gerasimenko shall be reached in 2014. From the comet orbit the spacecraft will provide detailed information on the properties of the comet nucleus to parameterize the Lander separation and descent

286

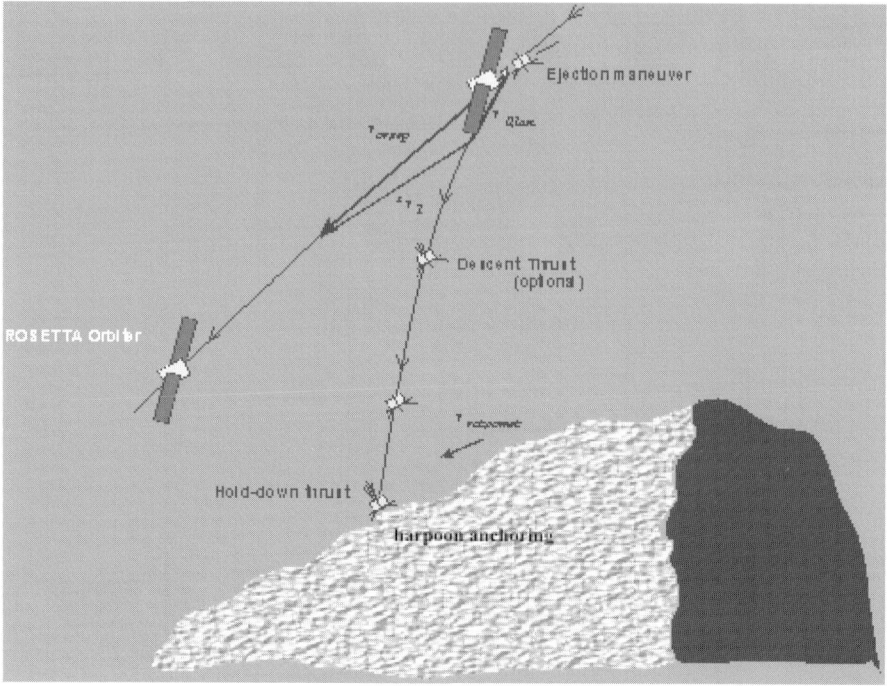

Figure 5. Rosetta Lander landing scenario. The harpoon anchoring is now a part of the Mission Analysis

sequence. The mapping time period must not be too short in order to construct detailed models of the nucleus, the coma and the landing site(s).

The delivery of the Lander to the surface of the comet is foreseen in November 2014 at a distance of 3 Astronomical Units (AU) to the Sun (Sánchez Pérez & Rodríguez Canabal, 2003).

As in the case of Wirtanen the Rosetta Lander will separate from the Orbiter with an adjustable velocity of 5 to 52 cm/sec and descend, stabilized by an internal flywheel to the comets surface. A cold gas system, originally foreseen to increase the descent and impact velocity may be used only in special cases and as hold down thruster immediately after touch down. Figure 5 illustrates the landing scenario.

The assumed properties of the nucleus of CG are summarized in table 2. The much larger radius (compared to Wirtanen) implies a higher risk for landing, see figure 6.

The landing gear was designed for impact velocities up to 1 m/s and optimized for a touchdown with about 50 cm/s, according to the Wirtanen mission (Rosetta Lander - Final Report, 1999). Since the new target comet is larger,

Table 2.   Assumed properties of nucleus. Comet 67P preliminary Engineering Model

	minimal	*typical*	*maximal*
radius (km)	1.5	2.0	2.5
density (g/cm^3)	0.2	0.5	1.3
rotation period (h)	12	12	240
surface compressible strength (Pa)	600	10^5	$2 \cdot 10^6$
Young modulus (Pa)	$7 \cdot 10^5$	10^8	$5 \cdot 10^9$
tensile strength (Pa)	10^3	10^5	10^6

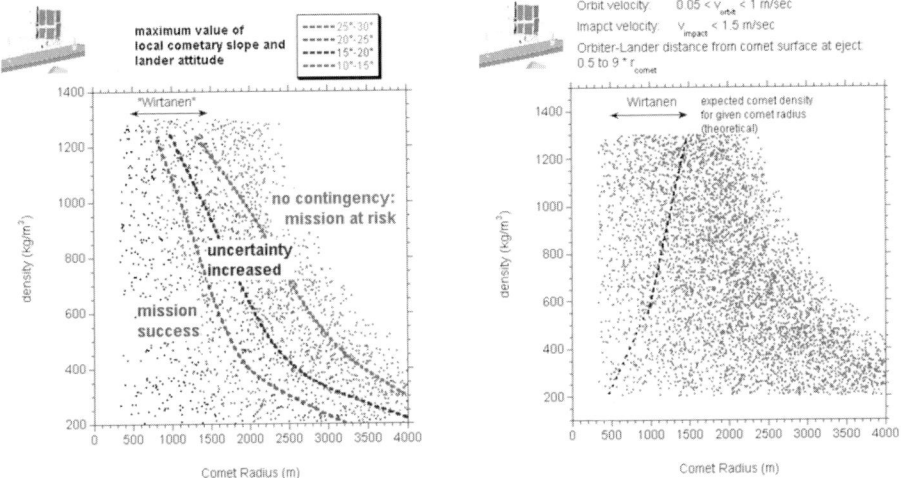

Figure 6.   Monte Carlo Analysis: Landing capability/risk as a function of radius and density of comet nucleus (Hilchenbach, 2003). Impact velocities up to 1.5 m/s and orbit velocity up to 1 m/s. Depending on the actual density (and partly on the rotation period) landing may even be impossible staying within the given requirements of maximum impact velocity (1 m/s) and delivery orbit (separation altitude >1 radius; no instable orbits).

depending on the actual density of the comet also higher touch down velocities must be covered.

A careful analysis revealed that the landing gear survives impacts of up to 1.5 m/s. However, in this case for local slopes greater than a few degrees, the dynamics at landing become very critical regarding the danger of capsizing. Consequently the subsystem has been mechanically modified. Satisfying results are provided by stiffening the cardanic joint and preventing it from any tilt greater than 5°.

288

4. Implications on science operations

All instruments will be able to operate during the new mission. The effects of the possibly different dust environment on science operation still need some investigation. The most important change concerns the descent: Due to the high mass of CG and the consequently very short descent time, some time-consuming experiment operations or operation of units with high data volumes (imagers) will very likely have to be shortened. This is relevant for the panoramic and descent cameras (CIVA/ROLIS) , thermal mapper (MUPUS-TM) and acoustic calibration (SESAME). In an emergency descent scenario (descent time $\lesssim$ 20 min) apart from the vital subsystems only few experiment operations can be foreseen. Since Churyumov-Gerasimenko seems to be rotating with a period of about 12 hours (compared to Wirtanen with about 7 hours), from a thermal point of view the new mission hardly affects the thermal behaviour.

5. Conclusions

Although the mission to comet Churyumov-Gerasimenko does imply an increased landing risk due to a higher impact velocity, the Rosetta Lander can cope with the expected range of possible nucleus parameters. However, depending on the actual density and shape (and partly on the rotation period) landing may be impossible staying within the given requirements of maximum impact velocity (1 m/s) and delivery orbit (separation altitude >1 radius; no instable orbits). Therefore a so-called "dive orbit" bringing the Rosetta S/C for a short while (during Lander separation) on a collision trajectory with the comet nucleus must be considered. Allowing this, worst case scenarios regarding density, size and shape could be covered. From a scientific point of view the new target seems to imply no obvious drawback compared to Wirtanen.

References

Biele, J., Ulamec, S., Feuerbacher, B., Rosenbauer, H., Mugnuolo, R., Moura, D. and Bibring, J.-P. (2002). Current Status and Scientific Capabilities of the Rosetta Lander Payload. *Advances in Space Research*, 29(8), 1199-1208.
Hilchenbach, M. (2003). Presentation at Mission Analysis Working Group, 31.03.2003.
Lamy, Ph. & Schulz, R. (2003). Presentations at the 13th Rosetta SWT, 16/17.06.2003 (covering observation results at ESO and with the Hubble Space Telescope).
Rosetta Lander Mission Analysis Working Group, Final Report; RO-ESC-RP-5003, April 1999
Sánchez Pérez, J. M. & Rodríguez Canabal, J. (2003). Rosetta: Consolidated Report on Mission Analysis. Churyumov-Gerasimenko 2004; RO-ESC-RP-5500, ESOC.

FIRST CONTACT WITH A COMET SURFACE: ROSETTA LANDER SIMULATIONS

M. Hilchenbach, H. Rosenbauer and B. Chares

Max–Planck-Institut für Aeronomie, 37191 Katlenburg–Lindau, Germany

Abstract   At the end of the year 2014 the Rosetta Lander will be ejected from the Rosetta orbiter and, after a descent time of about half an hour, touch the surface of 67P/Churyumov–Gerasimenko, the selected new mission target comet. This comet has a radius of about 2 km. The Lander impact velocity is increased as compared to the former mission target, 46P/Wirtanen, from about 0.5 m/s to 1.2 m/s. The 3-D simulations of the Lander touch–down on the comet surface are carried out in the frame of a multibody analysis and the new landing scenario with the higher impact velocity is analysed.

Keywords: Rosetta, comet, impact simulations

1.   Introduction

Rosetta is the corner stone mission of the European Space Agency (ESA) designed to rendezvous with a comet. The mission target was up to the year 2003 comet 46P/Wirtanen. In January 2003 the envisaged launch had to be postponed due to technical problems of the Ariane V rocket. A few months later the comet 67P/Churyumov–Gerasimenko was selected as the new target for the Rosetta mission. The Rosetta orbiter will release a surface science package or Lander which will land on the comet surface at the end of the year 2014. This Lander is prepared to carry out an extended science analysis on the comet nucleus surface. The new target comet has an observed nucleus radius of about 2 km (for 4% albedo, Lamy et al., 2003) and it is significantly larger than the former target comet. Assuming the same comet density range, the envisaged impact velocity of the Rosetta Lander is increased. The momentum of the Lander increases by a factor of up to 3 and the kinetic energy of the Lander by a factor of up to 10 as compared to the former target comet parameters. The impact of the Lander on the comet surface, as simulated in Hilchenbach et al. (2000), is revisited, the new landing scenario analysed and potential Lander adaptations considered.

L. Colangeli et al. (eds.), The New ROSETTA Targets, 289–296.
© 2004 *Kluwer Academic Publishers. Printed in the Netherlands.*

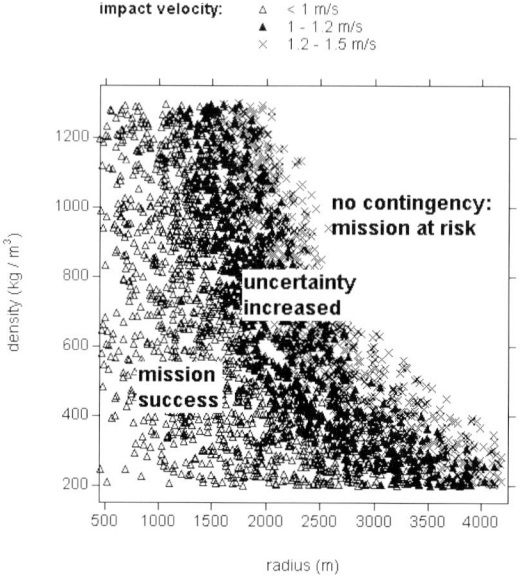

Figure 1. Monte Carlo calculation for target comets: Variation of radii and densities. The orbit is assumed to be circular and the Lander is ejected with a velocity in the range of 0.05 and 0.5 m/s. Minimum altitude of the Rosetta orbiter is 1 km, comet surface points are assumed to rotate with about 0.3 m/s.

2. Lander impact velocities as expected for new comet target comet

The principal limitation for the impact velocity from the Lander design point of view is the length of the damping mechanism of 20 cm. For the Rosetta Lander with a mass of about 96 kg and for the given damping constant of about 850 Ns/m (additional friction about 30 N), the maximum impact velocity with complete damping of the kinetic energy in the landing gear damping mechanism is about 1.5 m/s. This estimate only applies to the landing on a flat surface which is in reality extremely unlikely. For more authentic surface slopes on the Lander scale size of about 3 m, the Lander was designed for impacts with up to 1 m/s on slopes of up to 30°.

The orbiter is assumed to be in a circular equatorial orbit. The nucleus is assumed to rotate with a surface velocity of about 30 cm/s. The target comet is assumed to have a spherical shape, even so the real target is much more likely elongated with axis lengths of 3 km and 5 km (Lamy et al., 2003). The Lander eject velocity is up to 52.5 cm/s, however the rebound of the orbiter has to be taken into account, resulting in an effective orbital velocity between

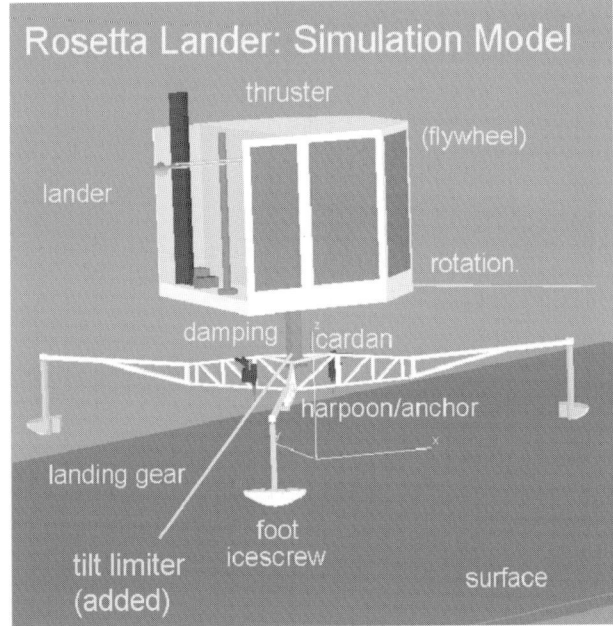

Figure 2. The Rosetta simulation model: A schematic view of the geometry and the dynamical elements are shown. The main elements are damping mechanism, cardan, thruster, and anchor or harpoon. The cardan and landing gear can store energy on impact due to their elasticity. The position of the tilt limiting device restricting the angular rotation of the cardanic joint to about 3° is shown. Its purpose is to prevent the Lander body touching a landing gear leg, i.e. for impacts on steep surface slopes with high impact velocities.

5 and 50 cm/s. The expected impact velocities are estimated via a Monte Carlo approach for varying cometary radii and densities and the results are shown in Fig. 1. For comet densities up to 1300 kg/m^3 and radii from 500 to 1500 m, the Lander was designed to land with impact velocities of up to 1 m/s on slopes of about 30°. Up to a comet radius of about 2000 m, the impact velocity increases to 1.2 m/s and even 1.5 m/s for high densities. Up to cometary densities of 800 kg/m^3, the impact velocity of 1 m/s is sufficient for landing on a comet with a radius of about 2000 m. The minimum goal is therefore the adaptation of the landing gear for impact velocities up to 1.2 m/s and inclusion of the new mission target in the engineering parameter frame of the landing gear.

3.   Modelling of the Rosetta Lander impact

The landing is modeled in a multibody system approach. Mass and inertia describe each body. They are connected by joints or force elements like springs and dampers as well as friction. The detailed model description of the Lander and the comet model surface can be found in Hilchenbach et al. (2000).

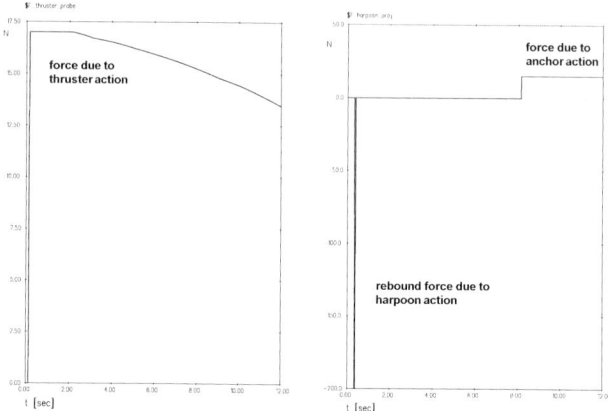

Figure 3.   Input data for the thruster and anchor forces. The timing is correlated to the impact of the landing gear on the surface. The latter triggers the touch down signal, which is used to switch on the thruster and fire the harpoon. 8 sec are assumed for rewinding the thread of the anchor.

The simulation is carried out with a commercial multibody simulation program package called SIMPACK (INTEC, 1996).

The geometry of the Lander, as used in the simulations, is shown in Fig. 2. The principal dynamic elements are the damping mechanism, cardan with cardanic joint and cardanic joint brake, thruster (cold gas), and anchor or harpoon system. Further elements are flywheel or icescrews in the Lander feet. The simulations can be carried out for different comet model surfaces, ranging from very soft to very hard surfaces within the engineering comet model parameter frame (Rosetta Lander Mission Analysis Working Group, 1999). In the following we concentrate on hard comet surfaces, i.e. the feet and icescrews penetrate only marginally into the comet model surface. This case is the worst case landing scenario. Even with the orbiter in the vicinity of the comet nucleus, the surface layer density parameters remain unknown until the actual landing.

In Fig. 3 the time–dependent forces of the thruster and harpoon or anchor are shown as function of time. These forces are included in the updated model based on the previous simulations (Hilchenbach et al., 2000). The cardanic joint friction is modeled as a friction force causing a torque of about 10 Nm. The lateral velocity components are set to 0.1 m/s and the angular rotation previous to impact to $0.4^o/s$. The gravity field acceleration is set to 10^{-3} m/s^2. The foot–soil interaction force due to surface hardness is set to about $2 \ 10^3$ N. The friction factor between foot/ icescrew and comet surface is set to 0.7. The rigidity of the landing gear is modeled via prismatic and torsion springs. The most important elasticity parameter is the elasticity of the gear along the damping axis. The damping factor and the vertical elasticity of the landing gear are designed as a matched system. If the damping factor would be increased by a factor of 1.4, the landing gear vertical elasticity has to be increased by a factor of about 2, otherwise the Lander would rebound on impact even on a flat surface. Alteration of the damping constant without adjustment of the matched landing gear parameter is not feasible. The cardanic joint–leg rotational stiffness is 200 Nm/rad. With the cardan tilt limiting device, the stiffness is increased to about 10^3 Nm/rad. Higher values would have been very much desirable to limit the energy storage. They are not achievable without interfering with the Lander body and landing gear interface, i.e. the landing gear release mechanism and launch locks.

4. Simulation results

In Fig. 4 and Fig. 5 the results of the mutibody simulations for the impact velocity of 1.2 m/s and with a cardanic joint without a tilt limitation are shown. This is the model of the Lander as envisaged for the landing on a comet, such as 46P/Wirtanen with impact velocities up to 1 m/s and slopes of up to 30°. At the higher impact velocity, the Lander touches the surface with one leg, the damping process via the damping mechanism is initiated. The touch down triggers the onset of the thruster action and the harpoon firing. Due to the 30° slope and the high impact velocity, the Lander body hits the landing gear leg, the damping process is stopped after about 9.5 cm and the Lander body rebounds. Up to impact velocities of 1 m/s, the cardanic joint device permits the Lander to impact on the comet with nearly no angular momentum exchange between comet and Lander. Due to the limited length of the damping device, at 1.2 m/s this is not the case anymore. Increasing the damping constant and therefore shorting the required damp length is not feasible as it requires a significant increase of landing gear stiffness. The Lander lifts off due to the exchanged angular momentum. After a few seconds the force due to the thruster pushes the Lander back towards the surface and the Lander touches down for the second time. After 8 sec the anchor motor should have rewound the anchor or harpoon thread and the anchor force grips the Lander on the surface.

Limitation of the cardanic joint freedom with a half cone angle of about 2.5^o was simulated and is shown in Fig. 6 and Fig. 7. All other model parameters are the same as in the previous simulation. The Lander touches the surface slope again with one leg, initiating the damping process, the thruster onset and the harpoon firing.

The damping proceeds up to 12 cm damplength, i.e. more energy is damped in this first impact than in the previous simulation without the tilt limiter. Due to the energy stored in the torsion spring of the cardan including the cardanic joint limitation, the 2 other legs of the Lander rebound and lift off while the

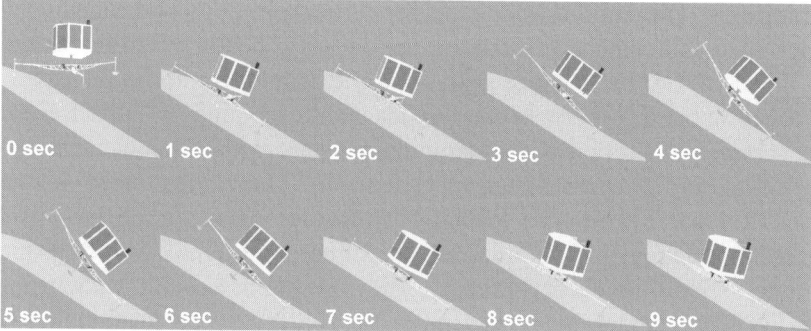

Figure 4.   Lander impacts on a model comet surface with a slope of 30° and a velocity of 1.2 m/s: Sequential impact time series, still frames taken in intervals of one second. The Lander touches the surface and rebounds due to the interaction between a Lander landing gear leg and Lander body. In spite of the cardanic joint operation, angular momentum is transferred to the Lander due to the body–leg interaction, short circuiting the damping mechanism. Only after about 7 to 8 sec, the Lander settles again (2nd touch down) due to the action of the thruster, and after further damping and action of the anchor force, the Lander comes to rest on the comet surface.

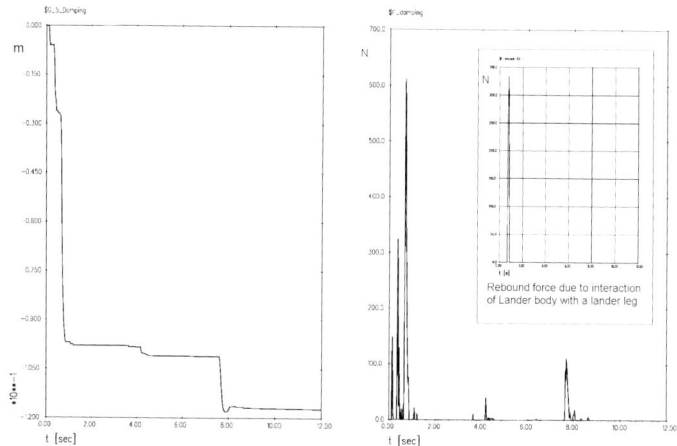

Figure 5.   Landing scenario as Fig. 4, the damping length and the damping force are plotted as a function of time. The damping is stopped some hundred milliseconds after impact due to the Lander body–leg interaction; i.e. a substantial part of the kinetic energy of the Lander is not damped, but transformed into rotational energy.

Lander slides along the slope. Due to the thruster action, the Lander is forced back towards the surface and finally settles after about 10 sec. The achievement of the tilt limitation is a landing scenario without discontinuities such as the one caused by the Lander body rebounding from a landing gear leg.

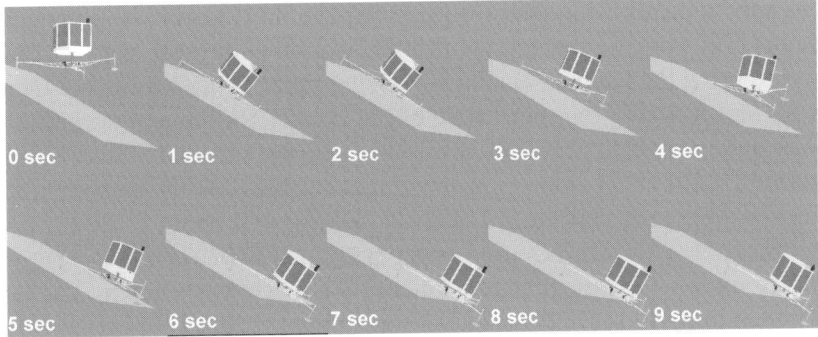

Figure 6.   Lander equipped with the cardan tilt limitation impacts on the model comet surface with a slope of 30° and a velocity of 1.2 m/s: Sequential time series, still frames taken in intervals of one second. The Lander touches the surface, rebounds due to the energy stored in the torsion of the tilt–limited cardanic joint and landing gear structure. The Lander lifts off, moves adjacent to the slope and settles after about 5 to 6 sec (2nd touch down) due to the action of the thruster. The Lander comes to rest on the comet surface after about 10 to 12 seconds due to the anchor and thruster action.

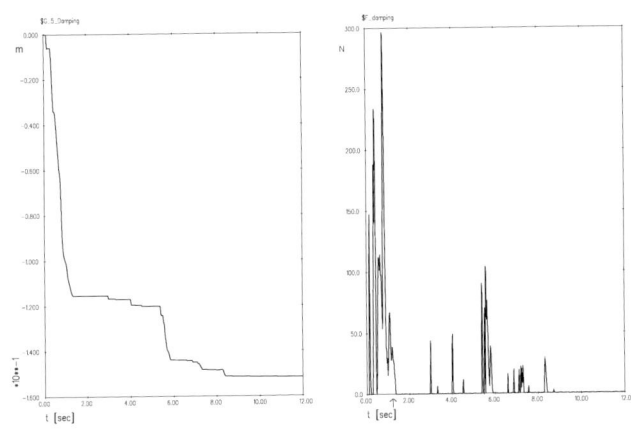

Figure 7.   Landing scenario as Fig. 6, the damping length and the damping force are plotted as a function of time. The damping stops after some hundred milliseconds after impact as the Lander lifts off due to the release of the energy stored in the torsion of the structure. On 2nd touch down, the kinetic energy is further damped and the Lander finally rests on the surface, secured by the anchor force.

5.   Summary

The implementation of the tilt limiter allows to set the upper limit for the Lander body impact velocity to 1.2 m/s for cometary surface slopes of up to 30°. A higher stiffness of the cardan including the limitation device would have been highly desirable resulting in less energy stored in this device on impact and in a dynamically smoother landing scenario. In the previous simulations for impact on a smaller comet, the thruster was an additional, not mandatory device securing the landing, having the main purpose of increasing the Lander velocity and reducing the travel time of the Lander. In the new scenario, the thruster action after first impact is essential for the successful landing. The duration of the Lander impact on the comet surface was in the scenario for comet 46P/Wirtanen about 1 - 2 sec. It took as well about 8 sec for tightening the anchor thread, but the Lander body was then already at rest. For 67P/Churyumov–Gerasimenko, the kinetics are much less smooth. The landing lasts up to 9 - 12 sec and the anchor force might even act before the Lander is completely at rest on the comet surface. For impact velocities independent of the comet size over a wide range the adjustment of the impact velocity via a cold gas system would have been required. The implementation of the limitation of the cardanic joint tilt was chosen to allow the Rosetta Lander to land successfully even on a larger comet with high density.

Acknowledgments

The authors wish to thank the organisers of the Rosetta Workshop on the island of Capri for their hospitality and the participants for many fruitful discussions.

References

Hilchenbach, M., Kuechemann, O., and Rosenbauer H. (2000). *Planet. Spa. Sci.*, 48:361–369.
INTEC GmbH (1996), Wessling, Germany.
Lamy, P. L., et al. (2003). Annual meeting of the Division of Planetary Sciences of the American Astronomical Society in Monterey, Calif., USA, September 2003.
Rosetta Lander Mission Analysis Working Group (1999). RO–ESC–RP–5006, ESOC/ESTEC.

THE ROSETTA LANDER EXPERIMENT SESAME AND THE NEW TARGET COMET 67P/CHURYUMOV–GERASIMENKO

K. J. Seidensticker, H.–H. Fischer, D. Madlener, S. Schieke
Institute of Space Simulation, DLR, Cologne

K. Thiel
Dept. for Nuclear Chemistry, University of Cologne

A. Péter
Atomic Energy Research Institute, KFKI, Budapest

W. Schmidt
Finnish Meteorological Institute, Helsinki

R. Trautner
RSSD, ESTEC, ESA, Noordwijk

Abstract   *SESAME* is an international instrument complex carried by the Rosetta Lander. Most of the instrument sensors are mounted within the soles of the landing gear feet in order to provide good contact with or proximity to the cometary surface. The main aim of these instruments is to measure physical properties of the cometary surface layer and of emitted particles by acoustic and electrical methods. The scientific goals, the measuring principles and performance parameters are described. In addition, we discuss the impact by selecting 67P/Churyumov–Gerasimenko as the new target comet.

1.   Introduction

The activity and evolution of comets is, apart from the solar radiation flux, strongly influenced by the properties of their surfaces. To understand the processes within and on comets, measurements of surface parameters like compo-

L. Colangeli et al. (eds.), The New ROSETTA Targets, 297–307.

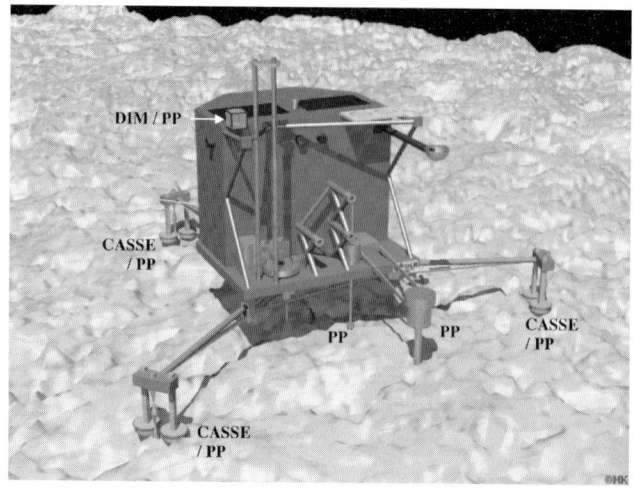

Figure 1. Positions of *SESAME* sensors on the Rosetta Lander (illustration: M. Kretschmer)

sition, structure, mechanical and electrical properties and their variation with rotational and orbital phase are of crucial importance. The Surface Electric Sounding and Acoustic Monitoring Experiment (*SESAME*) on the Lander of the ESA Rosetta Mission intends to determine some of these parameters as well as the properties of back–falling cometary particles.

These data are also necessary for modeling the gas transport in cometary surfaces. Describing this gas transport is a prerequisite for a solid interpretation of the gas measurements of elemental and molecular abundances. The diurnal variation of the gas flux is strongly governed by the diffusion properties of upper cometary surface layers. Vice versa, erosion by and recondensation of the gas flux steadily modifies the surface properties.

SESAME is a complex of three instruments: the Comet Acoustic Surface Sounding Experiment (*CASSE*), the Dust Impact Monitor (*DIM*) and the Permittivity Probe (*PP*). They share a Common Electronics (*CE*) and common control software. In order to save mass, all *SESAME* sensors except the *DIM* sensor, are mounted on other Lander instruments or subsystems (Fig. 1). The Central Electronics weighs 900 g and the sensors add up to a total mass of 1650 g. The typical power consumption of *SESAME* is about 2.5 W.

For the new target of the ESA Rosetta mission 67P/Churyumov–Gerasimenko Lamy et al., 2003 and Tancredi et al., 2000 derived radii ranging from 2.0 to 2.5 km based on the absolute magnitude of the nucleus, its geometric albedo and phase coefficient. The original target comet 46P/Wirtanen presumably has a much smaller radius of $\simeq 0.6$ km (Lamy et al., 1998). Taking into account the estimates for the bulk density of model comets of different size (ESA, 1999), the larger radius of the new target comet implies a mass of almost two orders of magnitude greater than that of 46P/Wirtanen.

The rotational period of 67P/Churyumov–Gerasimenko has been determined by means of Phase Dispersion Minimisation and Fourier techniques to be 12.30 ± 0.27 h (Lamy et al., 2003) compared to an estimate of 6–7 hours for 46P/Wirtanen.

The new target comet is reported to have suffered two Jupiter encounters in 1840 and 1959 that reduced the perihelion distance significantly (Kidger, 2003). Comets with such drops in perihelion distance in their recent dynamical history are believed to develop enhanced activity even at aphelion due to the removal of the thermally insulating dust mantle. Kidger, 2003 determined for the dust activity rate of 67P/Churyumov–Gerasimenko $af\rho$ values of 135 cm at perihelion to 290 cm 35 days after perihelion. The author pointed out that over several returns perihelic outbursts occurred that temporarily exceed the average dust emission near perihelion by a factor of $\simeq 3$. Such outbursts during the 2015 return could strongly affect *SESAME* dust measurements.

For comparison, the dust activity of comet 46P/Wirtanen was characterised by earth–bound observations 2.5 days before perihelion passage in 1997 yielding an $af\rho$ value of 157 cm (Jockers et al., 1998). No significant temporary outbursts near perihelion have been reported for Wirtanen.

2.   CASSE

CASSE will investigate the outermost layer of the cometary surface by transmitting and receiving acoustic waves in the frequency range from 100 Hz to 6 kHz. By measuring the velocities of the longitudinal (c_p) and transversal (c_s) acoustic waves, described as compressional p– and shear s–waves in seismic terms, the elastic parameters Young's modulus E and Poisson number ν can be determined:

$$E = \rho\, c_s^2\, \frac{4c_s^2 - 3c_p^2}{c_s^2 - c_p^2} \quad \text{and} \quad \nu = \frac{c_s^2 - \frac{1}{2}c_p^2}{c_s^2 - c_p^2}\,. \tag{1}$$

The density ρ of the cometary surface matter will be measured by other Lander instruments or estimated by using cometary models.

The acoustic transmitters (stacked piezoceramics), which can also be operated as receivers, and the receivers (triaxial piezoelectric accelerometers) have been integrated into the two soles of each of the Lander's three feet (Fig. 2). To guarantee sufficient ground contact at a possible rough surface the two soles can move against each other by about 10 cm in height. Sensor temperature can be measured down to $-105\ ^\circ C$. By switching between transmitters and receivers, an analysis of the surface material and an in–depth sounding for detection of a layered structure or embedded local inhomogeneities is possible.

300

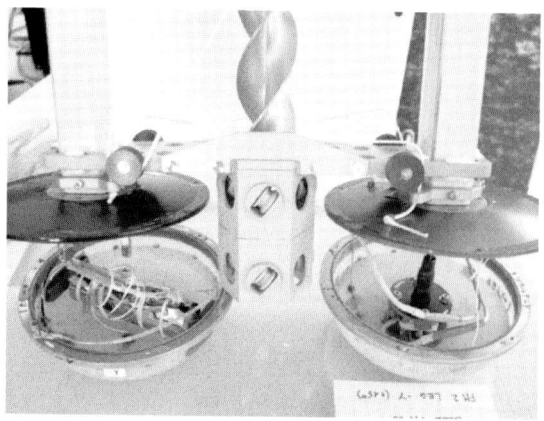

Figure 2. The *CASSE* sensors of the −Y foot of the Rosetta Lander before closing the covers during integration. The left sole contains the transmitter and the right sole holds the triaxial accelerometer fixed between two mounting plates made of glass fiber.

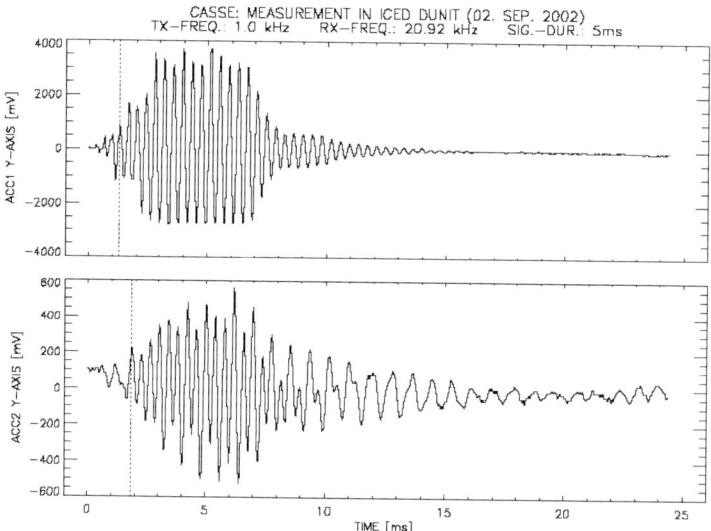

Figure 3. Sound transfer via a frozen Dunite sand bed. Laboratory models of the *CASSE* receiver soles were placed at a distance of 40 cm (upper panel) and 80 cm (lower panel) from a transmitter sole. A sound velocity of 700 ± 60 m s^{-1} has been derived from the arrival times of the p–wave (broken lines). One can also see the decrease of the signal amplitude with distance.

On the comet, sound will be transferred via the cometary surface as well as via the landing gear. The latter signal has to be removed from the first one during data analysis. Although first measurements of the sound transfer via the landing gear were done during the thermal–vacuum testing of the Rosetta Lander, it is necessary to repeat this calibration in free space during the descent

Figure 4.   *DIM* sensor with 3 piezo plates mounted on a cube

after landing gear foldout and before touchdown. The planned calibration sequence had to be drastically reduced due to the shorter descent duration (see Section 5).

Preparing for the analysis of the recorded signals, various experiments with cometary analogous materials have been conducted (Kochan et al., 2000). These studies of regolithic sand and hardened ice/dust mixtures showed that acoustic sounding could be applied to cometary surfaces (Fig. 3). Additionally, a numerical method has been applied to simulate the wave propagation in porous media.

3.   DIM

In active regions where ices are exposed to the solar radiation, intensively sublimating gas molecules lift up cometary dust grains ($\rho \simeq 1000$ kg m^{-3}) of different sizes. Depending on the combined action of gas drag and gravitational forces, the ice/dust particles are either ejected to the interplanetary space (smaller ones) or fall back onto the nucleus (bigger ones).

The Dust Impact Monitor (*DIM*), placed on the Rosetta Lander, will observe mainly those particles of the comet, which temporarily leave the surface but will fall back due to their insufficient velocity. *DIM* will sense the impacting particles with piezo plates mounted on 3 sides of a cube (Fig. 4).

The expected scientific results from the *DIM* instrument are statistics of impacting particles as a function of rotational and orbital phase, velocity and mass distributions of impacting grains and model parameters to describe mantle formation by back–fallen particles.

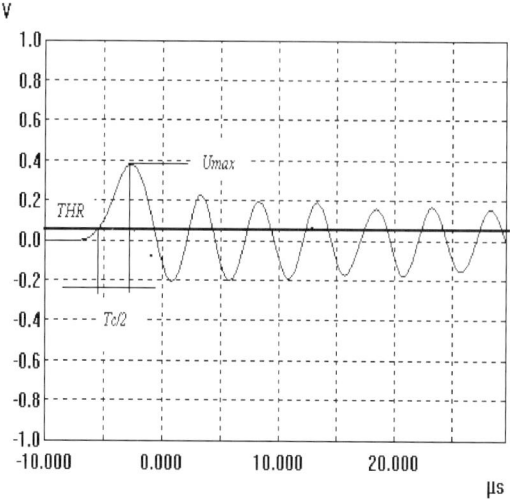

Figure 5. Typical impact signal and its measured parameters (impacting particle: mass 4 mg, velocity 0.44 m s^{-1}, energy 4×10^{-7} J)

Hertz theory was used to calculate the mechanical stress caused by the elastic impact of spherical particles and the time variation of the stress. This model gives the correlation between the measured peak voltage U_{max} and the duration T_c of the impact signal:

$$U_{max} = K_U(d_{33}, C, A)R^{0.2}E^{0.6} \quad \text{and} \quad T_c = K_T(A, \rho)R^{1.3}E^{-0.1} , \quad (2)$$

where d_{33} is the piezoelectric constant of the sensor plate and C its capacitance. R, E and ρ are the radius, the energy and the density of the impacting particle, respectively. A is a function of the elastic parameters Young's modulus and Poisson number of the piezo segment and the impacting material.

Figure 5 shows a typical signal with the measured parameters. The output of the sensor can be best described as a decaying sinusoidal electrical signal. The detection of an impact is based on an adaptive threshold crossing, where the threshold is the sum of the average of the signals and a margin. Changing the margin value can alter the sensitivity of the detection. As the rate of impacts increases, the average will be increased as well, and the sensitivity will be decreased. By this way there is no danger of saturation of the system: the higher the impact rate the lower will be the sensitivity and less impacts will be detected. In case of a very high impact rate the individual impacts cannot be distinguished and therefore just the average value of the signal will be measured. Based on these considerations and measurements, the expected performance of *DIM* is listed in Table 1.

Table 1.   *DIM* performance parameters (density: $1000 \ \mathrm{kg \, m^{-3}}$)

Quantity	Range
Energy	$2 \cdot 10^{-11} \ \ldots \ 3 \cdot 10^{-7} \ \mathrm{J}$
Radius	$5 \cdot 10^{-5} \ \ldots \ 6 \cdot 10^{-3} \ \mathrm{m}$
Velocity	$0.025 \ \ldots \ 0.25 \ \mathrm{m \, s^{-1}}$
Mass	$5 \cdot 10^{-10} \ \ldots \ 9 \cdot 10^{-4} \ \mathrm{kg}$

4.   PP

The Permittivity Probe (*PP*) is dedicated to the investigation of the low frequency electrical properties of the cometary surface material. It allows to determine the water ice content down to about 2 m below the surface and to
measure its diurnal temperature variation.

PP is based on the quadrupolar probe concept (Grard, 1990a; Grard, 1990b)
which employs a set of transmitter and receiver electrodes for emitting AC
currents into a medium of interest. The complex dielectric constant can be
determined by measuring the magnitude and phase shift of both the emitted
transmitter currents as well as the resulting potential difference at a pair of
receiver electrodes. Conductivity σ and dielectric constant ϵ_r of the medium
can be calculated according to:

$$\sigma = \frac{A_0}{A} \omega \, \epsilon_0 \sin(\varphi - \varphi_0) \quad \text{and} \quad \epsilon_r = \frac{A_0}{A} \cos(\varphi - \varphi_0) \,, \tag{3}$$

where A, φ are the receiver signal magnitude and phase measured in the medium
and the indexed quantities correspond to the signals observed in vacuum. ϵ_0 is
the permittivity of vacuum and $\omega = 2\pi f$ represents the angular frequency.

A quadrupolar probe has been implemented for space applications before
as a part of the Huygens Atmospheric Structure Instrument HASI (Trautner et
al., 2003). For the *PP* instrument, the receiver electrodes (RX in Fig. 6, left)
were integrated into the Lander +Y and $-$Y soles (Fig. 2) and the transmitter
electrodes (TX in Fig. 6, left) were added to the +X soles and the *APX* and
MUPUS Pen sensors (Fig. 6, right).

The *PP* instrument is based on a dedicated electronics board, which digitally
generates the desired transmitter frequency, and provides current, potential and
phase shift information by sampling the injected transmitter currents and receiver potentials simultaneously. The receiver electrode signals are picked up
via high impedance preamplifiers accommodated in the soles on two of the
three landing gear feet (RX in Fig. 6, left). They also provide guarding signals
for reducing parasitic effects caused by the presence of other sensors. Any two
of the three transmitter electrodes (TX in Fig. 6, left) can be selected in order

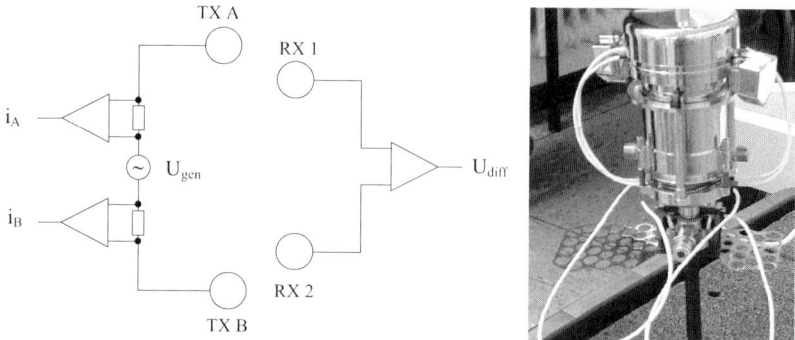

Figure 6. Left: Principle of the *PP* Active Mode measurement configuration. Right: *PP* sensor (mesh below hammer mechanism) on *MUPUS Pen*

to react to different deployment scenarios and for defining different quadrupole geometries. The operation parameters are given in Table 2.

Table 2. *PP* parameters for the Active Mode

Parameter	Range
Transmitter frequency	10 Hz ... 10 kHz sine
Max. peak–to–peak voltage	20 V adjustable
Phase resolution	1 deg
DAC / ADC resolution	8 bit

Without active transmitters, *PP* is able to measure the variation of potential differences between the receiver electrodes and detect plasma waves in the range of 10 Hz to 10 kHz. The difference signal is sampled with 20 kHz and analyzed with a simplified wavelet algorithm, providing the power distribution of plasma waves as a function of frequency.

An additional integrating spacecraft potential sensor attached to the *DIM* sensor cube (Fig. 4) generates alert information if the potential varies by more than 1 mV with frequencies above 10 Hz, indicating detectable plasma waves. This allows to distinguish electronic noise from real plasma effects in passive measurements.

5.   Operations

The selection of comet 67P/Churyumov–Gerasimenko as the new target has implications on the *SESAME* science operations sequences. The major impact will be for the Separation, Descent and Landing (S–D–L) phase including the touch–down on the cometary surface. Minor adjustments of instrument operations will be necessary for the on–comet phase. The major impacts on *SESAME* are summarized below:

- The greater mass via stronger gravity leads to a considerably shorter Separation, Descent and Landing phase (S–D–L) compared to Wirtanen ($\simeq$ 2 h). Presently two options are foreseen for Churyumov–Gerasimenko, related to as *basic descent* ($\simeq$ 25 min) and *fast descent* ($\simeq$ 8.5 min). The reduced S–D–L duration causes considerable restrictions in *SESAME* science operations and calibration measurements (cf. Table 3).

- A presumably higher dust emission activity of the new target comet may have consequences for the kind of information achievable by the *SESAME* dust detection instrument *DIM*. Particularly, measurements of single dust grain parameters may be prevented.

- The longer comet day requires only minor adjustments of *SESAME* post–landing measuring programs that are depending on sunrise and sunset phenomena or midday and midnight conditions.

- Whereas the first three items show the impact on operations and science data, the expected higher touch–down velocity up to (or even exceeding) $1.5 \ \mathrm{m\,s^{-1}}$ bears the risk of loosening or destroying the *CASSE* and/or *PP* transmitter and receiver fixings in the soles of the landing gear feet, originally designed for $0.5 \ \mathrm{m\,s^{-1}}$ landings.

6.   Conclusions

The situation of the Rosetta Lander experiment *SESAME* with respect to the change of the target comet can be summarized as follows:

- The *SESAME* instruments can operate on and deliver valuable scientific data from 67P/Churyumov–Gerasimenko.

- The amount of scientific data to be obtained about the dust and plasma environment between the Rosetta orbiter and the cometary surface will be significantly reduced due to the shorter descent duration. For the same reason, only a limited set of calibration data needed for *CASSE* data analysis can be collected.

- The higher touchdown velocity increases the risk for the mountings of the *SESAME* sensors and actuators in the soles of the Lander.

- The presumably higher dust emission activity can reduce the amount of information on single grains obtained by *DIM*.

Table 3. Comparison of *SESAME* science operations for old and new Rosetta target comet

Operation phase[a]	46P/Wirtanen Normal descent ($\simeq$ 2 h)	67P/Churyumov–Gerasimenko Basic descent ($\simeq$ 25 min)
S–F	• Continuous measurement of *DIM* and *PP*	• Limited data vs. height • Only *DIM* records dust impacts • Number of *PP* Langmuir probe data points reduced
F–T	• *CASSE* sound transfer calibration • Record of dust impacts on *CASSE* foot soles and *DIM* detector vs. height • Record of ambient electron density vs. height by *PP* Langmuir probe • *PP* plasma wave measurements vs. height	• *CASSE* sound transfer calibration reduced to few frequencies and amplifier gain values
T	• Record of touchdown by *CASSE* accelerometers • Record of surface approach by *PP* Active Mode	• *CASSE* touchdown measurement
T–C	• Normal measuring program	• Adjustment of operations time schedule to new comet day length

[a] S–F: Time from separation to foldout of landing gear; F–T: Time from foldout of landing gear to touchdown; T: Touchdown; T–C: On–comet time after touchdown

- On the other hand, the larger gravity of 67P/Churyumov–Gerasimenko could lead to a better contact of the Lander soles with the surface, what is advantageous for the *CASSE* acoustic measurements after touchdown.

References

Grard, R. (1990a). A quadrupolar array for measuring the complex permittivity of the ground: application to Earth prospection and planetary exploration, *Meas. Sci. Technol.*, 1:295–301.

Grard, R. (1990b). A quadrupole system for measuring in situ the complex permittivity of materials: application to penetrators and landers for planetary exploration, *Meas. Sci. Technol.*, 1:801–806.

Jockers, K., Credner, T. and Bonev, T. (1998). Water ions, dust and CN in comet 46P/Wirtanen, *Astron. Astrophys. Letters*, 335:L56 L59.

Kidger, M.R. (2003). Dust production and coma morphology of 67P/Churyumov–Gerasimenko during the 2002–2003 apparition, *Astron. Astrophys.*, 408:767–774.

Kochan, H., Feibig, W., Konopka, U., Kretschmer, M., Möhlmann, D., Seidensticker, K.J., Arnold, W., Gebhardt, W. and Licht, R. (2000). *CASSE* – The Rosetta Lander Comet Acoustic Surface Sounding Experiment – status of some aspects, the technical realisation and laboratory simulations, *Planet. Space Sci.*, 48:385 – 399.

Lamy, P.L., Toth, I., Jorda, L., Weaver, H.A. and A'Hearn, M. (1998). The Nucleus and Inner Coma of Comet 46P/Wirtanen, *Astron. Astrophys. Letters*, 335:L25–L29.

Lamy, P.L., Toth, I., Weaver, H., Jorda, L. and Kaasalainen, M. (2003). The nucleus of comet 67P/Churyumov–Gerasimenko, the new target of the Rosetta mission, *DPS 35th Meeting*, Monterey, USA.

ESA, Rosetta Lander Mission Analysis Working Group, Comet Surface Engineering Model (1999). RO–ESC–RP–5006, Issue 1.

Tancredi, G., Fernández, J.A., Rickmann, H. and Licandro, J. (2000). A catalog of observed nuclear magnitudes of Jupiter family comets, *Astron. Astrophys. Suppl. Ser.*, 146:73–90.

Trautner, R., Grard, R. and Hamelin, M. (2003). Detection of subsurface ice and water deposits on Mars with a mutual impedance probe, *J. Geophys. Res.*, 108 (E10):8047.

Astrophysics and Space Science Library

Volume 314: *Solar and Space Weather Radiophysics- Current Status and Future Developments,* edited by D.E. Gary and C.U. Keller, ISBN 1-4020-2813-X, August 2004

Volume 312: *High-Velocity Clouds,* edited by H. van Woerden, U. Schwarz, B. Wakker, ISBN 1-4020-2813-X, September 2004

Volume 311: *The New ROSETTA Targets- Observations, Simulations and Instrument Performances,* edited by L. Colangeli, E. Mazzotta Epifani, P. Palumbo, ISBN 1-4020-2572-6, September 2004

Volume 310: *Organizations and Strategies in Astronomy 5,* edited by A. Heck, ISBN 1-4020-2570-X, September 2004

Volume 309: *Soft X-ray Emission from Clusters of Galaxies and Related Phenomena,* edited by R. Lieu and J. Mittaz, ISBN 1-4020-2563-7, September 2004

Volume 308: *Supermassive Black Holes in the Distant Universe,* edited by A.J. Barger, ISBN 1-4020-2470-3, August 2004

Volume 307: *Polarization in Spectral Lines,* by E. Landi Degl'Innocenti and M. Landolfi, ISBN 1-4020-2414-2, August 2004

Volume 306: *Polytropes – Applications in Astrophysics and Related Fields,* by G.P. Horedt, ISBN 1-4020-2350-2, September 2004

Volume 305: *Astrobiology: Future Perspectives,* edited by P. Ehrenfreund, W.M. Irvine, T. Owen, L. Becker, J. Blank, J.R. Brucato, L. Colangeli, S. Derenne, A. Dutrey, D. Despois, A. Lazcano, F. Robert, hardbound ISBN 1-4020-2304-9, softcover ISBN 1-4020-2587-4, July 2004

Volume 304: *Cosmic Gammy-ray Sources,* edited by K.S. Cheng and G.E. Romero, ISBN 1-4020-2255-7, September 2004

Volume 303: *Cosmic rays in the Earth's Atmosphere and Underground,* by L.I, Dorman, ISBN 1-4020-2071-6, August 2004

Volume 302:*Stellar Collapse,* edited by Chris L. Fryer Hardbound, ISBN 1-4020-1992-0, April 2004

Volume 301: *Multiwavelength Cosmology*, edited by Manolis Plionis
Hardbound, ISBN 1-4020-1971-8, March 2004

Volume 300: *Scientific Detectors for Astronomy,* edited by Paola Amico, James W. Beletic, Jenna E. Beletic
Hardbound, ISBN 1-4020-1788-X, February 2004

Volume 299: *Open Issues in Local Star Fomation,* edited by Jacques Lépine, Jane Gregorio-Hetem
Hardbound, ISBN 1-4020-1755-3, December 2003

Volume 298: *Stellar Astrophysics - A Tribute to Helmut A. Abt,* edited by K.S. Cheng, Kam Ching Leung, T.P. Li
Hardbound, ISBN 1-4020-1683-2, November 2003

Volume 297: *Radiation Hazard in Space,* by Leonty I. Miroshnichenko
Hardbound, ISBN 1-4020-1538-0, September 2003

Volume 296: *Organizations and Strategies in Astronomy, volume 4,* edited by André Heck
Hardbound, ISBN 1-4020-1526-7, October 2003

Volume 295: *Integrable Problems of Celestial Mechanics in Spaces of Constant Curvature,* by T.G. Vozmischeva
Hardbound, ISBN 1-4020-1521-6, October 2003

Volume 294: *An Introduction to Plasma Astrophysics and Magnetohydrodynamics,* by Marcel Goossens
Hardbound, ISBN 1-4020-1429-5, August 2003
Paperback, ISBN 1-4020-1433-3, August 2003

Volume 293: *Physics of the Solar System,* by Bruno Bertotti, Paolo Farinella, David Vokrouhlický
Hardbound, ISBN 1-4020-1428-7, August 2003
Paperback, ISBN 1-4020-1509-7, August 2003

Volume 292: *Whatever Shines Should Be Observed,* by Susan M.P. McKenna-Lawlor
Hardbound, ISBN 1-4020-1424-4, September 2003

Volume 291: *Dynamical Systems and Cosmology,* by Alan Coley
Hardbound, ISBN 1-4020-1403-1, November 2003

Volume 290: *Astronomy Communication,* edited by André Heck, Claus Madsen
Hardbound, ISBN 1-4020-1345-0, July 2003

Volume 287/8/9: *The Future of Small Telescopes in the New Millennium,*
edited by Terry D. Oswalt
Hardbound Set only of 3 volumes, ISBN 1-4020-0951-8, July 2003

Volume 286: *Searching the Heavens and the Earth: The History of Jesuit
Observatories,* by Agustín Udías
Hardbound, ISBN 1-4020-1189-X, October 2003

Volume 285: *Information Handling in Astronomy - Historical Vistas,* edited
by André Heck
Hardbound, ISBN 1-4020-1178-4, March 2003

Volume 284: *Light Pollution: The Global View,* edited by Hugo E. Schwarz
Hardbound, ISBN 1-4020-1174-1, April 2003

Volume 283: *Mass-Losing Pulsating Stars and Their Circumstellar Matter,*
edited by Y. Nakada, M. Honma, M. Seki
Hardbound, ISBN 1-4020-1162-8, March 2003

Volume 282: *Radio Recombination Lines,* by M.A. Gordon, R.L. Sorochenko
Hardbound, ISBN 1-4020-1016-8, November 2002

Volume 281: *The IGM/Galaxy Connection,* edited by Jessica L. Rosenberg,
Mary E. Putman
Hardbound, ISBN 1-4020-1289-6, April 2003

Volume 280: *Organizations and Strategies in Astronomy III,* edited by André
Heck
Hardbound, ISBN 1-4020-0812-0, September 2002

Volume 279: *Plasma Astrophysics , Second Edition,* by Arnold O. Benz
Hardbound, ISBN 1-4020-0695-0, July 2002

Volume 278: *Exploring the Secrets of the Aurora,* by Syun-Ichi Akasofu
Hardbound, ISBN 1-4020-0685-3, August 2002

Volume 277: *The Sun and Space Weather,* by Arnold Hanslmeier
Hardbound, ISBN 1-4020-0684-5, July 2002

Volume 276: *Modern Theoretical and Observational Cosmology*, edited by Manolis Plionis, Spiros Cotsakis
Hardbound, ISBN 1-4020-0808-2, September 2002

Volume 275: *History of Oriental Astronomy*, edited by S.M. Razaullah Ansari
Hardbound, ISBN 1-4020-0657-8, December 2002

Volume 274: *New Quests in Stellar Astrophysics: The Link Between Stars and Cosmology*, edited by Miguel Chávez, Alessandro Bressan, Alberto Buzzoni, Divakara Mayya
Hardbound, ISBN 1-4020-0644-6, June 2002

Volume 273: *Lunar Gravimetry*, by Rune Floberghagen
Hardbound, ISBN 1-4020-0544-X, May 2002

Volume 272: *Merging Processes in Galaxy Clusters*, edited by L. Feretti, I.M. Gioia, G. Giovannini
Hardbound, ISBN 1-4020-0531-8, May 2002

Volume 271: *Astronomy-inspired Atomic and Molecular Physics*, by A.R.P. Rau
Hardbound, ISBN 1-4020-0467-2, March 2002

Volume 270: *Dayside and Polar Cap Aurora*, by Per Even Sandholt, Herbert C. Carlson, Alv Egeland
Hardbound, ISBN 1-4020-0447-8, July 2002

Volume 269: *Mechanics of Turbulence of Multicomponent Gases*, by Mikhail Ya. Marov, Aleksander V. Kolesnichenko
Hardbound, ISBN 1-4020-0103-7, December 2001

Volume 268: *Multielement System Design in Astronomy and Radio Science*, by Lazarus E. Kopilovich, Leonid G. Sodin
Hardbound, ISBN 1-4020-0069-3, November 2001

Volume 267: *The Nature of Unidentified Galactic High-Energy Gamma-Ray Sources*, edited by Alberto Carramiñana, Olaf Reimer, David J. Thompson
Hardbound, ISBN 1-4020-0010-3, October 2001

Volume 266: *Organizations and Strategies in Astronomy II*, edited by André Heck
Hardbound, ISBN 0-7923-7172-0, October 2001

Volume 265: *Post-AGB Objects as a Phase of Stellar Evolution*, edited by R. Szczerba, S.K. Górny
Hardbound, ISBN 0-7923-7145-3, July 2001

Volume 264: *The Influence of Binaries on Stellar Population Studies*, edited by Dany Vanbeveren
Hardbound, ISBN 0-7923-7104-6, July 2001

Volume 262: *Whistler Phenomena - Short Impulse Propagation*, by Csaba Ferencz, Orsolya E. Ferencz, Dániel Hamar, János Lichtenberger
Hardbound, ISBN 0-7923-6995-5, June 2001

Volume 261: *Collisional Processes in the Solar System*, edited by Mikhail Ya. Marov, Hans Rickman
Hardbound, ISBN 0-7923-6946-7, May 2001

Volume 260: *Solar Cosmic Rays*, by Leonty I. Miroshnichenko
Hardbound, ISBN 0-7923-6928-9, May 2001

Volume 259: *The Dynamic Sun*, edited by Arnold Hanslmeier, Mauro Messerotti, Astrid Veronig
Hardbound, ISBN 0-7923-6915-7, May 2001

Volume 258: *Electrohydrodynamics in Dusty and Dirty Plasmas- Gravito-Electrodynamics and EHD*, by Hiroshi Kikuchi
Hardbound, ISBN 0-7923-6822-3, June 2001

Volume 257: *Stellar Pulsation - Nonlinear Studies*, edited by Mine Takeuti, Dimitar D. Sasselov
Hardbound, ISBN 0-7923-6818-5, March 2001

Volume 256: *Organizations and Strategies in Astronomy*, edited by André Heck
Hardbound, ISBN 0-7923-6671-9, November 2000

Volume 255: *The Evolution of the Milky Way- Stars versus Clusters*, edited by Francesca Matteucci, Franco Giovannelli
Hardbound, ISBN 0-7923-6679-4, January 2001

Volume 254: *Stellar Astrophysics*, edited by K.S. Cheng, Hoi Fung Chau, Kwing Lam Chan, Kam Ching Leung
Hardbound, ISBN 0-7923-6659-X, November 2000

Volume 253: *The Chemical Evolution of the Galaxy*, by Francesca Matteucci
Paperback, ISBN 1-4020-1652-2, October 2003
Hardbound, ISBN 0-7923-6552-6, June 2001

Volume 252: *Optical Detectors for Astronomy II*, edited by Paola Amico,
James W. Beletic
Hardbound, ISBN 0-7923-6536-4, December 2000

Volume 251: *Cosmic Plasma Physics*, by Boris V. Somov
Hardbound, ISBN 0-7923-6512-7, September 2000

Volume 250: *Information Handling in Astronomy*, edited by André Heck
Hardbound, ISBN 0-7923-6494-5, October 2000

Volume 249: *The Neutral Upper Atmosphere*, by S.N. Ghosh
Hardbound, ISBN 0-7923-6434-1, July 2002

Volume 247: *Large Scale Structure Formation*, edited by Reza Mansouri,
Robert Brandenberger
Hardbound, ISBN 0-7923-6411-2, August 2000

Volume 246: *The Legacy of J.C. Kapteyn*, edited by Piet C. van der Kruit,
Klaas van Berkel
Paperback, ISBN 1-4020-0374-9, November 2001
Hardbound, ISBN 0-7923-6393-0, August 2000

Volume 245: *Waves in Dusty Space Plasmas*, by Frank Verheest
Paperback, ISBN 1-4020-0373-0, November 2001
Hardbound, ISBN 0-7923-6232-2, April 2000

Volume 244: *The Universe*, edited by Naresh Dadhich, Ajit Kembhavi
Hardbound, ISBN 0-7923-6210-1, August 2000

Volume 243: *Solar Polarization*, edited by K.N. Nagendra, Jan Olof Stenflo
Hardbound, ISBN 0-7923-5814-7, July 1999

Volume 242: *Cosmic Perspectives in Space Physics*, by Sukumar Biswas
Hardbound, ISBN 0-7923-5813-9, June 2000

Volume 241: *Millimeter-Wave Astronomy: Molecular Chemistry & Physics in
Space*, edited by W.F. Wall, Alberto Carramiñana, Luis Carrasco, P.F.
Goldsmith
Hardbound, ISBN 0-7923-5581-4, May 1999

Volume 240: *Numerical Astrophysics,* edited by Shoken M. Miyama, Kohji Tomisaka,Tomoyuki Hanawa
Hardbound, ISBN 0-7923-5566-0, March 1999

Volume 239: *Motions in the Solar Atmosphere,* edited by Arnold Hanslmeier, Mauro Messerotti
Hardbound, ISBN 0-7923-5507-5, February 1999

Volume 238: *Substorms-4,* edited by S. Kokubun, Y. Kamide
Hardbound, ISBN 0-7923-5465-6, March 1999

Volume 237: *Post-Hipparcos Cosmic Candles,* edited by André Heck, Filippina Caputo
Hardbound, ISBN 0-7923-5348-X, December 1998

Volume 236: *Laboratory Astrophysics and Space Research,* edited by P. Ehrenfreund, C. Krafft, H. Kochan, Valerio Pirronello
Hardbound, ISBN 0-7923-5338-2, December 1998

For further information about this book series we refer you to the following web site:
http://www.wkap.nl/prod/s/ASSL

To contact the Publishing Editor for new book proposals:
Dr. Harry (J.J.) Blom: harry.blom@springer-sbm.com